D1226417

DECISION MAKING, MODELS AND ALGORITHMS

Decision Making, Models and Algorithms

A First Course

SAUL. I. GASS

College of Business and Management
University of Maryland

A Wiley-Interscience Publication

JOHN WILEY & SONS

New York • Chichester • Brisbane • Toronto • Singapore

Library of Congress Cataloging in Publication Data:

Gass, Saul I.
Decision making, models and algorithms.

"A Wiley-Interscience publication."
Bibliography: p.
Includes index.
1. Decision-making—Mathematical models. I. Title

HD30.23.G37 1985 658.4'033 84-21996
ISBN 0-471-80963-2

Printed in the United States of America

10 9 8 7 6 5 4 3 2 1

To my parents

Bertha Gass
Louis Gass (in memoriam)

and

my friend and mentor
George B. Dantzig

The beginning is the most important
part of the work.

Plato, *The Republic*

Preface

My purpose in writing this book is to make available to undergraduate students a presentation of what I feel has been one of the most exciting advances in mathematics since the invention of the calculus. That is, the development and application of mathematical methods and analysis techniques that enable us to structure and solve important decision problems of business, industry, and government. This class of problems—decision problems—are different from the classical types in many respects. How they are different will become clear as you read the text.

It is my further belief that undergraduate training in mathematics must move away from the current required calculus-based program to a more varied one that stresses the important and modern applied areas contained in the mathematical decision sciences, as practiced by operations researchers and management scientists. There is a need to teach the fundamentals of the decision sciences to engineers, business majors, mathematicians, computer scientists, and just about anyone who wants to obtain an understanding of what mathematics and the decision sciences can do to solve real-world problems. To my mind, the finite mathematics books—as good as they are and as valuable as they have been to students and teachers—do not convey the excitement and structure inherent in decision-making models and the related mathematical content. Neither do the standard texts that explain management science/operations research. I have tried to develop a different approach. I emphasize the decision nature of problems. I bring the students into the problem-solving part by challenging them to use their training and basic knowledge. It is a book that will also challenge the instructors. They will have to teach model formulation and algorithm development, motivate students to study variations of a problem under changing assumptions, and encourage students to use their knowledge and intuition in developing solution techniques.

I believe the presentation and material in this text will assist both instructors and students to meet these important challenges. Thus, this is not a methodology book with which students just learn how to solve problems. Rather, I have attempted to present the material in a manner that would strengthen their abilities to analyze decision situations—situations that call for the selection of one possible solution over others.

Certain principles and ideas have been developed that help us organize our approach to analyzing decision problems. It is these principles and ideas—what I collectively term the *decision framework*—that are emphasized. This framework and the concept of mathematical decision models are described in Part I. In the succeeding Parts II–V, we introduce many problem situations and show how to formulate their mathematical statements. Depending on the resulting mathematical structures, the problems can be solved by obvious or not-so-obvious methods. Whenever possible, the student is asked to invent solution techniques, or is led to the development of a known procedure. As this is not a book on case studies, the full import and concerns of the decision framework cannot be demonstrated. For example, the problems of data collection, model validity, and model implementation can only be hinted at. The student is cautioned against indiscriminate use of a mathematical model.

Each of the five parts of the text ends with a chapter that combines further discussions, extensions, and exercises. These concluding chapters are designed to clarify and extend the previously discussed material, and to motivate and challenge the student. The teacher should use these chapters in planning the course content and include those sections that best meet the objectives of the course and the needs of the students. The student and teacher must treat the discussions, extensions, and exercises chapters as integral parts of the course and book.

A few teachers and researchers familiar with the totality of decision models might question my apparent emphasis on the linear-programming model. This emphasis is due to a number of reasons: (1) linear programming is the preeminent model of the decision sciences; (2) it is the most widely used modern mathematical optimization technique; and (3) its range of application and use in the practical worlds of business, industry, and government appears to have no bound. Further, as an instructional tool for linear algebra, optimization, and decision making, there is no competitor.

Algorithms imply computation. To learn just how an algorithm works you need to carry out the prescribed steps on a typical set of numbers. The problems you will encounter in this text are readily solved using the old reliable paper and pencil (with eraser) system, possibly augmented with a hand calculator. However, we need to recognize the pervasiveness of electronic computers and their use in solving decision problems. Thus, based on the backgrounds and needs of the students, we encourage the instructor to integrate, as much as possible, the course material with the local computer/software systems. This can be done in a number of ways. If the course involves computer programming (or requires it as a prerequisite), the students can test out their understanding of an algorithm and their programming abilities by writing codes for solving the related decision problems. The numerical exercises could then be solved using these codes. Software packages for some models may already be available, for example, in the case of linear programming. I have found the use of an interactive linear-programming system, such as LINDO (Linear, Interactive, and Discrete Optimizer), to be a valuable instructional aid. I would like to stress, however, that the complete reliance on computer software to solve problems without attempting to understand the associated algorithmic process limits one's future decision-making skills. Such reliance inhibits the ability to raise the right questions as to the structure and assumptions of the algorithm and related model, and leads to blind acceptance of computer printouts.

The level of mathematics required of the student is introductory mathematics, including linear equations and graphs, that is, deterministic mathematics at an elementary level. Introductory probabilistic notions are assumed, but they are not used extensively and can be introduced by the instructor. My target audience is undergraduates, certainly at the junior and senior levels, and advanced lower-division students. The text is for a one-semester course.

In sum, this book presents my approach to how undergraduate students in mathematics, business, computer science, and engineering should be introduced to the science of decision making. It is not a complete review of the subject matter. I have discussed those issues and problems that have been important to me as an analyst and appeal to me as a teacher. The material is designed to prepare the student for more advanced topics. No matter what career path is followed, the concepts of scientific decision making will prove to be of great value.

A text such as this is based on many sources. I would like to acknowledge my indebtedness to the many scientists whose works have formed the basis of my experiences and knowledge. I have tried to give complete references throughout. The explicit and subliminal use of their work is greatly appreciated. The student is encouraged to investigate the original sources to obtain a first-hand view of the ideas and models discussed in the text.

The ever-present encouragement and patience of my wife, Trudy, has always been the fuel that has enabled me to continue—there has never been any shortage here.

SAUL I. GASS

Potomac, Maryland
March 1985

Contents

PART I. A FRAMEWORK FOR DECISION MAKING 1

1. Decision Making **3**

1.1. The Science of Decision Making, 3
1.2. Problems and Decisions, 4
1.3. The Making of Decisions: You and the Experts, 7

2. Systems, Models, and Algorithms **9**

2.1. Systems and Decisions, 9
2.2. Everyday Models, 11
2.3. The Many Definitions of a Model, 13
2.4. The Types of Models, 14
2.5. Elements of a Model, 16
2.6. The Decision Problem, Models, and Algorithms, 18
2.7. The Abstract Decision Model, 20

3. The Decision Framework **23**

3.1. Decision Problems and the Scientific Method, 23
3.2. Newton as an Urban Planner, 24
3.3. The Decision Process Steps: The Decision Framework, 26

4. Our First Decision Model: The Transportation Problem **29**

4.1. A Slight Digression: On Getting Dressed, 29

4.2. The Statement of the Transportation Problem, 32
4.3. The Transportation Problem Model, 34

5. Part I. Discussion, Extensions, and Exercises 39
—which includes tomato sauce and pizzas; the general transportation problem and some variations; heuristic and optimizing transportation problem algorithms; measures of effectiveness; assignment of students to classes; urban planning and other gravity models; bounds for the number of guards in museums; when to trade in your automobile; on designing a model; heuristic methods; A Tale of Two Alternatives; and Spike's Choices.

Part I. References 60

PART II. THE LINEAR-PROGRAMMING MODEL: APPLICATIONS 63

6. What's for Breakfast?: The Diet Problem 67
6.1. Krispies and/or Crunchies, If You Must Know, 67
6.2. The Model and Its Selections, 69
6.3. Down on the Farm, 72

7. Mad Hatter, Inc.: The Caterer Problem 74
7.1. Two of a Kind: The Consultant and the Mad Hatter, 74
7.2. The Report, 77

8. I'll Take the Middle Slice: The Trim Problem 88
8.1. How to Make and Cut Cellophane, 88
8.2. Saving Trim the Mathematical Way, 90
8.3. Epilogue, 94

9. You're in the Army Now: The Personnel-Assignment Problem 95
9.1. Camp LP with Able, Baker, and Charlie, 95
9.2. The Assignment Model, 98

10. The Simple Furniture Company: The Activity-Analysis Problem 100
10.1. The Meeting of the Board, 100
10.2. One More Question, Please? 106

11. **Part II. Discussion, Extensions, and Exercises** **107**
—which includes the standard form of the linear-programming problem; the Junior Class's investment problem; mixing nuts; eating a better breakfast; nutritional recommended daily allowances; your all-day diet; competition from the Prince's laundry; making furniture and baked goods; scheduling police; cutting paper; planning menus; getting married; finding a job; greedy algorithms; fighting fires; outfitting a space laboratory; set-covering and set-partitioning problems; loading trucks and aiming missles; solving the museum director's problem; and the first linear program.

Part II. References **134**

PART III. SOLVING LINEAR-PROGRAMMING PROBLEMS: THE MODEL AND ITS ALGORITHM 137

12. **Single-Variable Problems (For the True Beginner)** **139**

12.1. How Simple Can We Get? 139
12.2. An Example: Solved by Inspection, 140

13. **Two-Variable Problems** **142**

13.1. Finding Our Way in 2-Space, 142
13.2. Examples: Solved by Enumeration, 143
13.3. Some Definitions, 145

14. **A Manufacturing Problem** **147**

14.1. Making Chairs and Tables: Graphical Techniques, 147
14.2. Should We Also Make Desks?: The Dual Problem, 152

15. **The Diet Problem (Again)** **156**

15.1. The Right Mix, 156
15.2. The Pill Peddler, 158

16. **The Simplex Algorithm** **159**

16.1. Inequalities to Equations, 159
16.2. Solutions and Extreme Points, 162
16.3. George B. Dantzig and His Simplex Algorithm, 163
16.4. Applying the Simplex Algorithm, 163

17. Part III. Discussion, Extensions, and Exercises **174**

—which includes more on the simplex algorithm; unbounded linear programs; minimizing; degenerate problems; cycling; multiple optimal solutions; finding a first feasible solution; artificial variables and artificial bases; the two-phased simplex algorithm; exercises and more exercises; finding a solution to the dual problem; nutrient pills for astronauts; sensitivity analysis; and the efficiency of the simplex algorithm.

Part III. References **203**

PART IV. NETWORK AND RELATED COMBINATORIAL PROBLEMS 205

18. Network-Flow Problems **207**

18.1. Basic Network Concepts, 207
18.2. Getting the Most: Maximal Flow, 208
18.3. Spending the Least: Minimal Cost, 212
18.4. From Here to There: Shortest Route, 214

19. The Traveling-Salesman Problem **218**

19.1. The Expense Account and the Horse, 218
19.2. Can We Solve It? The Problem, 220

20. The Transportation and Assignment Algorithms **224**

20.1. Properties of the Transportation Problem, 224
20.2. The Simplex Transportation Algorithm, 232
20.3. The Hungarian Method for the Assignment Problem, 243

21. Part IV. Discussion, Extensions, and Exercises **250**

—which includes a maximal-flow algorithm; networks with many sources and sinks; another shortest-route algorithm; acyclic networks; longest route; PERT; routing bombers; leasing trucks; traffic flow; transshipment of automobiles; a mathematical model of the traveling-salesman problem; random cities; heuristic algorithms for the traveling-salesman problem; Hamiltonian circuit; branch and bound for the traveling-salesman problem; Euler circuit; the Chinese postman problem; sensitivity analysis for the transportation problem; network models for the transportation problem and the caterer problem; production and inventory problems; the many-salesmen problem and algorithm; routing salesmen and trucks; minimal-spanning tree and greedy algorithms; traveling through the lower-48; and branch and bound for integer-programming problems.

Part IV. References 320

PART V. GAMES, TREES, AND DECISIONS 323

22. The Theory of Games 325

22.1. Basic Game Concepts, 325
22.2. How the Battle Got Its Name, 326
22.3. Matrix Games: The Model, 329
22.4. At the Carnival: The Skin Game, 336
22.5. Games and Life, 339

23. The Fruits of Decision Trees 341

23.1. The Rational Decision Maker: Utility Functions, 341
23.2. Uncertain Money: Expected Monetary Value, 344
23.3. Future Money: Expected Present Value, 345
23.4. Soybeans or Corn: A Decision Tree, 347
23.5. Business Investment: A Bigger Decision Tree, 350

24. The Analytic Hierarchy Process 355

24.1. Goals, Criteria, Alternatives: The AHP, 355
24.2. Will You Ever Buy That New Car? 364
24.3. Choosing the Right Job, 366

25. The Decision Framework, One More Time 368

25.1. Modeling Steps: An Expanded View, 368
25.2. Is the Model What We Think It Is? 370
25.3. Final Words, 372

26. Part V. Discussion, Extensions, and Exercises 373
--which includes game-theory primal and dual linear programs; solving 2 × 2's; "paper-scissors-stone" game; the farmer versus nature; marketing hi-fi's; negotiating union contracts and the price of wheat; the environmentalists versus the developers; randomizing the plays of the game; bimatrix games; "The Prisoner's Dilemma"; constructing utility functions; to fix or not to fix your old car; which horse to race; investment decisions; where to have

the cocktail party; expected values; game trees; The Petersburg Game; eternal happiness; the AHP consistency index; computing eigenvalues and eigenvectors; measuring areas and weighing bags with the AHP; conflict resolution and other uses of the AHP; establishing model validity; let's sing a song; and the last words.

Part V. References 398

Combined References 400

Index 407

DECISION MAKING, MODELS AND ALGORITHMS

PART I
A Framework for Decision Making

Chapter 1	Decision Making 3
Chapter 2	Systems, Models, and Algorithms 9
Chapter 3	The Decision Framework 23
Chapter 4	Our First Decision Model: The Transportation Problem 29
Chapter 5	Part I Discussion, Extensions, and Exercises 39

1 Decision Making

1.1 THE SCIENCE OF DECISION MAKING

All of us have encountered situations and problems that have required the making of a decision—the choosing between competing opportunities or alternatives. Most everyday problems are resolved without too much difficulty or without serious consequence if, from among the many available alternatives, the "correct" one is not selected. There are problems, however, for which we, as decision makers, want to do our utmost to ensure that the best possible solution is chosen. Such problems occur in most professions, in business, industry, and government, as well as in some personal situations. The medical doctor wants to prescribe the best treatment, the architect wants to produce the best design, the student wants to select the correct career, a company needs to decide where to build the new warehouse, a pension fund manager has to determine what stocks to have in the fund's portfolio and when to buy and sell, a city manager has to select the location of the new incinerator and sanitary landfill. These are hard choices and decisions.

Since the 1940s, our ability to understand, structure, and resolve decision problems has improved greatly. This is due to the increased study of applied problems by mathematicians and other scientists, the development of new mathematical techniques, and the power of the electronic computer. A new science of decision making has been evolving. This has produced a set of ideas, approaches, and procedures that can be considered to form a modern framework for decision making. In this book, we shall describe our view of this framework and focus attention on its centerpiece—the mathematical model. But first, some words of caution.

It is claimed that decision making is more art than science and that intuition and experience are the main resources of a decision maker. While we are in no position to refute this view, we do believe that many decision-making situations can be understood better by the application of the more disciplined approach to problem analysis that is imposed by the framework. With better understanding come better decisions.

Although the decision framework can be used in just about any decision situation, its abilities as an aid in solving certain types of problems are limited. This limitation is due to our inadequate knowledge of the underlying physical or behavioral phenomena.

The science of decision making is still developing. Opportunities exist for advancing and improving decision making in all fields of endeavor. However, we feel that students can benefit and learn from the present, incomplete state of decision analysis methodologies. We next discuss the general areas of problems and decisions, mathematical models and modeling, and a framework for decision making.

1.2 PROBLEMS AND DECISIONS

Let us consider the following situation. A museum director has been presented with two floor plans for a proposed art gallery. The first plan uses the available one-story rectangular site in the standard gallery form that has rooms interconnecting with one another (Plan One). The second plan makes

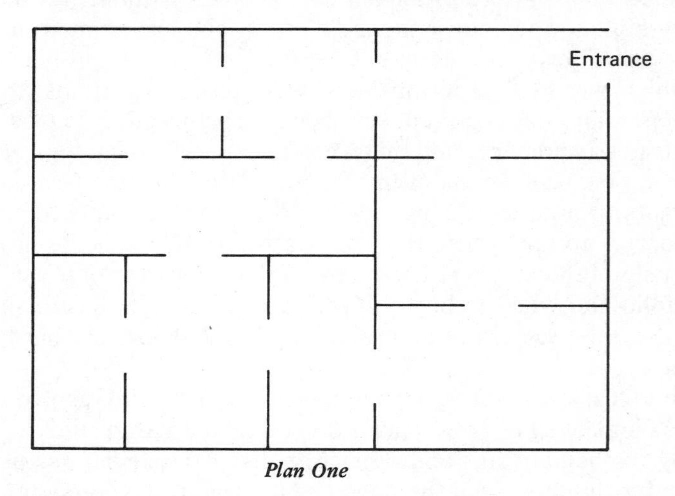

Plan One

a more daring use of the space and consists of one big room whose walls zig and zag to form an irregular polygon (Plan Two).

The competing architects have shown that each plan meets the exhibit wall space requirements and the budgets for construction and future maintenance. One other recurring cost must be considered: the payroll costs for guarding the exhibit space. The director would like to know the minimum number of guards required to safeguard the rooms of Plan One or the walls of Plan Two. All other things being equal, the plan with the fewest guards would then be chosen.

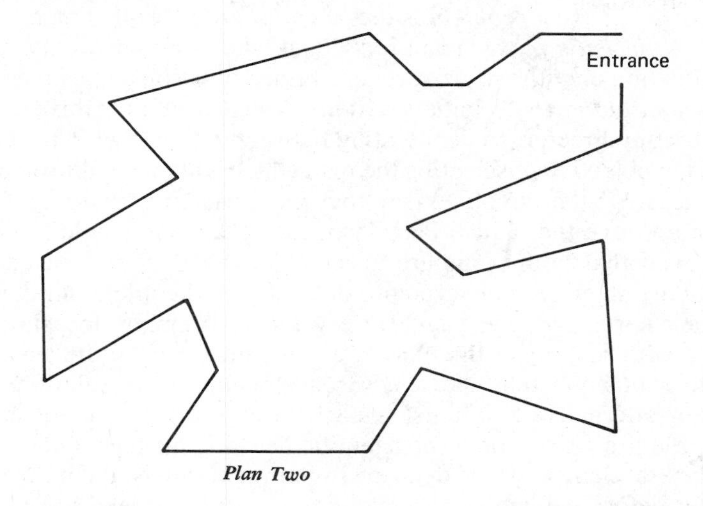

Plan Two

The museum director is faced with a basic situation which defines management: *decision making*. Decision making consists of recognizing that a problem exists, identifying possible causes, developing alternative solutions, choosing among alternative courses of action, and carrying out the action chosen.

The director has a problem to solve—a decision to make and to implement. But to do so, requires more information, that is, the cost of each plan's guards. By what means can the problem be analyzed and the information developed?

Any procedure should determine the minimum number of guards for each plan. Hopefully, the analysis can go further than just calculating these numbers. What if the director wants to add a room or a doorway in Plan One? What if a zig or zag is added or removed from Plan Two? How do such changes influence the number of guards required and the selection decision? If the framework of analysis can handle these variations, it would be of greater value to the questioning mind of the director.

The museum director's problem (which we will return to from time to time) illustrates the more general problem and decision-making concerns that are the subject of this book. *Problem solving* is that form of activity in which an individual or organization, wanting to achieve a desired objective, must make a selection from the set of alternative ways for accomplishing that objective. A decision problem is nonexistent if there is no way to accomplish the objective or if there is only one way. In either of these situations, there is no decision to be made—for, as we note from *Webster's Dictionary,* a decision is a conclusion arrived at after consideration. In the first case, no actions can be taken; while in the second, there is only one thing that can be done.

For our purposes, then, a problem is a problem because it has alternative solutions. The *decision problem* is the selection from competing alternative solutions. A *decision maker* is an individual who is dissatisfied with some existing situation or with the prospect of a future situation and who possesses the *desire and authority* to initiate actions designed to alter this situation.

The museum director, in deliberating between alternative Plans One and Two, has the objective of selecting the plan which will make future operating costs the least. If the director knew that the museum's board of directors would not approve the radical Plan Two, then Plan One would be the only choice. Given that both plans are acceptable, the selection based on the operating-cost objective converts into determining the minimum number of guards for each alternative plan. Here we have a simple measure for an alternative with respect to the objective. This measure can be used by the director to compare, rank, and select from the alternative solutions. These concepts of alternative solutions, measures, and alternative selection are central to the process of modern, scientific decision making. They combine to form the key element of the decision framework, that is, the mathematical model. But before delving into the definition, structure, and use of mathematical models, we look once more at the process of decision making.

"Now, that's a welcome sight! I was just beginning to miss decision-making."

1.3 THE MAKING OF DECISIONS: YOU AND THE EXPERTS

At this stage of your career, you probably have not been associated with a business enterprise in which you were required to make decisions. But as you read the following material, look back at your work experiences and home, school, and social activities that involved your making decisions. For example, how did you decide to go to college and on what basis did you select the school you are attending? You might not of thought of it in this way, but you are a decision maker. Like Monsieur Jourdain in Molière's play *Le Bourgeois Gentilhomme* who suddenly discovered that he had been talking prose for 40 years without knowing it, you have been a decision maker all of your life. So spend a few moments to review some of the decisions you effected and try to reason out why they were made and what makes a good decision and decision maker. Now let's see what some experts have to say about such matters.

In his book, Robert Townsend, the ex-president of Avis car rental, gives practical advice on decisions. His short chapter on decisions follows:[1]

> All decisions should be made as low as possible in the organization. The Charge of the Light Brigade was ordered by an officer who wasn't there looking at the territory.
>
> There are two kinds of decisions: those that are expensive to change and those that are not.
>
> A decision to build the Edsel or Mustang (or locate your new factory in Orlando or Yakima) shouldn't be made hastily; nor without plenty of inputs from operating people and specialists.
>
> But the common or garden-variety decision—like when to have the cafeteria open for lunch or what brand of pencil to buy—should be made fast. No point in taking three weeks to make a decision that can be made in three seconds—and corrected inexpensively later if wrong. The whole organization may be out of business while you oscillate between baby-blue or buffalo-brown coffee cups.

Townsend, in a sense, defines how an organization should delegate its decision-making authority and how it should operate when faced with a range of decision problems. This approach to organizational problem solving has been addressed more formally in the books by March and Simon, and by Gannon.[2] Organizational problems can be classified as structured or ill-structured. A *structured problem* can be solved by means of a *programmed decision* that applies a decision rule or routine. An *ill-structured problem* can-

[1] From Robert Townsend, *Up the Organization,* Knopf, New York, 1970. Copyright © 1970 by Robert Townsend. Reprinted by permission of Alfred A. Knopf, Inc., New York, and Michael Joseph Ltd., London, publishers.

[2] Book and other references are given at the end of each part, with all sources listed in the final Combined References Section.

not be solved readily by a programmed decision. The decision maker then must make a *nonprogrammed decision* in terms of a creative response. Most nonprogrammed decisions are made at the top echelons of an organization, with the programmed decisions usually a function of lower management. The museum director's decision problem is a structured one as there is a specific decision rule for selecting between plans, although how to determine the number of guards may not be clear, as yet. The director's original problem of designing a new gallery can be considered an ill-structured one. As exhibit space requirements, cost constraints, esthetic considerations, and so on, were addressed and resolved, the problem became more structured.

Whether a decision is the right thing to do cannot be determined unless it is carried out. But even so, as the future is uncertain no matter which alternative is chosen, what looks like a correct decision today may turn out to be the wrong one tomorrow. Alternatives not selected or not implemented cannot be evaluated as to their ultimate impact or correctness. What is the decision maker to do? "Once a decision was made, I did not worry about it afterward," is the advice from President Harry Truman. Professor Harvey Wagner states, "The only thing you control is your decision prior to the uncertain outcome." Implementation of a decision is certainly important. But in our effort to choose what appears to be the correct alternative, emphasis must be placed on the selection process—in the development and care taken in analyzing alternative solutions and the means for comparing them.

We will be concerned with problems, alternative solutions, decisions, and the process of what can be termed the *science of decision making*. Science here is used in the sense of knowledge or principles obtained by study and practice. Modern decision analysis applies this knowledge to provide decision makers with a quantitative basis for choosing among alternative solutions. This process is sometimes referred to as rational, objective decision making. However, we recognize that many decision problems and their resolution cannot be divorced from the biases, experiences, and prejudices of the decision maker. Also, not all facets of a problem can be quantified.

We hope that a decision maker would behave in a rational manner and correctly use the results of a decision analysis. This does not mean that a decision maker cannot or should not use intuitive, judgmental, and experiential factors in making a decision. It does mean that the decision maker should use all the information available and attempt to analyze it in an intelligent and sensible manner. Our approach to decision making is a rational one—but it is softened by the following quote by the psychologist Carl Jung:

> The great decisions of human life have as a rule far more to do with the instincts and other mysterious unconscious factors than with conscious will and well-meaning reasonableness. The shoe that fits one person pinches another; there is no recipe for living that suits all cases. Each of us carries his own life-form—an indeterminable form which cannot be superseded by any other.

2 Systems, Models, and Algorithms

2.1 SYSTEMS AND DECISIONS

A decision is never without consequences. Decision problems involve interactions between people, machines, organizations, raw and processed materials, buildings, and so on. The total scope and elements of a problem can be viewed as forming a system. A *system* is an arrangement or grouping of objects that operate together for a common purpose. Everything excluded from the system is called the *system environment*. In a very broad sense, a system's environment is the rest of the world. Most problems do not have such a grandiose geopolitical outlook; the environment is a local one. For example, the director's museum system includes the present resources (buildings, art, personnel, funds, etc.) and those additional resources associated with the new gallery. The system's environment can be defined to be the rest of the community. This interpretation of the museum system and its environment may be the correct one for the gallery-design problem. The resolution of the problem depends only on the information content of the museum system as defined. However, if the director wanted to study the problem of how to increase the number of visitors to the galleries, then the museum system would have to be redefined to include the local populace, tourists, and other elements. An important step in defining a problem is to determine the set of interacting elements that forms the system under investigation, that is, we need to determine the boundary between the system and its environment; we assume the boundary fixed for the problem setting.

An organization can be viewed as a system consisting of subunits that not only interact with one another, but also are heavily dependent upon one another. They range from simple repair shops to integrated petrochemical companies to governmental bureaucracies. Systems can be completely automated and include equipment and hardware (a ballistic missile system), or consist mainly of personnel (a university educational system), or a combination of human and nonhuman resources (a manned space system). We consider any element of a system (human or otherwise) to be a part of the system's resources.

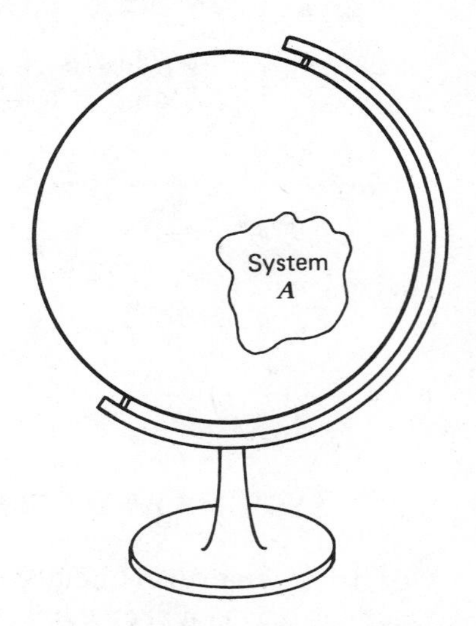

It is difficult to change an existing system. For example, we usually cannot experiment with or disrupt a working production line. In other instances, it is impractical or uneconomic to test alternative solutions. The museum director cannot build two art galleries; an incinerator location cannot be tried out. But to solve a problem, alternatives must be evaluated and compared, a choice must be made, and the selected solution must be implemented.

Some decision problems involve systems that are conceptual or nonexistent. They may or may not be built depending on the results of the analysis. Does the community need a new hospital (system), and if so, where should it be located? Can a proposed nuclear power plant be built that will serve the extended electric utility region and protect the local environment? Such decisions must be made prior to the construction of the proposed system.

For most decision problems, an experimental setting cannot be imposed upon the related system. We are constrained in any attempt to evaluate alternative solutions using the real-world system as a test bed. We must then make a deliberate effort to abstract the information that describes the problem and system. The information must then be organized to form a substitute for the actual or contemplated system. This abstracting, that is, the gathering and organizing of information, is the basic process for all decision making. If conducted properly, it will lead to a clear and concise statement of the problem; an understanding of what alternative solutions, if any, are possible; and a means of choosing among alternatives.

Information is used in a most general sense. For a decision problem it includes the system definition, resources, constraints (economic, political, organizational), and related data.

The hope is that once this information has been gathered (not an easy task for most problems), a structure or framework can be developed as an aid to the analysis. Certain principles for analyzing the information content of decision problems have evolved over the past few years. They combine to form the powerful concept of *decision-aiding models*.

2.2 EVERYDAY MODELS

The concept of a model is a familiar one. Most of us have played with toys and games that are models of the real thing: cars, airplanes, dolls, Monopoly. Depending on the purpose and the cost of a model, it more or less mimics or approximates the behavior of the real item. A rubberband airplane is a poor representation, while a radio-controlled one that can do loops is quite exciting and real to a 10-year-old child. A doll that crys and wets (ugh!) has proved to be an acceptable model of a live baby. And most of us have learned some truths about real-estate transactions and the capitalistic system by playing Monopoly.

When we leave the realm of toys, we find that models are a serious business. An architect builds a detailed scale model of a new museum to see how design ideas interact; an engineer builds a model airplane for wind-tunnel tests to determine the aerodynamic properties of a new wing; a clothes' designer tries out a concept by outfitting a fashion model with a new creation. Using a model, the architect can envision how the real museum will look and function in the true setting; the engineer knows the accuracy by which the wind-tunnel experiment predicts the real world of flight; and, hopefully, the future success of the clothes' designer can be measured by how the women shoppers, as idealized by the fashion model, will look in the new styles. These models, and the ones that we shall be interested in, are used as surrogate means for understanding how the real world works and as aids in predicting physical and human behavior.

Care must be taken to ensure that the type of model used is a valid one for the situation being studied. We do not have trouble accepting architect models or wind-tunnel experiments or fashion models. But what of a bridge builder? Would we give the go ahead if we were just presented with an erector-set model of how the bridge would look? In fact, we could dispense with a physical model of the bridge (a drawing would do), but we would insist upon a mathematical statement and calculations that showed how the stresses and strains were safely accommodated by trusses, cables, foundation, and so on. The mathematics of the bridge design, that is, its mathematical model, might be more difficult to appreciate, but it is the proper and correct type of model for predicting how the actual bridge would behave in its real-world setting.

Of course, we are not interested in only looking at models, just constructing them out of plaster or cloth or mathematical symbols. Models are used to try out new ideas and to measure design changes against each other. For example, the clothes' designer might be interested in knowing the material and labor costs of producing two ''great-looking'' lines of fashion. The architect can use the model of the museum to measure the esthetics of com-

peting designs, and to obtain some idea of how construction costs will vary. As usual, Shakespeare has the words for it:

> When we mean to build,
> We first survey the plot, then draw the model;
> And when we see the figure of the house,
> Then must we rate the cost of the erection.
>
> (*King Henry The Fourth, Part II*)

Models are used as aids in the understanding of a problem. As an aid to decision making, a properly constructed and valid model can predict the outcome of each possible solution. Thus, we can establish some scale that allows us to compare alternative choices and to select one to implement. The selection criterion can be cost, or some measure of efficiency, or even a judgmental and intuitive one.

2.3 THE MANY DEFINITIONS OF A MODEL

At this point, you might well ask, "But what is a model?" We have hinted that it is some representation of reality used to understand and study a complex problem. As this is a text that relies on the use of mathematics, we must be more precise. To accomplish this, we first present a short list of definitions of a model. We then indicate the definition that we think is most satisfactory for our purposes. We start with the following quote by the applied mathematician, Rufus Isaacs:

> Whenever an applied mathematician tackles a real-world situation, he makes simplifying assumptions. This simplified version of reality is called a *model*. It is a most convenient word, for it is his only recourse.
>
> *The human mind is incapable of thinking other than about models.*

Similar definitions are:

> A model is an idealized representation of reality describing some phenomenon whose behavior is to be highlighted. (Ackoff)
>
> A model is an abstract description of the real world. It is a simple representation of more complex forms, processes and functions of physical phenomena or ideas. (Rubinstein)

These definitions stress the simplifications required when we attempt to represent a complex situation; they imply that the representation—the model—should be accurate enough to be used instead of the real thing. As noted earlier, the real thing may be an ongoing system, but we cannot experiment with it; or it may be a proposed system. As we shall use models as aids to decision making, a more operative definition, and the one we like, is the following:

> Any subject using system A that is neither directly nor indirectly interacting with system B to obtain information about system B is using A as a model for B. (Apostel)

System B is termed the *reference system* and it is being modeled by system A. For example, the model airplane in a wind tunnel (system A) is used to understand the flight characteristics of the proposed aircraft (system B).

The models that we are concerned with are not constructed out of wood, concrete, or cloth. We use mathematics (and sometimes the languages of English and computer programming) to describe our models. To be more specific, we can alter the last definition to read as follows:

> A (formal) model of a given reference system is another system expressed in a (formal) language and synthesized from representations of selected elements of the reference system and their assumed interrelationships (Greenberger, et al.)

This last definition brings to the forefront a key aspect of model building when it stresses ". . . selected elements of the reference system and their assumed interrelationships." As we study a system—an ongoing one or a proposed one—we need to develop a model that is valid, that is, it must be capable of mimicking or behaving like the real reference system as we understand it. For example, the mathematical relationships that model a space rocket's flight to the moon are very exact in terms of telling us what decisions to make to control the flight. These relationships have been developed and refined by observations, experiments, and analyses of data. In contrast, the interrelationships of many systems of business and government are not that well understood. As we attempt to model these areas, we must be extra careful in stating and explaining our assumptions and in demonstrating that the assumed structure of the model is valid. An example is that of a model used to select from alternative advertising strategies. Such a model is usually based on relationships between exposures to an advertising medium and corresponding behavioral responses by the audiences. The forms of these relationships need to be hypothesized, tested by experiments or surveys, and validated to be correct for the purposes of the model.

The types and uses of models are manifold. We have already introduced the broad class of decision-aiding models. For completeness, we describe the following terms as they are used for models.

2.4 THE TYPES OF MODELS

Models have been classified into three basic types. The *iconic model* looks like what it is supposed to represent. It could be an architectural model of an apartment complex or a planetarium representing the celestial sphere. The *analogue model* relates the properties of the entity being modeled with other properties that are both descriptive and meaningful. For example, the concept of temperature is described by a graph in which a degree is equivalent to a specified unit of distance; or the concept of time as described by the hands and markings of a clock. Finally, the *symbolic model* or the *mathematical/logical model* represents a symbolic description of the process or problem under investigation. Einstein's famous equation $e = mc^2$ states, in symbols, that the energy e contained in a mass m is equal to the product of the mass and c^2, the square of the velocity of light. The mathematical model represents the translation of the statement of the problem into quantitative terms.

A model can be *predictive* in that it was developed for the purpose of forecasting the future. Many models that deal with national economic relationships are of this type, for example, predicting the gross national product or total unemployment assuming a certain level of federal expenditures or monetary policy. A model can be *normative* in that it can be used as an aid in identifying solutions that show what can be done to accomplish stated objectives or norms of operation. Such solutions represent baselines or standards against which proposed new solutions can be compared. For example, a manufacturer who maintains a single warehouse to serve the total country might wish to determine if customer service can be improved by a multiwarehouse, decentralized inventory system, and at what additional cost. To do this, a normative model or description would be required that defines the relationships that represent the manufacturer's inventory system and measures the current level of customer service and cost. Such models can also be *descriptive* or *procedural* as they describe the basic elements and interrelationships of a system. Descriptive models aid us in understanding the physical and behavioral aspects of the reference system. This type of model can also be used to try out proposed changes in the system, that is, as an experimental laboratory. Given a descriptive model of the manufacturer's centralized inventory system, proposals could be "modeled" to determine if improved customer service could be obtained without a new multiwarehouse system. Finally, a model can be *prescriptive* in that it can be used to prescribe a solution. If the manufacturer decides to decentralize the inventory operation, then choices must be made as to the locations of the new warehouses. A prescriptive model of this decision problem would have a means of comparing locations in terms of customer service, cost, and other measures, thus enabling the best configuration of warehouses to be selected. A solution would be best or optimal based on the measures that are important to the manufacturer.

For our decision situations, the model structures will be mathematical and prescriptive. However, we should not give too much importance to descriptive qualifiers. A model by any other name is still a model.

Some of you might still be asking "Why a model?"—to paraphrase the

Marx Brothers.[1] We need a basis upon which to make informed decisions. A properly conceived and clearly explained model affords us a consistent means of evaluating possible solutions. Our problem situations are complex. We need a way to sort out, arrange, and simplify the complexities to obtain a better understanding of the problem . As Isaacs emphasized, we can think only with models. We all have *mental models* or *internal models*. These are what we carry in our heads and are our implicit representations of real-world systems. Given nothing else to work with, decision makers must rely on their mental models. This is not to say that such models are not proper and cannot do the job. In fact, in many instances, we derive an explicit, external, mathematical model based on our ability to translate the decision maker's mental model. The mathematical representation, however, enables us to test the assumptions, vary data, and measure changes in an orderly manner. The aim here is to develop a decision-aiding model that would be of utility to the decision maker. The resulting model could replace the mental model or be used in conjunction with it. In either case, enough evidence must be presented to convince the decision maker that the external model can be used with confidence.

Many decision makers rely on their mental models and their intuition to make decisions. The human mind's ability to resolve situations by intuition based upon experience is not well understood, but is quite remarkable indeed. A mathematical model should try to encompass, explain, and extend intuitive concepts. A model might challenge our intuition about a system, but any counterintuitive results only mean that we did not understand correctly the problem complexities in the first place. We should be able to use the model to resolve any inconsistencies. With the above discussion in mind, we next review and extend the notion of a model (as described by the management scientist Roger L. Sisson).

2.5 ELEMENTS OF A MODEL

A model is a way of abstracting the real world so that not only the static picture of the world is obtained, but also the dynamic (time) interrelationships. With an appropriate model of a real-world situation, we should be able to predict certain outcomes or determine how the real world would behave if we implemented a particular alternative decision. In some instances, a model enables us to select the *best or optimal decision*.

Models used in decision making tend to be quantitative; the interrelationships are expressed in numerical terms, for example, the relationship between a coal-burning source and resultant pollution per ton of coal burned. However, techniques exist for representing some qualitative factors.

Models have two major components: *variables* and *relationships*. In many

[1] See their movie "The Cocoanuts."

real situations, it is possible to enumerate thousands of relationships and/or variables. The skill of the model builder enables the builder to capture the essence—only the important variables and relationships—so as to produce a meaningful and useful model. The creative input to the modeling process is in translating a perception of the world into the essential relationships and variables, and thus into a model which is tractable and, hopefully, computationally manageable.

In the model, the variables represent either (1) values (numbers which are real-world levels, counts, measurements, budget allocations, physical quantities, the results of survey measurements, etc.) or (2) codes (which identify projects, items, units, people, groups, proposals, etc.). The codes in turn can take on values such as zero if a project is not to be funded and a value of one if it is to be funded.

The relationships (equations, constraints) are expressed as procedures for computing the values of certain variables, given values of others; usually for computing the values of variables at a future time given the values at a present time. These computational processes represent the real interactions characterizing physical interconnections (the wear on a vehicle as it is used), economic relationships (the relationship between the demand for an item and its price), behavioral interactions (the effect of information produced by one group on the actions of another group) and many others.

In nearly all models it is found that the variables can be classified into four categories.

1. *Controllable or Decision Variables.* These are variables whose values can be determined by the decision process. (For example, in a vehicle replacement problem, the factor we can determine or control is when to trade in the old unit for a new one.)

2. *Uncontrollable Variables.* These are variables that are not under the control of the decision maker. Uncontrollable variables include physical phenomena (e.g., wear on a vehicle and, therefore, frequency of maintenance), variables which result from the actions of others, or certain economic parameters determined by the actions of the production and consumption processes in the country (e.g., price of a new vehicle). In a sense, the uncontrollable variables represent the "state of the world" as perceived by the model.

3. *Result or Output Variables.* These are variables characterizing the results of processes in the real world. They are usually defined by controllable variables. (In the vehicle replacement situation, a main output is the total cost of operation.)

4. *Utility or Value Variables.* The decision maker will set a utility or value on the results of the process. The value is a function of controllable and uncontrollable variables.

Uncontrollable variables are of two types: those whose values are computed in the modeling process and those which are inputs to the model. The latter represent the effect of the environment on the system—the action of the world outside penetrating the system's boundaries. For the model to operate, it is clearly necessary to obtain estimates of the values of these input variables over the time span of interest. In some cases, these are known from existing data (such as the average cost of a particular kind of maintenance operation). In other cases, data about the future are required (such as the amount of trash that will be generated in a given area per week over the next 5 years). Estimating values of these uncontrollable variables may be done by others, such as economic agencies or the Census Bureau, or the analyst may use forecasting processes subsidiary to the model itself. Some modeling theorists believe that every model should be self-contained and should not require subsidiary forecasts or inputs. Such models, which include their environments, are termed *holistic*.

With uncontrollable variables there are degrees of uncertainty; the estimates are never exact. Sometimes the estimates are sufficiently accurate and have a sufficiently small amount of variation. We can then assume that the uncontrollable variables take on specific values. This leads to what are called *deterministic models*; the uncontrollable variables are assumed to be determined.

More often some of the uncontrollable variables are difficult to estimate or there is a variability which exists in the real world. (The amount of refuse generated in an area is not constant but varies from week to week.) In these circumstances, the most accurate model (not necessarily the most useful) will represent the variables as statistical quantities and take the variations specifically into account. These are known as *probabilistic or stochastic models*.

The utility or value variables are computed by a formula from the result variables; the formula is called the *objective function* or *measure of effectiveness*. Note that the assumption is that there is *one* measure of value. In some studies this value is measured in dollars. However, in many situations, particularly in governmental applications, the measure of value is more complex. Indeed, in many cases there is no single measure of the utility of the outcomes (this is termed a *multiobjective* situation). No matter what is chosen as an indication of value, it is an important variable to be computed in the modeling process. The controllable variable values which give the best available utility are said to be the *optimum*.

2.6 THE DECISION PROBLEM, MODELS, AND ALGORITHMS

In most cases the model builder assumes the *decision problem* to be of the following nature: *Find the values of the controllable (decision) variables which produce the best utility (value) as measured by the utility variable(s), given the assumptions about the uncontrollable variables.* For example, what is the time to replace a vehicle to obtain the least total cost—the utility measure—given data about vehicle price and depreciation, maintenance requirements and cost, and vehicle usage? The utility is computed from the result or output variables (costs, in the example) estimated by the modeling process.

All modeling processes allow us to compute present and future values of result variables. Some models (and their associated computational processes) enable us to determine the utility-optimizing values of the controllable variables. The latter is known as *optimization* and, when considering situations over time, the former is called *prediction*.

An optimizing procedure has two important characteristics: (1) it is a well-defined series of computational steps, which, for real problems, can be executed in a reasonable amount of time; and (2) there is a mathematical proof that guarantees that the optimal solution is found at the end of the computational process. There are some intermediate analytic tools whose procedures do not satisfy characteristic number 2, but the procedures can be shown to yield good, nonoptimal values. These are called *heuristic methods* in that they involve special logical, mathematical, and sometimes intuitive steps that are based on the structure of a particular problem. The steps have not been shown to lead to an optimal solution. The mathematician George Polya in his famous book *How To Solve It* notes that heuristics is a branch of study dealing with the methods and rules of discovery and invention. As an adjective, heuristic means "serving to discover." In a similar vein, Pearl states that "Heuristics are criteria, methods, or principles for deciding which among several alternative courses of action promises to be the most effective in order to achieve some goal. They represent compromises between two requirements: the need to make such criteria simple and, at the same time, the desire to see them discriminate correctly between good and bad choices ."

Any explicit set of computational steps used to solve a particular problem is called an *algorithm*. The computational scheme you use to perform long division is an algorithm. In terms of a decision problem's measure of effectiveness, an algorithm, if it is based on an optimizing methodology, will find an optimal solution; if it uses heuristic ideas, an algorithm will usually produce a nonoptimal but acceptable solution.

Once a model has been shown to be an accurate representation of the problem situation, it becomes a powerful experimental device. A valid model enables us to measure the effect of changes to the problem structure without modifying the real-world system being modeled. Thus, models are used to answer "*What if" questions* of the following types: (1) "What if we set the decision variables at certain values?" (e.g., "What if we traded the vehicle every five years?"); and (2) "What if an uncontrollable variable takes on a different value?" (e.g., "What if maintenance rates increase at 10% per year instead of 6%?")

When analyzing a decision problem, the model builder must first identify the variables and then construct relationships (equations, computational processes) that interconnect the variables. In the most ideal case, the work is facilitated by (1) theories about the physical and behavioral phenomena of the problem situation and (2) theories about the structure of the specific operational or management process being studied. The theoretical results, if available, indicate to the model builder what the most important variables are, what the relationships between them are (in precise, computational form), and what procedures can be used to find values of the decision (controllable) variables which produce the best value of the measure of effectiveness.

Typical decision situations that exemplify the use of models are:

☐ A local sanitation agency operates vehicles to collect refuse. One operational decision it must make is when to replace a vehicle. If the vehicle is allowed to get too old, maintenance costs become high. But there is a real cost in purchasing a new unit in terms of both the purchasing process and early, large depreciation in value. Thus, there should be some best or optimal period after which the vehicle should be replaced. This is a relatively well-defined operational decision for which models are routinely used.

☐ A metropolitan air-conservation commission has the problem of setting air-pollution regulations affecting the sulfur content of coal burned in the area. Models are used to evaluate the economic feasibility of controls on coal pollutants in the area's airshed.

☐ A telephone company must plan the routes of its pay-telephone coin collectors to minimize the travel time (cost) of the collectors. This is a restatement of the *traveling-salesman problem* (how to route a salesman through the territory to minimize the total distance traveled). Problems with many locations can be solved only by heuristic procedures.

☐ A manufacturer wants to ship goods from warehouses directly to customers at minimum cost. This is known as the *transportation problem* and can be solved exactly for the optimal (least-cost) solution.

2.7 THE ABSTRACT DECISION MODEL

The general mathematical model that describes our decision problem can be stated in a succinct and formal fashion. Although this statement is quite abstract, it does represent the decision situations we will be concerned with. It offers us a way to summarize the above discussions, especially the verbal statement of the decision problem.

We define the set of controllable variables by the notation $\mathbf{X} = (x_1, x_2, \ldots, x_n)$; that is, there are n decision variables. For example, a manufacturer must decide how many units to produce each week for the next $n = 5$ weeks. We let x_j represent the production level for the jth week, where $j = 1, 2, 3, 4, 5$. The manufacturer must select values for these five variables. The uncontrollable variables are denoted by the notation y_j, where a specific y_j

represents a known uncontrollable input to the decision problem. For the manufacturer, a y_j could be the expected customer demand that must be satisfied by the jth-week's production. In general, we let $\mathbf{Y} = (y_1, y_2, \ldots y_m)$ represent the m different uncontrollable variables that need to be considered. The measure of effectiveness or objective function of the decision problem, denoted by E, is represented by some known or assumed relationship that combines the impact of the uncontrollable variables with the selected values of the controllable (decision) variables. This relationship is given by the formal equation

$$E = f(x_1, \ldots x_n; y_1, \ldots y_m) = f(\mathbf{X},\mathbf{Y})$$

The equation states that if we are given values of the uncontrollable variables and values of the decision variables, we can then determine the effectiveness of the solution. The aim of the decision problem is to determine those values of the decision variables that will yield the best—the optimal—value of E. For the manufacturer, E could represent profits and the values of the production variables x_j would be determined to maximize profits.

The selection of each x_j is restricted by the constraints of the decision problem. In the manufacturing situation, the manufacturer's daily production might be limited by available raw material, labor, warehouse space, and other resources. The conditions that restrict the possible choices of the decision variables x_j are represented by the notation $\{\mathbf{X}\}$; that is, the *decision space* available to the decision maker. The decision maker wants to determine the optimal (maximum or minimum) value of the measure of effectiveness, subject to choices of the decision variables that are available in the decision space. Letting $\overline{E}$ be the optimal value of the measure of effectiveness, the *decision problem model* is given by

$$\overline{E} = \underset{\{\mathbf{X}\}}{\text{maximum}}\ f(\mathbf{X},\mathbf{Y}) \quad \text{or} \quad \overline{E} = \underset{\{\mathbf{X}\}}{\text{minimum}}\ f(\mathbf{X},\mathbf{Y})$$

In the succeeding chapters we will state and formulate in mathematical terms a number of problems that use the general decision models given above. In some instances it will be an easy matter to determine the set of constraints that form $\{\mathbf{X}\}$ and the objective $f(\mathbf{X},\mathbf{Y})$, while for others we will have to struggle and compromise to obtain a model that can aid the given decision situation. Our modeling task will be to determine the form and substance of the $\{\mathbf{X}\}$, $\mathbf{Y}$, and E, and to demonstrate that the resulting model is an acceptable representation of the reference system.

The mathematical model is central to the decision framework discussed in the next chapter. But first, to put the formal aspects of our discussion of models in proper perspective, we offer the following anonymous "Declaration of Model Independence" and a quote by the mathematician Mark Kac:

> We hold these truths to be self-evident: that all models are not created equal; that they are endowed by their creators with certain inalienable traits; that among these are their assumptions, source of data, choice of methodology, and the pursuit of theory and behavioral laws; that in execution of their function, the models are asked to produce results, deriving their powers from the consent of the users; that whenever any model fails to properly serve its function, it is the right of the users to alter or abolish it and to institute new models laying a foundation on such principles and assumptions in such form as to them shall seem most likely to lead to greater understanding and better decisions. (Anonymous)

> Models are, for the most part, caricatures of reality, but if they are good, then like good caricatures, they portray, though perhaps in distorted manner, some of the features of the real world.
>
> The main role of models is not so much to explain and to predict—though ultimately these are the main functions of science—as to polarize thinking and to pose sharp questions. Above all, they are fun to invent and to play with, and they have a peculiar life of their own. The "survival of the fittest" applies to models even more than it does to living creatures. They should not, however, be allowed to multiply indiscriminately without real necessity or real purpose. (Mark Kac)

3 The Decision Framework

3.1. DECISION PROBLEMS AND THE SCIENTIFIC METHOD

The study of modern decision problems has led to the development of formal and informal procedures used to guide analysts in their quest of acceptable solutions. Today (as well as in the early days) most analysts have backgrounds in mathematics, physics, engineering, economics, or other technical disciplines; their basic education is that of a scientist. Hence, the *scientific method* of investigation has always been a part of the methodologies of decision science. As we shall see, these concepts have been redefined and restructured to meet the different needs of decision problems, as contrasted with those of pure-science problems.

When we think of the scientific method we think of the first modern philosopher of science, Francis Bacon, and the works of Galileo, Newton, and Einstein. These scientists were concerned with discovering new facts about the natural world. Their general approach to the study of physical phenomena is given by four interlocking steps that comprise the scientific method (as stated by the physicist E. M. Rogers):

1. Make observations and extract rules or laws.
2. Formulate a tentative hypothesis (a guess, which may be purely speculative).
3. Deduce the consequences of the hypothesis combined with known laws.
4. Devise experiments to test the consequences.

If the experiments confirm the hypothesis, adopt the hypothesis as a true law, and then proceed to frame and test more hypotheses. If the experiments refute the hypothesis, look for an alternative hypothesis. When the hypothesis answers suitably to repeated tests, the scientist has made a discovery. As Rogers points out, however, real scientific inquiry is not so "scientifically logical" or simple.

The processes of science and the qualities of its investigators are assumed to be present in the methodologies and traits of decision scientists. However, there is one big difference. Unlike most physical problems, decision science problems involve human beings as part of the reference system. This basic difference causes two concerns for the decision analyst: (1) How can we establish the validity of hypotheses that involve people interacting with one another and with manmade physical systems?; and (2) How can we erase or take care of biases, hidden desires, and lack of objectivity in the people who are part of the system? These are serious concerns that the analyst must address when attempting to hypothesize and determine the structure of the decision model.

There is a danger in assuming that we can extend the successes of the physical sciences into the world of decision making. The models of the physical sciences describe, explain, and can be used to predict. We strive to have our decision models do the same thing. Care must be taken not to have the mathematization of a problem cover our lack of knowledge of how the reference system really behaves. To illustrate these concerns, we next discuss Newton and urban planning.

3.2 NEWTON AS AN URBAN PLANNER

Newton's Law of Universal Gravitation states that each particle of matter attracts every other particle with a force which is directly proportional to the product of their masses and inversely proportional to the square of the distance between them. For two particles or bodies the mathematical model of this statement is given by the force formula

$$F = G\frac{m_1 m_2}{d^2}$$

where F is the force with which either body attracts the other (the attractive force between the particles), m_1 and m_2 are the masses of the two bodies separated by a distance d, and G is known as the gravitational constant. The number G has been established experimentally by physical experiments. The mathematical model for determining the force F has been shown to hold—to be valid—for the physical world.

Urban planners have attempted to extend this concept of attraction and gravity by hypothesizing a mathematical model that some planners claim can be used to determine the interaction between areas and populations. The urban-planning version of a gravity model for two areas can be stated mathematically as

$$I = H\frac{p_1 p_2}{d^b}$$

where I is the interaction between the two areas (populations), p_1 and p_2 are the populations of the respective areas, d is the distance between the areas measured from some center positions and b the exponent applied to the distance between the areas (to be determined experimentally), and H is a constant (similar in concept to the gravitational constant) used to relate the relationship to actual conditions. Urban planners use this type of formulation to estimate the number of trips (the interaction) between two areas.

The urban gravity model seems like a good idea. The form of the model captures our intuitive feeling that interaction between areas is inversely proportional to the distance between them and directly proportional to the size of the areas. But the model is not based on a proven, demonstrated theory of urban system behavior. The underlying hypotheses have not been shown to be valid. The urban planner Colin Lee notes in his book that these models are based on an analogy between the physical and social sciences which many people do not accept. Unlike Newton's law that describes, explains, and can be used to predict, the urban gravity analogy may describe past interactions and activity patterns between areas, but it does not explain them. For example, we may conduct a survey in a region to determine the flow of shoppers from their residences to their shopping places. Using the survey data we could then describe this type of interaction by calibration (determining H and b) of a gravity model. If a new shopping center is opened in the region, the question is whether or not the resulting gravity model can be used by the shopping-center planners to predict the trips from the region's areas to the new shopping center. Lee feels that the calibrated gravity model does not enable the planner to identify the complex chain of cause and effect which gives rise to urban activity patterns. Nevertheless, it does seem that many of the patterns of behavior, summarized by the calibrated constants of these models, may be expected to remain more or less stable over the short term. Lee notes that these models, inadequate though they may be, are among the best urban planners have and do give partial answers to some of the questions a planner needs to ask. If the limitations of these answers are borne in mind, the models can help to improve our understanding of the urban system's behavior and our ability to forecast and plan more effectively. The danger is that the assumptions and caveats of the model get lost and the answers are presented as facts.

We raise the above concerns now to guard against the teacher and student thinking that decision modeling in the real world is true science, and that our models are valid because they encompass aspects of science and mathematics. Our acceptance and use of a model to aid in resolving a particular decision must be demonstrated to the best of our abilities. If we can offer few or no tests of model validity, or the model is just our best guess of how the reference system behaves, we should not be afraid to say so.

3.3 THE DECISION PROCESS STEPS: THE DECISION FRAMEWORK

The decision process that has evolved over the past few years can be looked at as a model of the modeling process. We shall refer to it as the *decision framework*. It involves a series of interrelated steps or stages that can be viewed as the decision process adaptation of the scientific method.

For most purposes, the steps required in solving a decision problem are:

- ☐ Formulating the problem.
- ☐ Developing a mathematical model to represent the system under study.
- ☐ Deriving a solution from the model.
- ☐ Testing the model and solution.
- ☐ Establishing controls over the solution.
- ☐ Putting the solution to work.

These steps can be viewed as accomplishing the following: For any problem we need to define the broad objectives and goals of the system; examine the (possibly new) area we are working in; become familiar with the jargon, the people, and things associated with the problem; determine the alternative courses of action available to the decision maker; develop some statement, verbal or otherwise, of the problem to be investigated; translate the problem into a suitable logical or mathematical model which relates the variables of the problem by realistic constraints and a measure of effectiveness; find a solution which optimizes the measure of effectiveness; compare the model's solution against reality to determine if we have actually formulated and solved the real-world problem we started with; determine when the real-world situation changes and reflect such changes into the mathematical model; and, most important, implement the solution into operation (not just filing a report) and observe the behavior of the solution in a realistic setting. As our ability to develop precise mathematical models of operational problems is not a highly developed science, we must be sensitive to discrepancies in the solution and feedback to the model refinements that will cause future solutions to be more realistic and accurate.

The mathematical model is central to our decision-making methodology—it offers understanding of the process and problem under investigation; it provides a vehicle for the evaluation and comparison of alternative solutions; it enables us to evaluate the effects of a change of one variable on all the other variables; and finally, and somewhat mystically, it provides us with a quantitative basis to sharpen and evaluate our intuition of the process under investigation.

It should be stressed that the mathematical model is the prime distinguishing feature of mathematical decision making. Mathematical models enable us to bring some semblance of scientific methodology to areas of decision making heretofore characterized by intuition and experience.

The role of the mathematical model in decision making can be summarized diagrammatically. After the statement of the problem, which includes the

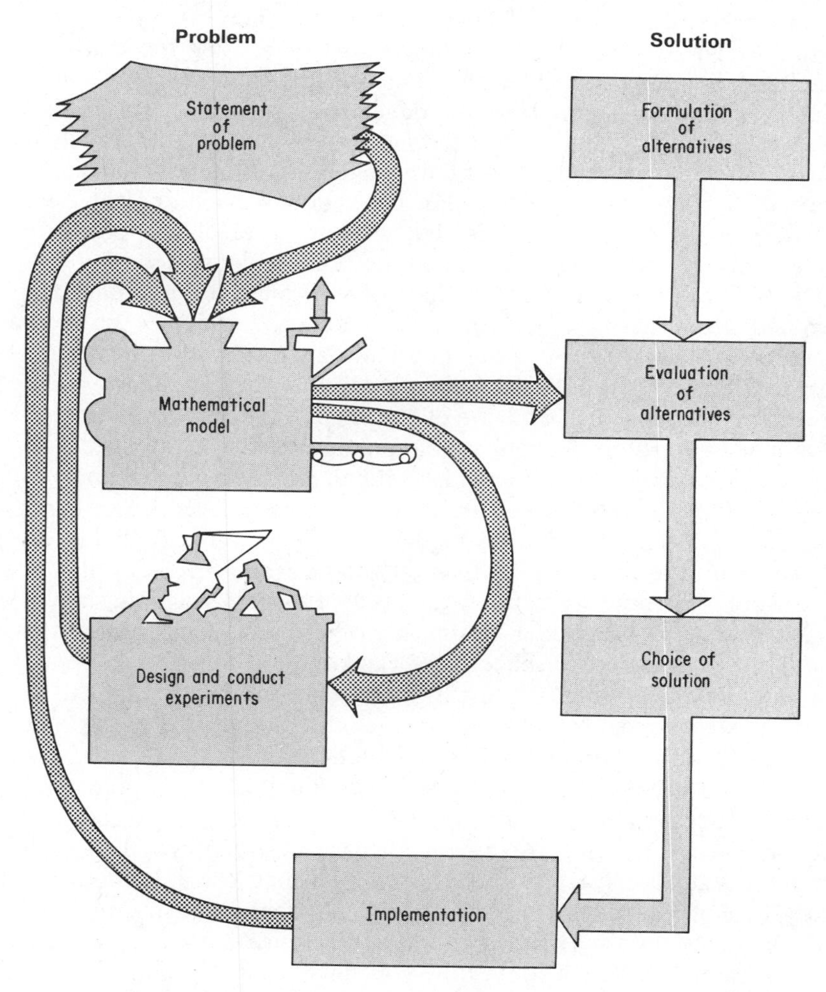

The role of the mathematical model in the decision framework[1]

[1] S. I. Gass, *An Illustrated Guide to Linear Programming,* McGraw-Hill Book Co., New York, N.Y., 1970. Reproduced with permission.

choice of the all-important measure of effectiveness, the functional form of the mathematical model is determined. As this requires specifying how the variables are related with associated data, certain experiments designed to aid the structuring of the correct form must be carried out. In some instances, this experimentation could be just the opening of the accounting ledger to gather the needed information; in others, it might call for complex and expensive efforts. In any event, the results are fed back into the structure of the model as shown in the diagram.

It is by means of the mathematical model that the problem is connected with its proposed solution. The major activities here are the devising of alternative ways of solving the problem and then using the mathematical model and its measure of effectiveness to evaluate and choose a solution to implement. For some problems the development of a set of feasible alternative solutions is automatically accomplished by computation involving the related mathematical model. For others, ingenuity and innovation are a must in proposing alternatives. As the implementation of a solution can affect the structure of the mathematical model, we have a feedback loop from the solution side to the problem side, and the process continues.

The above decision framework includes the analyst's hypothesizing the form and structure of the mathematical model used to depict the problem's reference system. Part of the testing and experimentation of the process must be devoted to establishing the model's validity. *Validation* tests the agreement between the behavior of the model and the real-world system being modeled. The process of validation attempts to state how closely the model mirrors reality. This is a most difficult task and one the analyst must face even though tests for validity may not be easy to devise or carry out. The validity of the urban gravity model is a case in point. In the last chapter we shall return to the concern of validity and a broader view of the decision framework. For now we caution the student against accepting any of the models that are discusssed in this or any other book unless recognition and resolution is given to the concerns of validation.

The age-old adage of "learning by doing" is certainly true for the student and analyst of decision problems. In a text we are limited in our ability to relate and portray the interactions and subtleties of the real-world environment. We usually can offer a succinct statement of the problem, but the detailed process implied by the application of the decision framework is a difficult one to describe. An involved, case-study approach to particular problems would detract from our accomplishing the basic objective of strengthening the student's ability to analyze varied decision situations using mathematical models. However, to illustrate some aspects of the decision framework and to demonstrate how your level of mathematical knowledge can be used to formulate and study important decision problems, we next discuss the classical and important transportation problem.

4 Our First Decision Model

The Transportation Problem[1]

4.1 A SLIGHT DIGRESSION: ON GETTING DRESSED

The transportation problem and other problems discussed in this text are examples of a widely used class of decision problems known as mathematical programs.[2] In a most general sense, programming problems are concerned with the efficient use or allocation of limited resources to meet desired objectives. Such allocation problems are central to the field of economics. However, they are found not only within industrial and corporate entities, but also arise in many guises during an individual's daily activities.

The first thing each morning we all face the programming problem of getting dressed. We must select a program of action which enables us to become dressed in a manner which meets the constraints or accepted fashion rules of society—socks are not worn over shoes, but socks are worn. Our basic resource is time, and the selected program must be best in terms of how each individual interprets his or her expenditure of early morning time.

[1] The material in this chapter is from S. I. Gass, *An Illustrated Guide to Linear Programming*, McGraw-Hill Book Co., New York, N.Y., 1970. Reproduced with permission.

[2] As the initial developments in the fields of electronic computers and mathematical programming evolved at about the same time period—the late 1940s—and since these developments have been mutually beneficial, it is unfortunate that a serious confusion in terminology occurred. We have computer programming, which deals with the logical analysis and the set of machine instructions to solve a given problem. Thus, a computer program can be used to solve a mathematical programming problem.

From a man's point of view, ignoring the "bare" essentials, a program of action involves the putting on of six items of clothes: shoes, socks, trousers, shirt, tie, and jacket. A program of action is any order in which the clothes can be put on. There are $6 \times 5 \times 4 \times 3 \times 2 \times 1 = 720$ different orderings. Many of these are not *feasible programs* as they do not meet the constraints of society (socks over shoes) or are impractical (tie on before shirt). Even after eliminating these infeasible solutions from consideration, there still are a number of feasible programs to contend with.

How is the final selection—the optimal decision—made? The dressing problem, like all those to be considered, has some measure of effectiveness—some basic objective—which enables us to compare the efficacy of the available feasible programs. If, in some fashion, we can compare the measures for each program, we can select the optimal one. For the dressing problem, we are concerned with minimizing the time it takes to get dressed. This is the measure of effectiveness—the *objective function* in programming terminology—with which we can compare the various feasible orderings of clothes. From the author's perspective, the optimal solution has been the following ordering: socks, shirt, trousers, tie, shoes, jacket. It minimizes the time to get dressed within the fashion constraints of society. Someone else with a different objective function—minimize the opening and closing of drawers, that is, minimize the early morning noise—might select a different optimal solution.

The dressing problem typifies decision problems in that it has many possible solutions. If there was only one feasible solution, there would really not be any problem or any fun in solving it. There is also some objective to be optimized that enables us to select at least one of the feasible solutions to be the optimum. The finding of the feasible solutions and the determination of an optimal one represents the computational aspects of programming problems that are discussed in later sections.

A very special subset of programming problems is called *linear programs*. The mathematical description of a linear program, that is, the model, can be written in terms of linear or straight-line relationships. The concept of a linear relationship is illustrated by the following example. If one pound of candy costs \$1.00, and the seller offers no quantity discount, the total cost is directly proportional to the amount purchased as shown in Fig. 4.1. The algebraic description is given by the equation

$$TC = (\$1.00)n$$

where TC is the total cost and n is the number of pounds purchased.

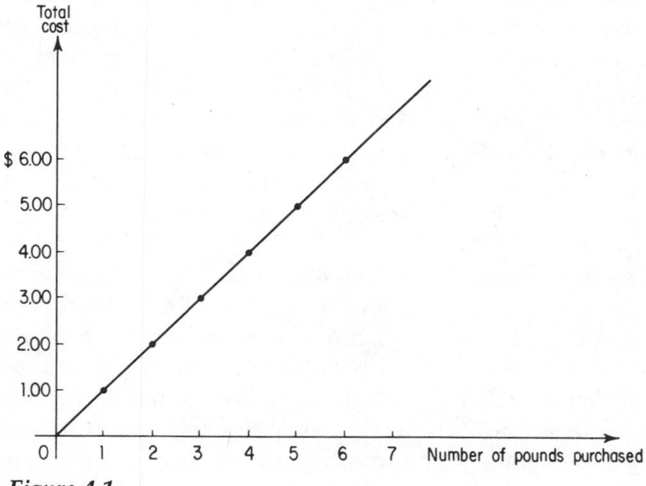

Figure 4.1

On the other hand, if the seller allowed 10 cents off for the second pound purchased, 20 for the third, and so on, up to the fifth pound, and 50 cents per pound afterwards, the cost curve would be nonlinear, as shown in Fig. 4.2. Although the assumption of a mathematical description in terms of linear relationships appears to be quite restrictive, many important problems, including the transportation problem, have this simplifying feature.

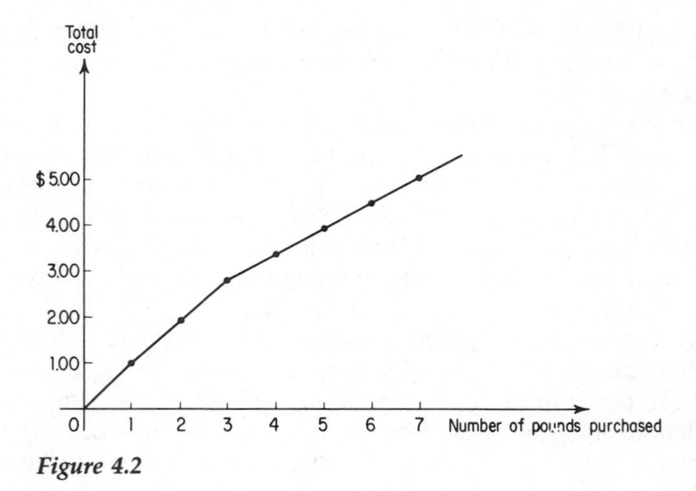

Figure 4.2

4.2 THE STATEMENT OF THE TRANSPORTATION PROBLEM

In a competitive, industrial society, such as the United States, manufacturers of finished goods strive to produce their products at the lowest cost consistent with quality and design standards. A not so insignificant part of this cost is the expense of shipping finished products to customers. We all have seen the latest automobiles loaded and moving on freight trains or have had a giant automobile transporter whiz by us on the highways. Ultimately, these movements are translated into a dealer destination charge that shows up in the fine print of the new car ''window sticker''—a cost that must be borne by the customer.

As the United States and most developed nations have an integrated network of highways, waterways, and air and rail lines, manufacturers have many alternatives to consider when deciding on how to ship their products from factories to final destinations. The decision process may be further compounded as a manufacturer might have, for example, the choice of contracting out the transportation activity and selecting from competing bidders, or operating a company-owned transportation facility. Persons in charge of shipping must determine which alternatives are available to the company and their costs.

This type of decision problem has been well studied. The basic problem, termed the *transportation problem,* has a mathematical model and associated optimizing algorithm that generates a minimal-cost solution. We describe below a simple transportation problem and develop the corresponding mathematical model. As the problem situation unfolds, you should attempt to identify the corresponding steps of the decision framework.

A refrigerator manufacturer has two factories that supply three retail stores. At the beginning of each month, the manufacturer receives from each store manager a list of unfilled sales that must be filled in the coming month by new production. This set of requirements represents the total number of new refrigerators that must be produced by the two factories. To simplify the discussion, we assume the manufacturer has enough resources—labor, raw materials, and so on—to fulfill the requirements, and in this instance, does not have any overproduction; that is, there is no facility for storage. (The production process itself could possibly be treated by means of a mathematical model, but here our interests are in a different area.) The manufacturer's store one, denoted by S_1, requires 10 refrigerators, S_2 requires 8, and S_3 requires 7, for a total of 25 refrigerators. The manufacturer has decided to produce 11 at factory one, F_1, and the remaining 14 at F_2. The problem we wish to consider is how many refrigerators should be shipped from each factory to each store so as to minimize the total cost of transporting the refrigerators from the factories to the stores.

We need additional information concerning transportation restrictions and costs. It is assumed that it is possible to ship any specified number of refrigerators from each factory to any store; that is, a transportation link—rail, truck, or other mode—connects any factory to all stores. We also assume knowledge of the costs of shipping one refrigerator from a factory to a store. Here we must make a linearity assumption about these costs which is quite critical—and in some instances quite debatable. This linearity or proportionality assumption requires that if the cost of shipping one refrigerator from F_1 to S_1 is \$10, then the cost of shipping two refrigerators is \$20, for three the cost is \$30, and so on. We can argue about this assumption based on real-world experience. If it costs \$100 to hire a truck to deliver one refrigerator, the unit cost for two refrigerators (excluding handling costs) would be \$50, for three it would be $\$33\frac{1}{3}$—a nonlinear cost relationship. In most transportation situations, if the cost is not linear, a good approximation can be obtained by averaging costs used by previous solutions.

For this problem then, we assume that the costs of transportation for a refrigerator between each factory and store are known and are linear. These costs are shown on the price tags of Fig. 4.3. For example, the cost of shipping a refrigerator from F_2 to S_3 is \$7. Fig. 4.3 illustrates the ways in which the refrigerators can flow from the factories to the stores and presents all the information of the problem. In a sense, it is an iconic model. Although such a depiction does not solve the problem, it can aid us in the development of a suitable mathematical model. Transportation problems have, in general, many possible feasible solutions. We shall develop some below, along with the linear-programming model of the problem.

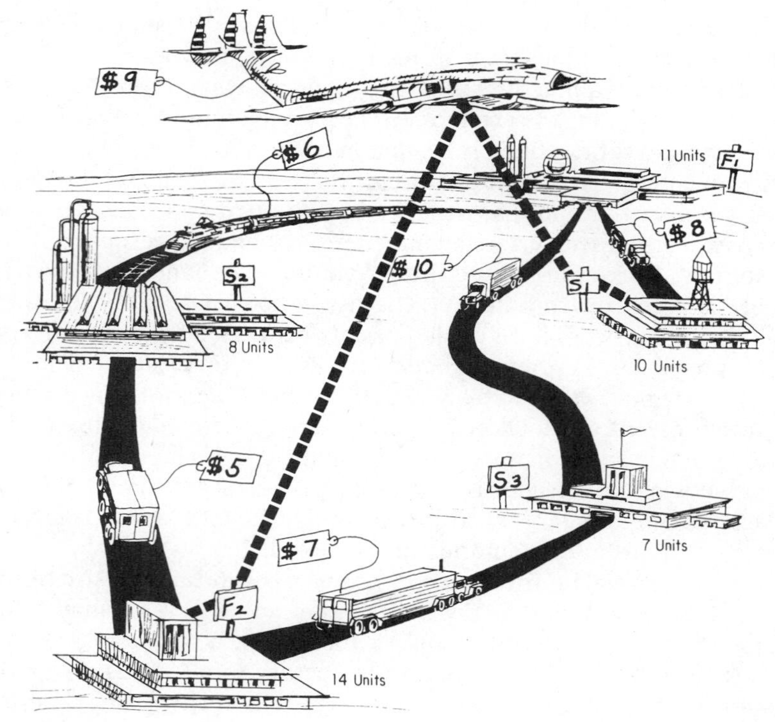

Figure 4.3

4.3 THE TRANSPORTATION PROBLEM MODEL

We are looking for a solution which meets the constraints of the problem—that is, ship 11 units from F_1, ship 14 units from F_2, S_1 receives 10 units, S_2 receives 8 units, and S_3 receives 7 units—and, at the same time, minimizes the measure of effectiveness, the total transportation cost.

In order to proceed with the formulation of the model, we rearrange the given information in a tableau:

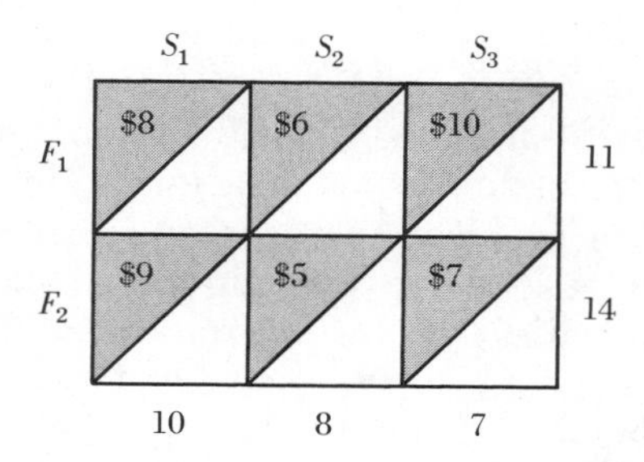

	S_1	S_2	S_3	
F_1	$8	$6	$10	11
F_2	$9	$5	$7	14
	10	8	7	

The blank triangles correspond to the unknown number of refrigerators to be shipped from the corresponding factory and store. To demonstrate that an experienced—or even inexperienced—shipping clerk (or vice-president in charge of shipping refrigerators) has no difficulty in coming up with a solution, we exhibit two possible solutions:

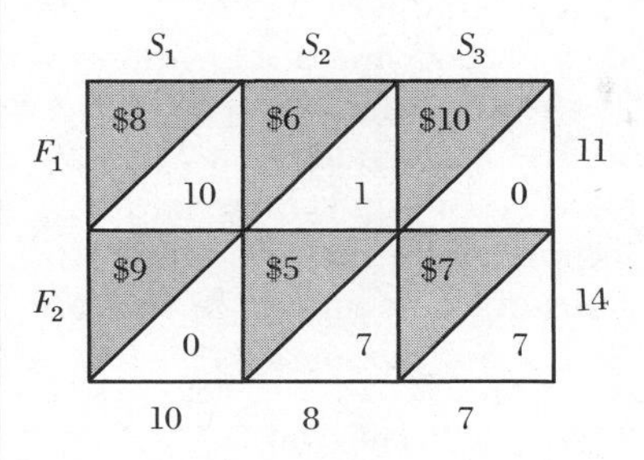

and

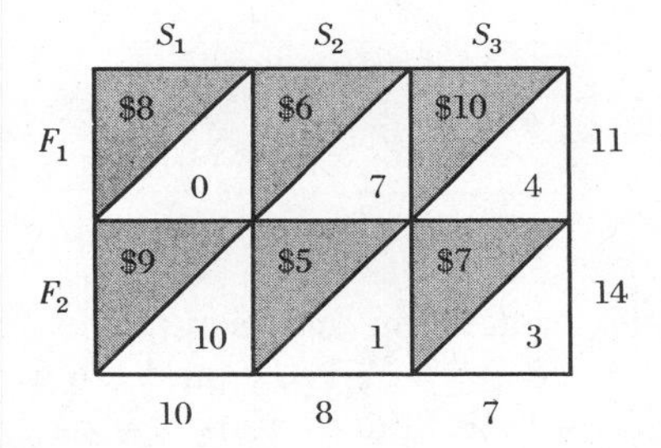

You should have no difficulty writing down others. The numbers written in each box of the first solution tableau constitute a solution with the total number shipped from F_1 to each store being $10 + 1 + 0 = 11$; from F_2 we have $0 + 7 + 7 = 14$. Also, the total amount received by S_1 from both factories is $10 + 0 = 10$; S_2 receives $1 + 7 = 8$; and S_3 receives $0 + 7 = 7$. It is similar for the second solution. Assuming linearity of the cost of transportation, the total cost of the first solution is given by the expression

$$\$8 \times 10 + \$6 \times 1 + \$5 \times 7 + \$7 \times 7 = \$170$$

and the cost of the second solution is

$$\$6 \times 7 + \$10 \times 4 + \$9 \times 10 + \$5 \times 1 + \$7 \times 3 = \$198$$

For the solutions exhibited, the first has a lower cost, but the question remains as to whether there is another feasible solution that is cheaper. A shipping clerk who does not use linear-programming techniques as an aid in solving the problem must rely heavily on experience and intuition. The clerk does not compare exhaustively all possible solutions—neither does the linear-programming procedure. The shipping clerk does select a particular solution to implement, but in general, cannot guarantee that it is the absolute minimum. The linear-programming approach offers an unconditional guarantee that the minimum will be determined. For the refrigerator problem, the first solution is the minimal solution.

To proceed with the development of the mathematical model of the transportation problem—and to simplify the discussion—we shall introduce some necessary mathematical shorthand. Let x_{11} be the number—the unknown number—of refrigerators to be shipped from F_1 to S_1, x_{12} the number to be shipped from F_1 to S_2, and so on, and, in general, x_{ij} the unknown number of refrigerators to be shipped from factory i to store j. We enter these notations into the tableau structure as shown in the tableau:

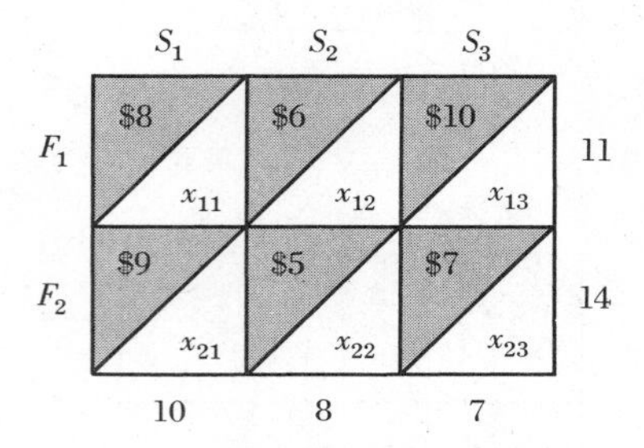

It is now a simple matter to develop the form of the mathematical model.

The total amount shipped from F_1 is 11, and the amounts shipped from F_1 are x_{11}, x_{12}, and x_{13}. Similarly, the total from F_2 is 14, and the shipments from F_2 are x_{21}, x_{22}, and x_{23}. Since the manufacturer requires a total of 25 units (10 + 8 + 7) and since the combined factory production is exactly 25 units (11 + 14), the total made at each factory must be shipped to the stores. Thus, the total amount shipped from F_1 is given by the equation

$$x_{11} + x_{12} + x_{13} = 11$$

the total shipped from F_2 is

$$x_{21} + x_{22} + x_{23} = 14$$

These sums are obtained by adding across the rows.

As each store must get exactly the amount asked for, the amounts shipped to each store—found by adding down the columns—are given by the equations

$$x_{11} + x_{21} = 10 \qquad \text{for } S_1$$

$$x_{12} + x_{22} = 8 \qquad \text{for } S_2$$

and

$$x_{13} + x_{23} = 7 \qquad \text{for } S_3$$

For any set of numbers x_{ij}, where again i represents one of the factories and j represents one of the stores, the total cost—the sum of the costs of individual shipments—is

$$\$8x_{11} + \$6x_{12} + \$10x_{13} + \$9x_{21} + \$5x_{22} + \$7x_{23}$$

These equations represent the basic constraints of the mathematical model. The only element missing is that we must limit the possible values of x_{ij} to positive values or zero. A negative x_{ij} would represent a shipment of refrigerators from a store to a factory; that is, a source of refrigerators other than those manufactured at the factories would be introduced. We disallow this possibility by restricting $x_{11} \geq 0$ (x_{11} is greater than or equal to 0), $x_{12} \geq 0 \ldots, x_{23} \geq 0$, or in general notation, $x_{ij} \geq 0$. These are called the *nonnegativity restrictions* of linear programming. As we wish to determine the set of numbers x_{ij} satisfying the equations, the nonnegativity restrictions, and which minimizes the total cost, we have the following mathematical decision model—the linear-programming model of this transportation problem:

Find the set of numbers $x_{ij} \geq 0$ which minimizes

$$\$8x_{11} + \$6x_{12} + \$10x_{13} + \$9x_{21} + \$5x_{22} + \$7x_{23}$$

subject to the constraints

$$\begin{aligned}
x_{11} + x_{12} + x_{13} \phantom{+ x_{21} + x_{22} + x_{23}} &= 11 \\
x_{21} + x_{22} + x_{23} &= 14 \\
x_{11} + x_{21} &= 10 \\
x_{12} + x_{22} &= 8 \\
x_{13} + x_{23} &= 7
\end{aligned}$$

The first solution given above satisfies these equations, where $x_{11} = 10$, $x_{12} = 1$, $x_{13} = 0$, $x_{21} = 0$, $x_{22} = 7$, and $x_{23} = 7$, and as noted before, this solution minimizes the objective function with a value of \$170.

Each factory and store contributed an equation in terms of the decision variables related to the corresponding factory or store. These equations, as well as the objective function, are linear equations, since they are simple sums of the variables. The total number of variables is the product of the number of factories and the number of stores; in this case $2 \times 3 = 6$. Also, the number of equations is the sum of the number of factories and the number of stores; here it is 5. Transportation problems can become quite large, but computational procedures—algorithms—for solving rather large-sized problems are available for most electronic computers.

From a mathematical standpoint a number of interesting items should be noted about the above system of equations. First, there is one equation too many, in that any one of them is implied by the remaining ones. For example, if we drop the first equation, it can be found by adding the last three and then subtracting the second. A more important point, however, is that if we solve the problem using the standard computational procedures of linear programming, we will determine an optimal solution whose values of the variables, the x_{ij}'s, are in terms of whole numbers. It was tacitly assumed that the x_{ij}'s must be in integers—we cannot ship $3\frac{3}{4}$ refrigerators! We can prove mathematically that we will obtain an optimal integer solution for the transportation problem, given that the amounts available at the factories and required by the stores are integers. This is not the case for the general linear-programming problem. For the transportation problem, it is due to the special structure of the equations of the corresponding mathematical model.

Although we described this linear-programming model in terms of factories, stores, and refrigerators, it is quite important to recognize that we could have cast the discussion into a more general format dealing with origins (factories), destinations (stores), homogeneous units to be shipped (refrigerators), and some measure to be minimized (total transportation cost).

In terms of the abstract decision model of Section 2.7, the above mathematical model of the transportation problem can be interpreted as follows. The decision variables are the x_{ij}; the uncontrollable variables are the supplies (production) at the factories and the demands at the stores; the decision space is defined by the equations and the nonnegativity conditions; and the measure of effectiveness is the cost function to be minimized. In some instances, a combined production–transportation problem can be considered in which the production at the factories and the amounts to be shipped are decision variables. (This type of problem is discussed in Chapter 18.)

5 Part I Discussion, Extensions, and Exercises

5.1. Most decision problems encountered in this text (and the real world) can be represented by mathematical models that have the characteristics of the abstract decision model discussed in Section 2.7, that is, of the form

$$\bar{E} = \underset{\{\mathbf{X}\}}{\text{maximum (minimum)}}\ f(\mathbf{X},\mathbf{Y})$$

where $\bar{E}$ is the optimal value of the measure of effectiveness $f(\mathbf{X},\mathbf{Y})$, $\mathbf{X}$ is the set of decision (controllable) variables, $\mathbf{Y}$ is the set of uncontrollable variables, and $\{\mathbf{X}\}$ is the set of alternative solutions (the decision space). As you probably noticed, the model of the transportation problem is more or less in the form of the abstract decision model; we did not express the transportation problem's measure of effectiveness (the cost function) as a composite function $f(\mathbf{X},\mathbf{Y})$ of the controllable and uncontrollable variables (see Section 4.2). The relationships between these variables, the decision space, and the measure of effectiveness is implied by the statement and assumptions of the transportation problem model. To show that there are some decision situations for which the form of the abstract model does occur, we develop the decision model for the following problem.

The local 24-hour pizza shop uses quite a lot of tomato sauce in its operation. If it runs out of tomato sauce, it is out of business until it can be resupplied. As the storage space for the cans of tomato sauce is limited, the owner worries a bit about running out. The owner has made an agreement with a local food supplier that whenever the sauce is running low (say the chef is starting on the last can), a call can be made to the warehouse and the supplier would immediately deliver an order to the pizza shop (guaranteed to be there before the chef emptied the last can).[1] The warehouse and its delivery operation also operate 24 hours and offer such a service to beat out the competition.

The owner is, of course, interested in minimizing the total cost of the tomato sauce inventory system. This cost is the sum of three types of cost:

1. The actual purchase cost c for a can of tomato sauce.
2. The cost k of placing an order.
3. The holding cost h of keeping a can in storage.

The cost of placing an order, termed the setup or fixed cost, takes into account the owner's expenses for billing and telephoning for an order. These costs tend to be the same no matter how big the order is. The holding cost, sometimes referred to as the inventory carrying cost, takes into account the value of the money tied up in inventory (the owner could invest the money in a bank if it wasn't being used for inventory), the cost of storage space, insurance on the inventory, and pilferage and damage to the inventory.

One extreme solution is to have only one can in inventory, and whenever the chef opens it, order another one from the supplier. Our intuition tells us that for this to be a minimizing solution, the setup cost would have to be quite low or the holding cost quite high. Obviously, this solution is impractical as it would be rather difficult to find a cooperative supplier; also, the owner would be on the telephone continually.

Another extreme inventory policy is to have the owner place orders that always fill the pizza shop's storage space. The difficulty here is that the cost of buying and maintaining the cans in inventory may put a strain on the shop's cash flow. Depending on the actual values of c, k, and h, one of these extreme solutions may prove to be the minimal-cost solution. But, in general, we would not expect this to be the case. Let's analyze the situation to determine what is the best ordering policy.

To simplify the ordering and handling of the inventory, the owner and supplier agree that whenever an order for cans is placed it will always be for the same amount. We denote this amount by the decision variable Q. Q is a positive number. The owner has collected data on the use of cans of tomato sauce and has determined that the pizza shop usage rate is constant from month to month. This uncontrollable factor (it depends on customer demand) is denoted by r, that is, an average of r cans of tomato sauce are used each month. With the constant order quantity and usage rate assump-

[1] This type of response to inventory depletion has been given the Japanese name of *kanban* or "just in time" system of inventory control. The system requires suppliers to deliver goods exactly when needed.

tions, we can study the inventory of cans over time with the following "sawtooth" graph. The figure is drawn on the basis that we start with Q items

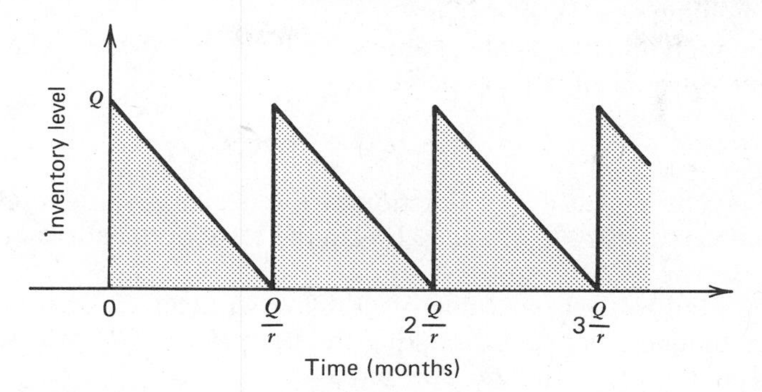

in inventory and the usage (demand) rate is smooth and constant over time. We see that the Q cans will run out after Q/r months. For example, if Q = 100 cans and r = 40 cans per month, then an order of Q cans will last 100/40 = $2\frac{1}{2}$ months. The time Q/r is called a cycle, that is, at the end of a cycle of length Q/r months, the inventory level is zero and an order is placed for Q cans which are delivered immediately (instantaneous—or close to it—replenishment of the inventory occurs). We need to develop a decision model that enables us to determine the value of Q that minimizes the average cost of operating the inventory system. This is done in the following manner.

The measure of effectiveness of the inventory system that we use is the average monthly cost. (Can you think of any others?) This cost, which we denote by C, is the sum of: (1) the monthly purchase cost of cans, (2) the monthly setup cost, and (3) the cost of holding inventory for a month. These individual costs are as follows:

1. *Monthly Purchase Cost.* As r cans are used per month, this cost is simply cr.

2. *Monthly Setup Cost.* A starting inventory of Q cans will last Q/r months. At the end of Q/r months, an order of Q cans is placed and a setup cost is incurred. This cost must be allocated on a monthly basis. For example, for Q = 100 and r = 40 there will be one setup in $\frac{5}{2}$ months. Thus, $\frac{2}{5}$ of this setup can be thought of as occurring in one month. For Q = 50 and r = 100, we have $Q/r = \frac{1}{2}$ month and we have $\frac{2}{1}$ setups per month. In general, we see that r/Q setups occur in one month. Thus, the average monthly setup cost is given by $(r/Q)k$.

3. *Monthly Inventory Holding Cost.* To determine the average monthly inventory, we note from the sawtooth inventory graph that the shaded area of a cycle (say from time 0 to Q/r) represents the total inventory during the cycle. This area is calculated by the right-triangle area formula given by $\frac{1}{2}(Q/r)Q = Q^2/2r$. Since this amount of inventory lasts Q/r months, the average monthly inventory is given by $(Q^2/2r)/(Q/r) = \frac{1}{2}Q$. [Another way of

looking at this is to note that the maximum inventory is Q and it decreases steadily to zero, with the average inventory being $\frac{1}{2}(Q + 0)$.] The monthly inventory holding cost is then $\frac{1}{2}Qh$.

From the individual costs (1–3) above we have that the average monthly cost (the measure of effectiveness) is

$$C = cr + (r/Q)k + \tfrac{1}{2}Qh \tag{1}$$

Here C is given explicitly as a function of the decision variable Q and uncontrollable variables c, k, h, and r. It is in the form of the abstract decision model.

We now need to find the value of Q that minimizes C. This is done by calculus techniques that require taking the derivative of C with respect to Q and setting the result to zero. We have

$$dC/dQ = -(r/Q^2)k + \tfrac{1}{2}h = 0 \tag{2}$$

We solve for Q to obtain the basic inventory formula of

$$Q^* = \sqrt{2rk/h} \tag{3}$$

where Q^* is the value of Q that minimizes C, that is,

$$\text{minimum } C = \overline{C} = cr + (rk)/Q^* + \tfrac{1}{2}Q^*h \tag{4}$$

Note that Q^* varies directly with the usage rate r and the setup cost k, and indirectly with the holding cost h. Q^* is independent of the actual product cost c. Does this agree with your intuition? The formula for Q^* is known as *Wilson's formula,* named for R. H. Wilson, a pioneer in inventory control procedures. The formula (3) is also called the simple lot size or economic order quantity (EOQ) formula.

The pizza shop's data are $r = 100$, $k = \$2.00$, $h = \$0.50$, and $c = \$10$. Find Q^* and $\overline{C}$ using formulas (3) and (4). You don't need calculus to come up with the value of Q that minimizes C, as given by Eq. (1). For a given set of data, all you need to do is to plot C as a function of Q and to find the point Q that yields the lowest point on the curve C. Your approximate answer should be good enough. Do this for the pizza shop data. What form of quadratic (conic section) is Eq. (1)?

We have tacitly assumed that the pizza shop could store the optimal order quantity Q^*. How would we change the solution if this was not the case? If Q^* required more storage space, how would you try to convince the owner to enlarge the shop's storage area?

How would the solution change if the pizza shop's insurance rate increased and caused $h = \$1.00$? How small must k be to make $Q^* = 1$? How large must h be to make $Q^* = 1$? Discuss how the owner should change the inventory policy based on seasonal factors, for example, higher sales during the summer. What would Q^* be if the owner started a year-long sales campaign and sold two pizzas for the price of one?

Can you think of other situations for which the assumptions of the simple lot size model with no inventory shortages apply?

5.2. In developing the transportation problem model for the refrigerator manufacturer, we assumed linearity for the measure of effectiveness, that is, for the cost function to be minimized. We did not make such an assumption for the constraints (equations) that defined the shipments from the factories (the row equations) and for the shipments to the stores (the column equations). These type of equations are just statements that reflect bookkeeping or accounting records for what goes out of a factory or what comes into a store. No need for simplifying assumptions or magic here—it is quite straightforward.

Attempt to generalize the two-factory and three-store example in Section 4.2 by writing the constraints for a four source (row) and five destination (column) transportation problem. How many variables will you have and how many equations?

The most general statement of the transportation problem assumes that there are m origins and n destinations. Using standard algebraic summation notation, the transportation problem, as a linear-programming model, can be written as: Find values of the variables x_{ij} that minimize

$$x_{00} = \sum_{i=1}^{m} \sum_{j=1}^{n} c_{ij}x_{ij}$$

subject to

$$\sum_{j=1}^{n} x_{ij} = a_i \qquad i = 1, \ldots, m$$

$$\sum_{i=1}^{m} x_{ij} = b_j \qquad j = 1, \ldots, n$$

$$x_{ij} \geq 0$$

where x_{ij} = the unknown number of items to be shipped from origin i to destination j (decision variables)

a_i = the amount to be shipped from origin i (uncontrollable variables)

b_j = the amount to be shipped to destination j (uncontrollable variables)

c_{ij} = the (linear) cost of shipping one unit from origin i to destination j

$x_{00} = \sum_{i=1}^{m} \sum_{j=1}^{n} c_{ij}x_{ij}$ = the cost function (the measure of effectiveness)

Test your understanding of this notation by writing out the equations for ($m = 3, n = 2$) and ($m = 3, n = 3$).

5.3. We noted, but did not prove, that the optimal solution to the refrigerator problem was given by

$$x_{11} = 10, \quad x_{12} = 1, \quad x_{13} = 0$$

$$x_{21} = 0, \quad x_{22} = 7, \quad x_{23} = 7; \quad x_{00} = \$170$$

At this point we have not discussed how you go about solving such problems for the optimum. An algorithm needs to be developed. However, for this (2×3) problem, you should be able to demonstrate that there are no better solutions than the one given above. (*Hint:* Any different solution must involve shipments along routes that have no shipments in the above solution, i.e., from F_1 to S_3 and from F_2 to S_2. Let $x_{13} = 1$ and show that this results in two new solutions whose total transportation costs will increase by \$2.00 or \$4.00, respectively. Do the same for x_{21}. Can you make a statement that describes the conditions that must hold for a solution to be optimum?)

5.4. Do you have any decisions to make if $m = 1$ or $n = 1$ in the transportation problem?

5.5. You probably noted that in the refrigerator problem the total amount to be shipped, $25 = 11 + 14$, is equal to the total requirements from the stores. Show that this always has to be true, that is, $\sum_{i=1}^{m} a_i = \sum_{j=1}^{n} b_j$, otherwise the mathematical model would contain conflicting mathematical statements. [*Hint:* Add together all the origin (row) equations and then all the destination (column) equations. Also, any one origin or destination equation is implied by the remaining equations.]

5.6. To get a feel on why the transportation problem can always be solved for an optimal solution with integer values, write out the equations of the refrigerator problem, let $x_{13} = 0$ and $x_{21} = 0$, and solve for the other variables. As noted in Sections 4.3 and 5.5, we have one too many equations, so drop the last one before you start solving the equations. Now you have four equations in four variables. Why does the solution of this set of equations appear to be easier to compute than others you might have solved?

5.7. Reanalyze the refrigerator problem and indicate what was done that corresponds to the steps of the decision framework; also indicate any omissions.

5.8. It should be rather easy to develop a heuristic algorithm for the transportation problem. Such an algorithm should be based on finding a feasible solution to the constraints, that is, a set of nonnegative values of the decision variables x_{ij} that satisfy the equations and keep the total cost as small as possible. The rationale for choosing which x_{ij} are in the solution should be related to the values of their corresponding costs. (*Hint:* Why not use shipments that involve the smallest costs?)

Try out your heuristic algorithm on the following problems:

(a)

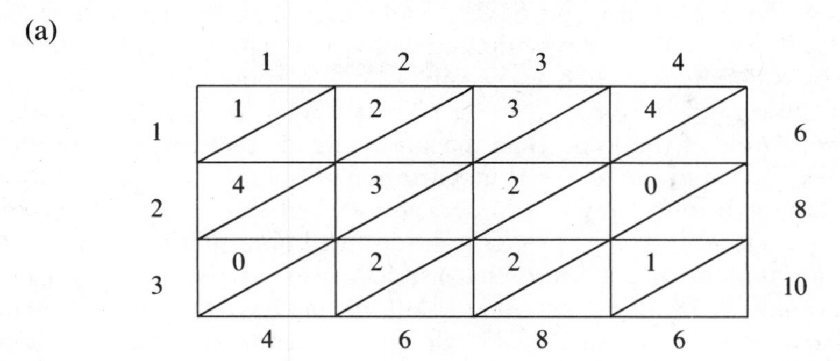

The minimum cost is 28.

(b)

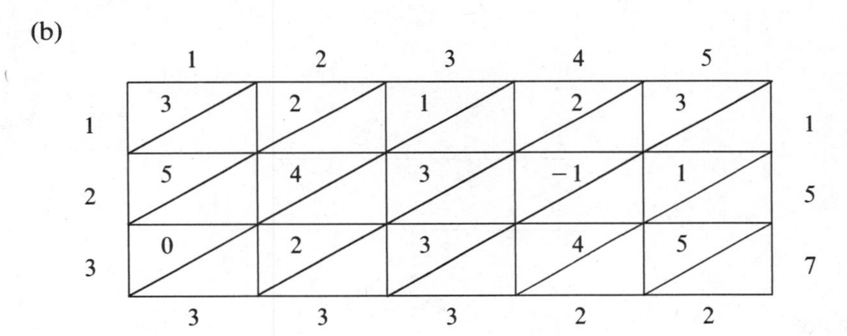

The minimum cost is 13.

(c) The refrigerator problem.

Make up a transportation problem for which your heuristic algorithm generates a poor solution. (*Hint:* You can select the magnitudes of the c_{ij} to meet your needs.)

5.9. By now you should have enough experience with the transportation problem to develop a part of the algorithm for solving it for the minimum cost. Optimization algorithms come in many forms. A basic approach for constructing such an algorithm involves the steps of: (1) finding a first feasible solution, that is, some set of the decision variables that satisfies the problem constraints; (2) applying a mathematical test to determine if this initial solution is optimum; and, if it is not, (3) applying a procedure for changing the solution while improving the value of the measure of effectiveness. Not all algorithms apply these three steps in a sequential manner. Some algorithms systematically search for a solution in a manner that guarantees that when one is found it is also optimal. An optimizing algorithm, unlike a heuristic one, is based on corresponding mathematical theory that enables a test for optimality to be applied to a proposed solution.

For the transportation problem, we saw how easy it is to come up with a solution. What we want to do next is to develop a mathematical test for determining if such a solution is optimal. Here we have to deal with solutions obtained by shipping as much as possible along a route. The refrigerator problem solution of $x_{11} = 10$, $x_{12} = 1$, $x_{13} = 0$, $x_{21} = 0$, $x_{22} = 7$, $x_{23} = 7$, with $x_{00} = \$170$ is of this type. The reason for this is that it can be proved mathematically that for an $m + n$ transportation problem an optimal solution exists that involves only $m + n - 1$ decision variables. For the refrigerator problem, $m + n - 1 = 2 + 3 - 1 = 4$. You will note that this number is equal to the number of equations that are left after we get rid of an extra one. You should determine some other solutions that have only four decision variables and ship as much as possible along a route, for example, start with $x_{21} = 10$ or with $x_{12} = 8$.

Let us work with the supposedly optimal solution to the refrigerator problem given in the tableau below:

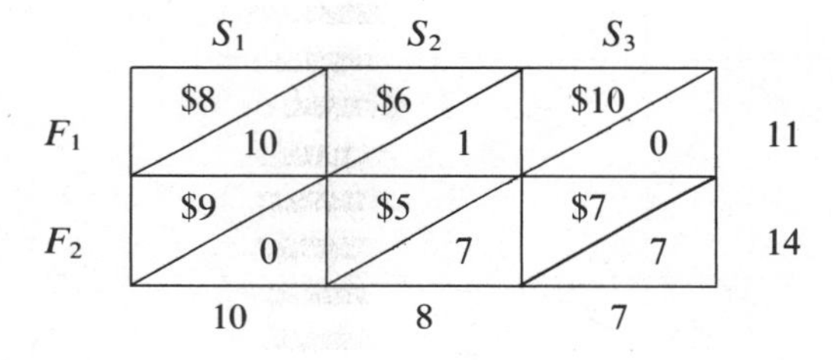

	S_1	S_2	S_3	
F_1	\$8 / 10	\$6 / 1	\$10 / 0	11
F_2	\$9 / 0	\$5 / 7	\$7 / 7	14
	10	8	7	

Here $x_{00} = \$170$.

The basic mathematical model for this problem is to find a set of $x_{ij} \geq 0$ that minimizes

$$x_{00} = 8x_{11} + 6x_{12} + 10x_{13} + 9x_{21} + 5x_{22} + 7x_{23}$$

subject to

$$\begin{aligned}
x_{11} + x_{12} + x_{13} \qquad\qquad\qquad\qquad &= 11 \\
x_{21} + x_{22} + x_{23} &= 14 \\
x_{11} \qquad\qquad + x_{21} \qquad\qquad &= 10 \\
x_{12} \qquad\qquad + x_{22} \qquad &= 8
\end{aligned}$$

Here we have left out the last equation $x_{13} + x_{23} = 7$ because it is implied by the others. You can show this by adding the first two equations and subtracting the third and fourth from this sum. We want to test the given solution to see if it is optimal. If it is not optimal, then there is another solution that has one of the routes (variables) that is now zero at a positive level. Explain why this must be the case. What we want to do is to see if it makes any sense to find a solution with either x_{13} or x_{21} positive; that is, can we find a solution with either x_{13} positive or x_{21} positive that has a value of the objective function x_{00} less than \$170? There is an easy algebraic way of testing this, but you have to keep your wits about you to make sure you do not make any computational errors. Can you think of how such a test can be constructed?

As the direct way is often the best way, let us do the following. We first rewrite the four equations so that only the positive variables are on the left-hand side, as follows:

$$
\begin{aligned}
x_{11} + x_{12} \qquad\qquad\qquad &= 11 - x_{13} \\
x_{22} + x_{23} &= 14 - x_{21} \\
x_{11} \qquad\qquad\qquad\qquad &= 10 - x_{21} \\
x_{12} + x_{22} \qquad\quad &= 8
\end{aligned}
$$

If we let $x_{21} = 0$ and $x_{13} = 0$, these equations are the ones solved in Section 5.6. Since we are interested in whether x_{21} or x_{13} should be different than zero, we now want to solve these equations keeping x_{21} and x_{13} in as symbols. First, notice what happens as we allow x_{13} and/or x_{21} to be positive, for example, if $x_{13} = 4$ and $x_{21} = 10$, then we are forced to have $x_{11} = 0$, $x_{12} = 7$, $x_{22} = 1$, and $x_{23} = 3$. This is the solution given in Section 4.3 that had a value of the objective function $x_{00} = \$198$. You should solve the equations yourself to understand how these calculations are done.

Coming back to the four equations, we see from the third equation that any solution must have

$$x_{11} = 10 - x_{21}$$

Substituting this value of x_{11} into the first equation, we have that any solution must have

$$x_{12} = 11 - x_{13} - x_{11} = 11 - x_{13} - 10 + x_{21} = 1 - x_{13} + x_{21}$$

Then, using this value of x_{12} in the fourth equation, we have

$$x_{22} = 8 - x_{12} = 8 - 1 + x_{13} - x_{21} = 7 + x_{13} - x_{21}$$

Finally, substituting this value of x_{22} in the second equation, we have

$$x_{23} = 14 - x_{21} - x_{22} = 14 - x_{21} - 7 - x_{13} + x_{21} = 7 - x_{13}$$

The above relations are summarized as

$$
\begin{aligned}
x_{11} &= 10 - x_{21} \\
x_{12} &= 1 + x_{21} - x_{13} \\
x_{22} &= 7 - x_{21} + x_{13} \\
x_{23} &= 7 \qquad\quad - x_{13}
\end{aligned}
$$

Of course, if here we let $x_{21} = 0$ and $x_{13} = 0$, we obtain the original solution we are testing for the optimal. But, what about the objective function? Its explicit form is

$$x_{00} = 8x_{11} + 6x_{12} + 10x_{13} + 9x_{21} + 5x_{22} + 7x_{23}$$

However, we now have new expressions for the positive solution variables x_{11}, x_{12}, x_{22}, and x_{23} in terms of the variables x_{21} and x_{13} that are now zero. Let us take these expressions and use them as substitutes for the solution variables in the objective function:

$$x_{00} = 8(10 - x_{21}) + 6(1 + x_{21} - x_{13}) + 10x_{13} + 9x_{21} + 5(7 - x_{21} + x_{13}) + 7(7 - x_{13})$$

Performing the multiplications and collecting terms, we obtain

$$x_{00} = \$170 + \$2x_{21} + \$2x_{13}$$

Here, if $x_{21} = 0$ and $x_{13} = 0$ then we have $x_{00} = \$170$, the correct value of the objective function for the solution we are testing. But, what are the implications of this new expression of the objective function? It tells us that for a unit increase of x_{21} the value of the objective function will *increase* by \$2.00 and similarly for a unit increase of x_{13}. In fact, if either x_{21} or x_{13} are anything but zero, the value of the objective function will be greater than \$170. For example, the solution discussed above with $x_{13} = 4$ and $x_{21} = 10$ has

$$x_{00} = \$170 + \$2(10) + \$2(4) = \$198$$

We must restrict $x_{13} = 0$ and $x_{21} = 0$, and hence, our test solution is optimal!

This test for optimality was rather straightforward. You should reproduce it for the solutions ($x_{21} = 10$, $x_{22} = 4$, $x_{12} = 4$, $x_{13} = 7$; $x_{00} = ?$) and ($x_{11} = 3$, $x_{12} = 8$, $x_{21} = 7$, $x_{23} = 7$; $x_{00} = ?$). As these solutions are not optimal, you should try to arrive at a procedure that enables you to change these solutions and determine a different one with a lower value of the objective function. If you can do that step, you would have completed the optimization algorithm for the transportation problem.

For longer problems, this form of the algorithm is rather cumbersome for hand calculations and costly to apply on a computer. A more routinized procedure is available based on the mathematical structure of the transportation problem and theoretical results of linear programming. We shall discuss it in Chapter 20.

Since it can be proved that an optimal solution to a transportation problem contains no more than $n + m - 1$ variables that are positive with the rest being equal to zero, you might have wondered why we just do not explicitly calculate all such solutions and then select the one with the lowest transportation cost. Any algorithm that finds all possible solutions to a problem is termed *complete enumeration*. For problems with many possible solutions, complete enumeration can be quite inefficient and costly. An efficient algorithm has a means of limiting its search to a small subset of possible solutions, that is, *partial enumeration* takes place with a guarantee that an optimal will be included in the search.

You might try to find all $n + m - 1$ variable solutions to the refrigerator problem. Can you state an upper bound on the number of such solutions? For a 10 origin and 10 destination (10×10) transportation problem, there is an optimal solution with nine variables. But there are $10 \times 10 = 100$ variables. How many combinations of nine from the 100 are there?

5.10. In the statement of the refrigerator problem, it was assumed that there was no warehouse facility and thus, the total refrigerators produced at the factories was equal to exactly the number required by the stores. This assumption is, of course, fairly unrealistic in an industrial setting. Manufacturers try to meet demand by efficient, rather level (smooth) production runs that might be above or below the demand for a particular time period. Demand is met by current production and previously produced units that have been placed in storage. Adapt the transportation model so that it can handle the situation where more units are shipped from the origins than are required by the destinations, that is, $\sum_{i=1}^{m} a_i > \sum_{j=1}^{n} b_j$. (*Hint:* Assume that the manufacturer has a warehouse. Illustrate your approach for the refrigerator problem if $F_1 = 15$ and $F_2 = 20$.)

There could be situations where the manufacturer finds that total production is less than the demand, that is, $\sum_{i=1}^{m} a_i < \sum_{j=1}^{n} b_j$. What can be done in this case to utilize the transportation model to describe how the available units can still be shipped to minimize the total transportation cost? (*Hint:* Assume that the shortage can be obtained from a competitor or that some destinations will not get what they want. Illustrate your approach for the refrigerator problem if $F_1 = 9$ and $F_2 = 12$.)

5.11. In some cases, we express a nonlinear relationship by a suitable approximation consisting of linear relations. The nonlinear curve of Fig. 4.2 is just a sequence of linear segments. Determine these segments and express the complete curve as a set of relationships involving TC (total cost) and n (the number of pounds purchased). Instead of expressing the costs of Fig. 4.2 by line segments, attempt to find a single linear equation of the form $y = ax + b$, where y is cost, x is pounds, and b is the cost when $x = 0$ (the y-axis intercept). For any x, y will then approximate the true cost. Such linear approximations can be used, but with care.

Some models have a measure of effectiveness that can be approximated mathematically as a quadratic equation with the general form of $y = ax^2 + bx + c$. For example, profit may be a quadratic function of the money spent on advertising; here $a < 0$ as the curve must have a finite maximum (check your calculus text). Total cost for some inventory systems (see Section 5.1) can be expressed as a quadratic function of the number of items ordered to replenish the stock; here $a > 0$ as the total cost must have a finite minimum.

Plot the curves of the following quadratic functions and check the corresponding high or low points:

(a) $y = x^2 - 6x$ $\quad (x = 3, y = -9)$

(b) $y = -2x^2 - 4x + 1$ $\quad (x = -1, y = 3)$

(c) $y = x^2 - 2x + 6$ $\quad (x = 1, y = 5)$

(d) $y = -x^2 + 6x + 6$ $\quad (x = 3, y = 15)$

Why would equations (a) and (b) not be suitable measures of effectiveness?

Approximate the graphs of equations (c) and (d) by two and then four straight-line segments. For (c), let $-1 \leq x \leq 3$; for (d), let $0 \leq x \leq 6$.

Quadratic profit measures of effectiveness illustrate the economic *law of diminishing returns*. This law usually states the relation between an input of production and the resulting output of a good: Each added unit of production produces less extra output. For a profit measure, it states that as we keep adding extra units of say advertising dollars, the extra profit we get per unit decreases. There is some point beyond which the extra contribution to profit is negative (oversaturation of advertising). For equation (d), determine the extra contribution to profit for each unit of x as x takes on the values $1, \ldots, 6$.

5.12. A powerful tool of analysis is that of developing analogies. The urban gravity model is analogous to Newton's Universal Law of Gravitation. Polya in his book *How to Solve It* stresses heuristic reasoning that is expressed by the simple but incisive questions: "What is the unknown? What are the data? What is the condition? Do you know a related problem?" Can you think of a problem analogous to the transportation problem? (*Hint:* There are many such situations at school, e.g., assigning professors or students to classes or sections.)

5.13. An assignment problem faced by colleges involves determining which students should be let into a class that is oversubscribed. For example, if there is one class with 40 seats and 60 students have preregistered for the class, on what basis should 40 students be selected? Should it be alphabetically; first come, first served, if who is first can be determined; or how? Think of an algorithm and an appropriate measure of effectiveness by which an assignment (selection) of the students can be compared to other assignments. You might want to check your college's registration office to see if they will tell you how it is done.

A certain college has an assignment algorithm that has been programmed on its computer. The algorithm was devised by one of the computer-center's analysts based on the procedures and constraints put forth by the administration. What this heuristic algorithm does for oversubscribed sections is to first assign athletes, then band members, then seniors, juniors, sophomores, and finally, freshmen. You should discuss what the implied measure of effectiveness is and whether it is an acceptable one for your school. There is a mathematical model for this problem, but it is difficult to solve due to its size and complexity (because of many students and many sections and the need to develop a suitable measure of effectiveness). Try constructing a mathematical model for a small example.

5.14. In his book on urban planning models, Professor A. G. Wilson describes the following retail-sales gravity model. Assume a geographical area is divided into n zones, where a zone can be a residential area, shopping area, or combined residential–shopping area. In general, we can have the residents in zone i shopping in zone j. We want a model to aid in predicting total dollar sales in shopping zone j by the residents from zone i. We make the following "reasonable" assumptions:

1. The flow of cash from i to j is proportional to the total amount of cash available for shopping in zone i.
2. The flow of cash from i to j is proportional to the "attractiveness" of the shopping center in zone j.
3. The flow of cash from i to j decreases in relation to increasing travel cost from zone i to zone j; that is, the flow is inversely proportional to travel cost.

To structure the model, we define the following terms:

S_{ij} = sales in zone j made by residents of zone i
P_i = population of zone i
e_i = average expenditure on shopping per person in zone i
e_iP_i = money spent on shopping by residents of zone i
W_j = size (attractiveness) of the shopping area in zone j, measured in square feet
c_{ij} = cost of travel from zone i to zone j as a function of the cost of the travel time and the transportation cost

Based on the assumptions, the simple retail-sales gravity model is given by

$$S_{ij} = K\frac{(e_iP_i)W_j}{c_{ij}}$$

where K, a constant of proportionality, needs to be determined by empirical studies.

This model implies that you can predict total dollar value of sales S_{ij} between zone i and zone j by knowing the K, e_i, P_i, W_j, and c_{ij}. How would you go about determining these data? Discuss the reasonableness of the assumptions in terms of the metropolitan area near your school. How would you test the ability of this model to predict sales in your area? You need to be concerned with how your zones are formed (you would like residential zones to have households with similar incomes and shopping zones to have one shopping center), the definition of attractiveness (it is also a function of the types of shops, restaurants, theatres, etc. in the shopping zone), and the definition of travel cost (we really have $c_{ij} = c(t_{ij}) + m_{ij}$, where t_{ij} is the travel time from an assumed center of zone i to assumed center of zone j, c is a constant that converts travel time to money, and m_{ij} is the cost of transportation from zone i to zone j). What is the formula for determining the total sales made from all zones i to a particular shopping zone j?

Professor Wilson also describes the following modification of the retail-sales model:

$$S_{ij} = K \frac{(e_i P_i) W_j^a}{c_{ij}^b}$$

where the exponents a and b are determined by empirical studies so that the model's predictions (and ability to replicate the past) are more accurate. How would you go about calibrating this model?

A builder is considering using the modified model to aid in determining where to build a shopping center with F square feet. The new center is to be located in one of three possible zones. Discuss how you would perform the analysis and aid the builder in making the selection. What is an appropriate measure of effectiveness? (*Hint:* Consider sales per square foot of total shopping area.)

5.15. The astrophysicist John Q. Stewart has written about the concept of "social physics." Its principal thesis is that the behavior of people in large numbers may be predicted by mathematical rules. His reasoning is based on an analogy between distribution and movement of a large number of molecules in a gas and large populations in a social setting. Much of statistics, surveys and polls, and economic theory and forecasts are based on behavior of populations, that is, collective individual behavior.

Dr. Stewart notes that the number of telephone calls between any two cities in the United States is roughly proportional to the product of their populations divided by the distance between them. This is similar to the gravity model with the constant $H = 1$ and the exponent $b = 1$; $I = p_1 p_2 / d$. He also notes that such a relationship appears to hold for the number of bus passengers between cities and the number of railroad tickets sold. Can you hypothesize other activities that might fit this model?

A related model approximates the number of students from a state attending a particular private college campus. This number tends to be proportional to the population of the state divided by the distance from the campus. Can you think of an experiment that involves your class by which a gravitylike model can be hypothesized and tested? Describe how you would go about collecting data and calibrating a gravitylike model for an investor who wants to build a new movie house and has to decide between two locations in a city. How would the investor use the model? What assumptions have to be made?

5.16. When we encounter a new problem and do not have a method of solution, it is sometimes helpful to develop bounds for the answers. Returning to the museum design problem, we see that Plan One (the standard gallery design) would require a maximum of nine guards, one for each room. This is a feasible solution, but is certainly too high a number. Show how you can reduce this number by placing guards in the doorways connecting the rooms. Start thinking about a mathematical model for this problem that would involve linear constraints, but has some complications. What would be the set of decision variables? Assume a room is protected by a guard standing in the doorway. The measure of effectiveness to be minimized is the number of guards. What would be an upper bound for the number of guards for Plan Two (the irregular polygon design)? Certainly a guard facing each of the 19 walls would be too many. We could place a guard at every other point where two walls meet and require only 10 guards, with the tenth guard assigned to watch one wall. This approach yields a bound of $\frac{1}{2}n$ if n is even and $\frac{1}{2}(n + 1)$ if n is odd, where n is the number of walls. Can you combine the 10 guard positions so that fewer guards are required? This type of adjustment is a hit-or-miss approach. You cannot demonstrate that you have the minimum number unless all possibilities have been investigated. A more systematic approach would involve the partitioning of the interior space into guard areas. Assume a guard remains fixed at a post, but is able to turn around on a spot. (*Hint:* Try to block out the Plan Two gallery space in nonoverlapping triangles with the walls forming part of each triangle. You might want to experiment with figures that have fewer walls, e.g., a three-, four-, or five-sided room with zigs and zags. Do you need more than one guard for rooms with three, four, or five walls? We would like to obtain a formula for the number of guards that is a function of the number of walls.)

5.17. A decision problem faced by most automobile owners is when to trade in the old car for a new one. This type of replacement problem is encountered by transportation-fleet owners, heavy-equipment users, and, in general, by anyone who invests in capital equipment that wears out and needs to be serviced to keep it going. The older a piece of equipment is the more costly it is to maintain; also, the salvage, or trade-in value is less. A suitable measure of effectiveness must be related to the cost of owning and operating the equipment as a function of time. What data would you need to collect or assume for such an analysis?

We next illustrate a simple replacement decision model by considering the problem of when to trade in an automobile. (The problem is taken from one of the first books on operations research written by the trio of authors

Sasieni, Yaspan, and Friedman.) The basic data that can be readily obtained is the initial cost of the automobile (given to be $6000) and the annual operating costs. We would like to obtain information—a decision rule—that tells us when to replace the automobile. A simple measure of effectiveness is the average yearly cost of owning and operating the car. What components of cost go into this measure?

We assume the car's owner keeps a record of costs and at the end of the first year we find that the running cost was $1000 (this includes cost of gasoline, oil, repairs, etc.). The automobile "blue book" of car values indicates that the car could be resold for $3000. Thus, at the end of the first year, the capital cost of the car is $6000 − $3000 = $3000 and the total (average) cost for the first year is $3000 + $1000 = $4000. We would expect this average cost to change over time and how it changes, that is, when it decreases or increases, yields some clue as to when to unload the car. Would you expect the average yearly cost to exhibit a pattern of changes that increases or decreases in a random manner? Remember that the running costs will steadily increase from year to year while the trade-in value will decrease.

The replacement decision cannot be made until the data are collected for each year. But to illustrate the analysis, we assume that the owner can extrapolate the running costs for the next eight years, as given in the following table (The table also gives the expected trade-in value by year.):

	Year							
	1	2	3	4	5	6	7	8
Running costs (in dollars)	1000	1200	1400	1800	2300	2800	3400	4000
Trade-in value (in dollars)	3000	1500	750	375	200	200	200	200

As we are studying the average cost per year, we form the next table in which the first row is the cumulative running costs by year and the second row is the capital cost by year obtained by subtracting the trade-in value from the initial cost of $6000:

	Year							
	1	2	3	4	5	6	7	8
Cumulative running costs (in dollars)	1000	2200	3600	5400	7700	10,500	13,400	17,900
Capital costs (in dollars)	3000	4500	5250	5625	5800	5800	5,800	5,800

Next, we add the cumulative running costs and capital costs to obtain the total costs to the owner at the end of each year:

	Year							
	1	2	3	4	5	6	7	8
Total costs (in dollars)	4000	6700	8850	11,025	13,500	16,300	19,700	23,700

If we then divide these costs by the corresponding number of years, we obtain the desired measure of effectiveness, the average cost per year:

	Year							
	1	2	3	4	5	6	7	8
Average cost per year (in dollars)	4000	3350	2950	2756	2700	2717	2814	2963

We see that the average cost per year decreases steadily to year five and then starts to increase. This analysis suggests that the owner should consider trading in the car at the end of the fifth year, and do so no later than the end of the sixth year.

It is clear that the above analysis is quite dependent on the assumptions made as to future running costs and trade-in values. An important part of any analysis is determining how sensitive the answer (here the decision rule) is to the data. We always have uncertainty in our data due to measurement problems or assumptions, or because much data are "guesstimates." All studies should include *sensitivity analyses* to indicate whether the results are stable as we change the given values of the data, including the values of the uncontrollable variables. Sensitivity analyses are made by varying one set of data while keeping all other data at their given values.

For the above automobile replacement problem, perform a sensitivity analysis on the running costs by assuming that these costs, starting with year two, can be $100 above or below the given costs. Perform one analysis with all the running costs increased by $100, and a second with all the running costs decreased by $100.

A third sensitivity analysis should be made to determine when the time for trading in the car changes as a function of the assumed trade-in value. Would you expect the time to decrease or increase if the trade-in value was less each year? What if the trade-in value was more each year? Recalculate the average cost per year if all the trade-in values given in the first table are increased by $400. Then do it for a $500 and $600 increase. (If you have a programmable hand calculator, or are adept at programming a computer, you might want to automate the calculations.) What happens between the shift from $500 to $600?

The understanding and acceptability of decision information is enhanced by exhibiting it in a clear fashion. Draw a graph or barchart figure that compares your sensitivity analysis results with the initial solution. You might want to prepare more than one graph and use overlays.

For the decision framework, describe the steps and procedures used to define and solve the auto replacement problem. Make sure you define the system environment. Can you think of any other models for this problem?

5.18. The automobile replacement problem of Section 5.17 used average yearly cost as the measure of effectiveness. Can you think of other measures that might be more appropriate? How does the factor of safety enter the picture, especially as the car gets older? Can you develop a measure of safety that can be combined with the cost measure? How about the image of you driving in an old car?

Most decision problems do not have a single measure of effectiveness, but, usually, one is more overriding than the others. An open research problem is how to make a decision based on *multicriteria objective functions* or measures of effectiveness. Part of the problem is due to the way the criteria are evaluated. Can you combine average costs in dollars with a measure of safety? Can safety be given in terms of dollars? (We discuss multicriteria problems in Chapter 24.)

The transportation problem minimal-cost solution might be in conflict (different) with one that minimizes the time the goods are in transit or the minimal distance traveled. Discuss how you might want to analyze and select a solution to the transportation problem that has two measures of effectiveness; also do the same for the automobile replacement problem. What type of graphical presentation material would you prepare to aid in the analysis?

5.19. The science of decision making is the subject of two early, but important, books. You should try to obtain copies to read along with this text. The first by I. D. Bross titled *Design for Decision* is a nontechnical discussion of decisions, models, and the area of statistical decision theory. He describes decision making as follows:

1. There are two or more alternative courses of action possible. Only one of these lines of action can be taken.
2. The process of decision will select, from these alternative actions, a single course of action which will actually be carried out.
3. The selection of a course of action is to be made so as to accomplish some designated purpose.

The second book is by R. L. Ackoff, one of the pioneers in the field of operations research. His book titled *Scientific Method: Optimizing Applied Research Decisions* develops some of the same material as Bross's book (and more), but in greater technical detail. Ackoff's book is "a book on planning or designing the use of science in the pursuit of objectives" and is a "how it ought to be done" book with respect to rational planning. His paradigm for the existence of a problem is given by the following:

1. An *individual* who has the problem—the *decision maker.*
2. An *outcome* that is desired by the decision maker (an objective). An objective is an outcome which has positive value for the decision maker. Every problem situation must involve at least two possible outcomes.
3. At least *two unequally efficient courses* of action which have some chance of yielding the desired objective.
4. A state of doubt in the decision maker as to which choice is "best."
5. An *environment* or *context* of the problem.

To *solve* a problem is to make the best choice from among the available courses of action.

5.20. In his book, the mathematical biologist Anatol Rapoport discusses "Metaphors and Models." A metaphor is a figure of speech in which a word or phrase denoting one kind of object or idea is used in place of another to suggest a likeness or analogy between them, for example, "The ship spread its wings to the breeze." Our vocabulary is largely built on metaphors. To quote Rapoport:

> Scientific metaphors are called "models." They are made with the full knowledge that the connection between the metaphor and the real thing is primarily in the mind of the scientist. And they are made with a clearly definable purpose—as starting points of a deductive process. . . . Like every other aspect of scientific procedure, the scientific metaphor is a pragmatic device, to be used freely as long as it serves its purpose, to be discarded without regrets when it fails to do so.

Of course, there are good and bad metaphors as there are models. Metaphors and models must at least pass what some researchers call the "laugh test," that is, not be obviously implausible. The following mixed metaphor does not pass: "The King is leading his people over the cliff with his head in the sand." Can you describe a model that would produce answers that fail the laugh test? The statement of a model may show that it is obviously incorrect, but sometimes we might have to wait for the answer before we discover that the model is wrong.

5.21. In his book on urban planning, Professor Wilson suggests the following rules for model design:

1. What is the *purpose* behind the particular model-building exercise?
2. What should be represented as *quantified variables* within the model?

3. Which of these variables are under the *control* of the decision maker?
4. How *aggregated* a view can be taken?
5. How should the concept of *time* be treated?
6. What *theories* are we trying to represent in the model?
7. What *techniques* are available for building the model?
8. What relevant *data* are available?
9. What methods can be used for the *calibration* and *testing* of the model?

Discuss each rule for predicting retail sales by the gravity model and the shipping of goods from origins to destinations by the transportation problem model.

5.22. In succeeding chapters, we will encounter many decision problems which we cannot solve optimally. This is because we will not be able to devise an algorithm for finding the optimal solution and/or because any such algorithm would take an inordinate and costly amount of (computer) time. But, these real-world problems are "solved" every day, that is, we humans are adept at choosing a "reasonable, good, and acceptable" solution that makes the system work rather well. These situations require the managers or analysts to develop and use heuristic algorithms. The more experienced and successful managers combine an ability to understand complex decision situations with a sharp intuitive sense to produce winning solutions.

You will be called upon to design heuristic methods for solving decision problems. In doing so, you should keep in mind the following aspects of heuristic methods (as given in the paper by Silver et al.). A definition of a heuristic method (from Nicholson) is a procedure ". . . for solving problems by an intuitive approach in which the structure of the problem can be interpreted and exploited intelligently to obtain a reasonable solution." We assume that we have a mathematical model of the problem with which we can demonstrate that the heuristic solution (the values of the decision variables) is a feasible solution with a specified value of the measure of effectiveness. A good heuristic algorithm should have the following features:

1. Realistic computational effort to obtain the solution,
2. The selected solution should be close to the optimum on the average, that is, we want good performance on the average.
3. The chance of selecting a very poor solution (i.e., one far from the optimum) should be low.
4. The heuristic algorithm should be as simple as possible for the user to understand, preferably explainable in intuitive terms, particularly if it is to be used manually.

5.23. *A Tale of Two Alternatives.* It was the worst of times. Two operations research analysts were shipwrecked on a tropical island. Like true analysts, they defined their problem—how to get off the island and make their way safely to the mainland—and listed all the alternative solutions. They ranked the solutions using a measure that combined risk with the chance of success. They chose the solution of building a raft.

They scoured the island looking for wood and came upon an old, but usable wooden lifeboat. They dismantled the boat and built a raft.

(*MORAL:* Your decision framework should not be made of wood—you need to construct it out of flexible material.)

5.24. "Action is the proper fruit of knowledge." (from Chinese fortune cookie)

5.25. *Spike's Choice.* Snoopy's brother Spike is about to make a decision. What measure(s) of effectiveness do you think he will use?

PEANUTS® by Charles M. Schulz

Part I References

Ackoff, R. L.: The Development of Operations Research as a Science, *Journal of the Operations Research Society of America*, Vol. 4, no. 3, June 1956.

Ackoff, R. L.: *Scientific Method: Optimizing Applied Research Decisions*, Wiley, New York, 1962.

Apostel, L.: Towards the Formal Study of Models in the Non-Formal Sciences, chapter in *The Concept and the Role of the Model in Mathematics and Natural and Social Sciences*, Gordon & Breach, New York, 1961.

Bartlett, J.: *Bartlett's Familiar Quotations*, 14th Edition, E. M. Beck (Ed.) Little, Brown, Boston, 1968.

Baumol, W. J.: *Economic Theory and Operations Analysis*, 2nd Edition, Prentice-Hall, Englewood Cliffs, New Jersey 1965.

Bross, I. D.: *Design for Decision*, Macmillan Co., New York, 1953.

Bullock, A., and O. Stallybrass (Eds.): *The Fontana Dictionary of Modern Thought*, Harper & Row New York, 1977.

Churchman, C. W., R. L. Ackoff, and E. L. Arnoff: *Introduction to Operations Research*, Wiley, New York, 1957.

Cleland, D. J., and W. R. King,: *Systems Analysis and Project Management*, McGraw-Hill, New York, 1968.

Emshoff, J. R., and R. L. Sisson: *Design and Use of Computer Simulation Models*, Macmillan Co., New York, 1970.

Flanders, H., and J. J. Price: *Elementary Functions and Analytic Geometry*, Academic Press, New York, 1973.

Gannon, M. J.: *Management: An Organizational Perspective*, Little, Brown, Boston, 1977.

Gass, S. I.: *An Illustrated Guide to Linear Programming*, McGraw-Hill, New York, 1970.

Gass, S. I.: *Linear Programming: Methods and Applications*, 5th Edition, McGraw-Hill, New York, 1985.

Gass, S. I., and R. L. Sisson (Eds.) *A Guide to Models in Governmental Planning and Operations*, Sauger Books, Potomac, Md., 1975.

Goode, H. W.: An Application of a Highspeed Computer to the Definition and Solution of the Vehicular Traffic Problem, *Journal of the Operations Research Society of America*, Vol, 5, no. 6, December 1957.

Gordon, G.: *Systems Simulation*, Prentice-Hall, Englewood Cliffs, N.J., 1969.

Greenberger, M., M. A. Crenson, and B. L. Crissey: *Models in the Policy Process*, Russell Sage Foundation, New York, 1976.

Hadley, G., and T. M. Whitin: *Analysis of Inventory Systems*, Prentice-Hall, Englewood Cliffs, N.J., 1963.

Hille, E., and S. Salas: *First-Year Calculus*, Blaisdell, Waltham, Mass., 1968.

Hogan, W. W.: Energy Modeling: Building Understanding for Better Use, *Proceedings of the Second Lawrence Symposium on Systems and Decisions*. Lawrence Energy Laboratory, Livermore, Calif. 1978.

Honsberger, R.: *Mathematical Gems*, Vols. I and II, The Mathematical Association of America, Washington, D.C., 1973 and 1976,

Isaacs, R.: On Applied Mathematics, *Journal of Optimization Theory and Applications*, Vol. 27, no. 1, January 1979.

Kac, M.: Some Mathematical Models in Science, *Science*, Vol. 166. no. 3906, November 1969.

Kemeny, J. G., A. Schleifer, Jr., J. L. Snell, and G. L. Thompson: *Finite Mathematics*, 2nd Edition, Prentice-Hall, Englewood Cliffs, N.J., 1972.

Lee, C.: *Models in Planning*, Pergamon Press, Elmsford, N.Y., 1973.

March, J., and H. Simon: *Organizations*, Wiley, New York, 1958.

Miser, H. J.: Operations Research and Systems Analysis, *Science*, Vol. 209, July 4, 1980.

Nicholson, T.: *Optimization Techniques*, Vol. I, Longman Press, London, 1971.

Pearl, J.: *Heuristics*, Addison-Wesley, Reading, Mass., 1984.

Polya, G.: *How to Solve It*, 2nd Edition, Doubleday, New York, 1957.

Rapoport, A.: *Operational Philosophy*, Harper & Brothers, New York, 1954.

Richmond, S. B.: *Operations Research for Management Decisions*, Ronald Press, New York, 1968.

Rogers, E. M.: *Physics for the Enquiring Mind*, Princeton Univ. Press, Princeton, N.J., 1960.

Rubinstein, M. F.: *Patterns of Problem Solving*, Prentice-Hall, Englewood Cliffs, N.J., 1975.

Samuelson, P. A.: *Economics*, 5th Edition, McGraw-Hill, New York, 1961.

Sasieni, M., A. Yaspan, and L. Friedman: *Operations Research: Methods and Problems*, Wiley, New York, 1959.

Silver, E. A., R. V. V. Vidal, and D. de Verra: A Tutorial on Heuristic Methods, *European Journal of Operational Research*, pp. 153–162, Vol. 5, 1980.

Sisson, R. L.: Introduction to Decision Models, Chapter 1 in *A Guide to Models in Governmental Planning and Operations*, S. I. Gass and R. L. Sisson (Eds.), Sauger Books, Potomac, Md., 1975.

Stewart, J. Q.,: Concerning "Social Physics," *Scientific American*, Vol. 178, May 1948.

Strauch, R. E.: *A Critical Assessment of Quantitative Methodology As A Policy Analysis Tool*, P-5282, The Rand Corporation, Santa Monica, Calif., August 1974.

Townsend, R.: *Up the Organization*, Fawcett Publications, Greenwich, Conn., 1970.

Vajda, S.: *Problems in Linear and Nonlinear Programming*, Hafner Press, New York, 1975.

Wagner, H. M.: *Principles of Management Science*, 2nd Edition, Prentice-Hall, Englewood Cliffs, N.J., 1975.

Wilson, A. G.: *Urban and Regional Models in Geography and Planning*, Wiley, New York, 1974.

PART

II The Linear-Programming Model

Applications

Chapter 6 What's for Breakfast?: The Diet Problem 67

Chapter 7 Mad Hatter, Inc.: The Caterer Problem 74

Chapter 8 I'll Take the Middle Slice: The Trim Problem 88

Chapter 9 You're in the Army Now: The Personnel-Assignment Problem 95

Chapter 10 The Simple Furniture Company: The Activity-Analysis Problem 100

Chapter 11 Part II Discussion, Extensions, and Exercises 107

In this part we shall discuss some of the more important and basic problems that can be formulated as linear programs.[1] Of all the decision-aiding models, the linear-programming structure is the most versatile and adaptable, as well as the easiest to explain. To set the stage, we first offer the following thoughts concerning problem solving in the real world versus textbook discussions—we find our abilities limited in what we are able to convey to the student relative to what really happens on the firing line. Although these comments refer to the linear-programming model, they apply to all models used as decision aids.

The basic aim in developing a linear-programming model of an operational problem is to be able to predict what the optimal solution should be, given the initial conditions of the problem. In making this statement, we assume we have been able to capture the real-life situation—that is, the actual problem we wish to solve—by proper definition and manipulation of variables and constraints. In many instances, in fact, in most instances, our ability to portray the true, the genuine, *the problem,* is open to question. As in all other areas of human endeavor, we find that compromises, resourcefulness, and a little finagling help to obtain a better understanding of the process in question and, we hope, lead to a mathematical model which yields an improved solution that *can be put to work.* A model which claims to portray the production capacity of a manufacturing plant can only do so within certain limits. The predicted production, which is used to plan transportation, storage, and related needs, is based on assumed or measured processing rates, availability of labor, resources, and so on. What actually happens—a slowdown here, a shortage there—molds and forces the events which define the true production. If the model is a "reasonable" mathematical representation of the real problem, the plans based on the predicted production have not led us too far astray and, in fact, have allowed us to plan a more efficient and thus a more profitable operation.

In developing the formulation of problems in terms of a linear-programming model, we must guard against being accused of having a tool for a job and, if the tool does not fit, reshaping the job to fit the tool. This *caveat* does not and should not rule out our being allowed to make proper simplifying assumptions in light of the desire to develop models which capture the

[1] The material in Chapters. 6–10 is from S. I. Gass, *An Illustrated Guide to Linear Programming,* McGraw-Hill Book Co., New York, N.Y., 1970. Reproduced with permission. We also wish to acknowledge the excellent illustrations by W. F. McWilliam that originally appeared in the cited book.

essentials of the problem. Our mathematical models must yield answers that can be understood by the people responsible for the process being studied. These people—the decision makers—must be able to put the results to work. These results must cause the operation to improve in terms of the agreed-upon measure of effectiveness.

How do we go about formulating the mathematical model of a linear-programming problem? What are the danger spots? How can we get results that work? The answers to these and related questions cannot be delineated in clear terms. At best, we can talk around the answers and impart their meaning by the discussions and illustrations that follow.

As noted earlier, although we first introduced the transportation problem as one in which we shipped goods from factories to stores, the general transportation problem can involve generic origins and destinations. We emphasize this point. We shall illustrate a number of linear-programming problems by imbedding the problem in a specific environment. The student should realize that the resulting mathematical model has a use beyond the given setting. The following fable illustrates the point.

In the early days of linear programming, circa 1953, one of the few formal publications in the field described the essentials of problem formulation in terms of determining an optimal product mix. A processor of nuts wished to mix three grades of nuts, each mix consisting of cashew nuts, hazel nuts, and peanuts, in such a fashion as to meet certain specifications. For example, one mixture had to contain not less than 50% cashews and no more than 25% peanuts. The processor wished to combine the available resources of nuts, subject to the capacity restrictions of the manufacturing plant, so as to maximize profits. The problem was explained in complete detail, the mathematical model developed, and a numerical example was solved. The example and the publication had a wide audience.

Included in this audience was one of the pioneering operations-research (OR) consultants who worked in Detroit. The consultant was approached by the production manager of a large automobile manufacturing company who inquired how he could learn about this new technique called linear programming. The OR man briefed him on the subject and recommended the nut-problem publication for further information.

A few weeks later he called the production manager to inquire about the progress of the self-study course. The manager was perplexed. He chastised the OR consultant and mentioned something about wasting his time. Undaunted, the consultant pressed on to determine the trouble. Finally it came out. The manager had read and reread all about the nut problem. He felt that he might now be able to become a big production man in a nut factory—but he made automobiles. The manager was unable to transfer the concepts of the product-mix, resource-allocation nut problem to his own environment. He could not "relate." Today's students are more astute. At least you have been forewarned.

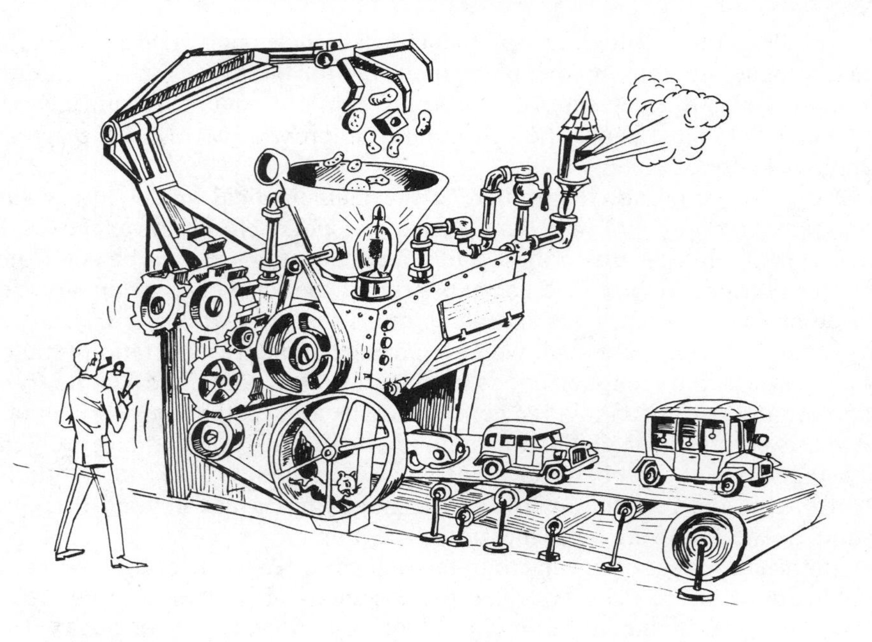

6 What's for Breakfast?

The Diet Problem

The proof of the pudding is in the eating.

CERVANTES

6.1 KRISPIES AND/OR CRUNCHIES, IF YOU MUST KNOW

Anyone attempting to live within a budget has a number of alternative ways to allocate his or her limited funds. A budget-minded housewife has to set aside so much for rent, clothes, food, entertainment, transportation, and so on. The fixed costs like rent are easy to allocate, while the split of money between food and entertainment is usually made based on past experience, with occasional spur-of-the-moment fluctuations. The specific allocations, however, are made based on the housewife's measure of effectiveness, evaluated with respect to the expenditure of her funds in all the budget areas. In these types of decision problems, it is rather difficult to optimize or even to determine an overall measure of effectiveness, and thus one attempts to *suboptimize*—break up the problem into tractable subproblems, each subproblem consisting of an applicable measure of effectiveness and associated constraints. For example, let us consider the subproblem faced by the housewife as she attempts to feed her family. In fact, we shall greatly simplify the problem—as shall be our approach to such matters—by considering only her plans for getting her children to eat a proper breakfast.

Forgetting the budget constraint for the moment, the harried housewife wishes to feed her children a breakfast menu which contains a specified level of nourishment. After consultation with her vitamin/calorie counter, she decides that her children should obtain at least 1 milligram of thiamine, 5 milligrams of niacin, and 400 calories from their breakfast foods. The children have a choice of eating the latest in dry cereals—Krunchies, the old standby Crispies, or, as children are wont to do, a mixture of the two. The fine print on the side of each cereal box contains, among other things, the fact that 1 ounce of Krunchies contains 0.10 milligram of thiamine, 1 milligram of niacin, and 110 calories; while 1 ounce of Crispies contains 0.25 milligram of thiamine, 0.25 milligram of niacin, and 120 calories.

We see that it is quite easy to find menus, that is, solutions, to this problem. The desired level of nutrients can be reached or exceeded by eating 10 ounces of Krunchies or 20 ounces of Crispies. But what of the cost of such a diet and what of the budget? The objective of the housewife is to plan a breakfast menu such that her children obtain at least the specified nutrients at the minimum cost. If Krunchies cost 3.8 cents per ounce and Crispies 4.2 cents per ounce, we see that the two suggested solutions would cost 38 cents and 84 cents, respectively. The first solution, to eat 10 ounces of Krunchies, contains 1 milligram of thiamine, 10 milligrams of niacin, and 1100 calories.

The second solution, to eat 20 ounces of Crispies, contains 5 milligrams of thiamine, 5 milligrams of niacin, and 2400 calories. The first solution gives exactly the right amount of thiamine and too much of the other nutrients, and the second solution contains exactly the right amount of niacin and too much of the others. If the children wish to eat only Krunchies, the housewife cannot lower the amount of Krunchies below 10 ounces, as a lower quantity would not contain at least 1 milligram of thiamine. Similarly, if the children wish to eat only Crispies, they must eat 20 ounces, for otherwise they would not get enough niacin. The only alternative solution left is a mixture of the two cereals—the housewife must determine if there exists a combination of the two which contains enough of the nutrients and is cheaper than the 10-ounce Krunchies solution. In fact, she wants the cheapest mixture which satisfies the nutrient constraints. How can she determine the minimum cost diet? Linear programming to the rescue!

6.2 THE MODEL AND ITS SELECTIONS

During the above discussion we have tacitly assumed certain linear relationships with respect to combinations of the two cereals. First, we assumed that the amount of nutrients ingested is proportional to the amount of food eaten; 1 ounce of Krunchies contains 110 calories, while 2 ounces contains 220 calories, and so on. Also, the amount of nutrient contained in one food is additive to the amount contained in the other; 1 ounce of Krunchies and 1 ounce of Crispies contain 110 + 120 or 230 calories. The constraints of the problem state that the children must take in at least so much of a nutrient, and if a solution causes them to have an excess of a nutrient, this is acceptable. The 10 ounces of Krunchies solution contained 1100 calories, while only 400 are required. To formulate the linear-programming model let us rearrange the data of the problem in the following tableau. We denote the amount of Krunchies to be eaten by K and the amount of Crispies by C:

Nutrients	Amount of each nutrient (in mg) in 1 ounce of Krunchies (K)	Amount of each nutrient (in mg) in 1 ounce of Crispies (C)	Requirement of each nutrient (in mg)
Thiamine	0.10	0.25	1
Niacin	1.00	0.25	5
Calories	110.00	120.00	400
Cost	3.8 cents per ounce	4.2 cents per ounce	

To construct the constraints of the problem we note that the total amount of each nutrient in the left-hand columns must be greater than or equal to ($\geq$) the numerical amount shown in the right-hand column. Thus, for each nutrient we have a constraint. For thiamine, K ounces of Krunchies contain $0.10K$ milligram, and C ounces of Crispies contain $0.25C$ milligram. A solution to the problem must have

$$0.10K + 0.25C \geq 1$$

Similarly, for niacin, we must have

$$1.00K + 0.25C \geq 5$$

and for calories,

$$110.00K + 120.00C \geq 400$$

The amount of each food eaten must be zero or a positive amount; that is, $K \geq 0$ and $C \geq 0$. Finally, the total cost of the breakfast menu is given by

$$3.8K + 4.2C$$

Putting the above together, the housewife's breakfast-menu linear-programming problem is to find values of K and C which minimizes the total cost

$$3.8K + 4.2C$$

subject to

$$\begin{array}{rcrl} 0.10K & + & 0.25C & \geq 1 \\ 1.00K & + & 0.25C & \geq 5 \\ 110.00K & + & 120.00C & \geq 400 \\ K & & & \geq 0 \\ & & C & \geq 0 \end{array}$$

You can readily check that our two previous menus $K = 10$, $C = 0$ and $K = 0$, $C = 20$ satisfy the above inequalities. The minimum solution is when $K = 4\frac{4}{9}$ ounces and $C = 2\frac{2}{9}$ ounces, with a total cost of $26\frac{2}{9}$ cents. This solution will give the children 1 milligram of thiamine and 5 milligrams of niacin, which are the exact requirements, and $755\frac{5}{9}$ calories, an overage of $355\frac{5}{9}$ calories.[1]

This simple diet or menu-planning model can be extended to include the problem of determining a three-meal diet such that minimum daily require-

[1] How the solution to this diet problem was determined is shown in Chapter 15.

ments for all the basic nutrients are met in a manner which minimizes the total cost of the foods used in making the meals. This type of diet problem was first formulated in the early 1940s—before the discovery of the mathematics and solution procedures of linear programming. At that time, the economist, George J. Stigler, formulated a 77-food, nutrient-diet problem, using the 1939 costs of the foods. Stigler's approach to solving his problem was by trial and error. By diligent analysis and keen insight he determined the types and amounts of each food which would satisfy the daily minimum nutrient requirements for a very low-cost, but not minimum, diet. His solution called for the use of only *five* foods at a total yearly cost of $39.93! The foods were wheat flour, evaporated milk, cabbage, spinach, and dried navy beans. The true minimum cost diet obtained by linear-programming methods called for nine foods—wheat flour, cornmeal, evaporated milk, peanut butter, lard, beef liver, cabbage, potatoes, and spinach, with a slightly better yearly cost of $39.67. Such diets, although quite inexpensive, are certainly unpalatable over any period of time, and the selection of foods would do justice to the chief dietician of a slave-labor camp.

As Stigler points out, "No one recommends these diets [i.e., true minimum cost diets] to anyone, let alone everyone." He also cites a low-cost diet for 1939 that was constructed by a dietician and cost $115. The difference in cost was attributed to the dietician's concern with the requirements of palatability, variety of diet, and prestige value of certain foods. More recent attempts at constructing diets for human beings using linear programming have met with acceptance due to the problem formulator being able to express, via linear constraints, the dietician's concern for flavor, taste, and variety. Such menu planning is now being done for large institutions, with a reasonable cost savings over the standard dietician's approach. But, even if the success of the diet problem for people has been limited, the diet problem for chickens, cattle, and pigs has been a notable linear-programming example.

6.3 DOWN ON THE FARM

Diet problems, or, more generally, *blending problems,* arise in a number of manufacturing activities. We have problems of minimal-cost feed mixtures for farm animals (chickens and cattle), or the mixing of various elements, (chemicals or fertilizers) to meet requirements at least cost. The mathematical model of these problems is an extension of the simple problem faced by our housewife. Even for animals we have such restrictions as palatability, but they can be taken care of in a straightforward manner. Although the mathematics of such applications are standard, the ability to determine the full set of proper restrictions depends on the analyst's understanding of the problem area. In one attempt to develop a feed mix for cows, a restriction limiting the total amount of molasses was included to ensure that the resultant feed could be processed by the manufacturing equipment which pressed the feed into pellets. It turns out that molasses is cheap and reasonably high in calcium and protein, and the cows love it. Thus, the optimal solution included as much molasses as possible, but the optimum feed was rejected by the manufacturer. Although the manufacturer could produce it in pellet form, the molasses content would make the manure too soft to use as fertilizer.

This rather earthy example illustrates how an analyst must go about developing a mathematical model which produces solutions that can be put to work. It is an evolutionary process and calls for the close interactions of the people with the problem and the analyst with the model. We must try to include the whole system, for example, the internal workings of the cow!

7 Mad Hatter, Inc.

The Caterer Problem

While we wait for the napkin, the soup gets cold,
While the bonnet is trimming, the face gets old,
When we've matched our buttons, the pattern is sold,
And everything comes too late—too late
FITZHUGH LUDLOW

7.1 TWO OF A KIND: THE CONSULTANT AND THE MAD HATTER

As it must happen to all good things, a management science consultant was called in to "unmadden" the Mad Hatter's tea parties. For all these many years, the Mad Hatter, his friends, and guests have been going around and around the table encountering dirty place settings. The table was by now becoming rather unsightly.

"If you plan to entertain each day, you must plan ahead," expounded the consultant to the Mad Hatter.

"But I have no time to plan," complained the Mad Hatter, "as I must entertain my friends with poems and riddles and giving them more or less tea, as the case may be."

"Leave it all to me," said the consultant. "I have observed your affairs for some time now, and with a little cooperation on your part and my model, we shall be running bigger and better and cheaper parties before you know it."

"My parties have been everyone's model for years. I would like to learn about yours. You're hired," said the Mad Hatter.

The guests at the table all moved up one and the consultant joined them, pen and contract in hand. He and the Mad Hatter signed, and it was duly witnessed by the March Hare. The contract called for the consultant to study the problem in depth and to submit his report and recommendations in 30 days. The report follows.

7.2 THE REPORT

AN ANALYTICAL ANALYSIS OF
INTERACTIVE ACTIVITIES AS RELATED TO
THE ECONOMICS OF FUNCTIONAL GATHERINGS

A Preliminary Linear Programming Model
of a Tea-party Subsystem

by

Super Management Consultants

Introduction

A brief study of tea-party operations as manifested by the present management, Mad Hatter, Inc., led us to the immediate and conclusive conclusion that things were in a terrible state of affairs. Any attempt to bring some semblance of analytical methodology to the present operations would meet with disaster. Thus, we recommend that tea-party operations be stopped as soon as present invitations are honored and a program of action, as outlined below, be initiated as soon as possible. Although the suggested program considers only one aspect of the total tea-party operations, it is felt that if we can accomplish a cost-effective program for this pivotal part of the operations, we can then push on to full tea-party service with a high probability of success. More specifically then, we recommend that all of the present stock of dirty napkins be destroyed and a purchase-laundering programming for new napkins be instituted.[1] We have attacked this element of the total problem using the powerful tools of modern mathematical decision-making. Our approach follows.

---

[1]Controlled experiments showed that the napkins were in such a deplorable condition they could not be cleaned.

The Mathematical Model

Based on our extensive experience in mathematical modeling, we were able to structure within the total tea-party system a critical subsystem which lends itself to a complete mathematical analysis. This analysis is based on the well-known caterer problem, in which a caterer wishes to determine how many napkins he must purchase and how many dirty ones he must send to the laundry in order to have enough for his customers. He wants to achieve a proper balance between purchases and laundry so as to minimize the total cost of the napkin subsystem. Our plan is to adapt this linear-programming model to the tea-party problem and, once we have optimized the flow of napkins, to extend our analysis to the other elements of tea-party operations. We feel confident that within time we shall optimize the flow of food, the flow of guests, and, finally, the flow of tea.

Analysis

To illustrate the application of the caterer's problem model to the operations under consideration, we undertook a data-collection phase of our project and determined for a typical week that the following number of persons attend your tea parties:

1. Monday--5
2. Tuesday--6
3. Wednesday--7

4. Thursday--8
5. Friday--7
6. Saturday--9
7. Sunday--10

Any purchases of new napkins can be made on the desired day and delivered in time at a cost of 25 cents per napkin. (Delivery is free.) There are two laundries in the area which have been tested and approved. The King's Laundry can clean a napkin in two days and charges 15 cents per napkin; while the Queen's Laundry takes three days and charges 10 cents per napkin. Assuming we have burned all the old napkins and have no new ones on hand, we next construct the linear-programming model for your data.

First, let us state our notation. We let n_1, n_2, n_3, n_4, n_5, n_6, n_7 be the number of new napkins purchased to be used on the corresponding day of the week. Similarly, we let k_1, k_2, k_3, k_4, k_5, k_6, k_7 and q_1, q_2, q_3, q_4, q_5, q_6, q_7, be the number of napkins sent to the King's and Queen's laundries, respectively. Finally, let d_1, d_2, d_3, d_4, d_5, d_6, d_7, be the number of dirty napkins which are not sent to a laundry on the corresponding day. As we want to minimize the total cost of maintaining the proper inventory of clean napkins, it does not pay to buy more napkins than needed for the current day, or to send out a napkin to be laundered unless it will be used at a future time.

For the first day of operation, i. e., on Monday, we must buy exactly the number needed. As we expect five guests, we conclude that

$$n_1 = 5.$$

At the end of Monday's tea party, we have a choice of sending all or some of the five dirty napkins to the fast King's Laundry, to the slower Queen's Laundry, or letting them stay dirty in the laundry room. What happens to these five napkins can be represented by the equation

$$k_1 + q_1 + d_1 = 5$$

The total cost for the first day's operation is $(25n_1 + 15k_1 + 10q_1)$ cents. Of course, our problem is to determine exactly what numerical values to give to the variables of the problem—so far we have n_1, k_1, q_1, and d_1 as variables, with n_1 equal to exactly 5. Once we have all the equations of our problem, the computational procedures of linear programming can be put to work to find the minimum cost solution. We next develop the rest of the model's constraints, remembering that laundered napkins will be back in the system in 2 or 3 days, depending on the service. The k_1 napkins sent on Monday will return in time to be used for Wednesday's tea party, while the q_1 napkins will be ready for Thursday. Also, the d_1 napkins not sent out on Monday can be sent out the next day.

For Tuesday's tea party, we again must buy the required number of napkins, i.e.

$$n_2 = 6$$

After they have been used, the actions taken to dispose of these six napkins, and the dirty pile of d_1 napkins, is given by the equation

$$k_2 + q_2 + d_2 = 6 + d_1$$

This last equation states that we now have a dirty stockpile of d_1 from Monday and six more from Tuesday that can be either laundered or kept in a dirty pile. The cost of Tuesday's operation is

$$(25n_2 + 15k_2 + 10q_2) \text{ cents}$$

For Wednesday's affair, we need seven napkins. This is the first day in which laundered napkins from the King's fast service can be put into use again. Hence, the required seven clean napkins can be obtained either by purchases and/or from the amount sent to the fast laundry on the first day. We then have

$$n_3 + k_1 = 7$$

Also, as before

$$k_3 + q_3 + d_3 = 7 + d_2$$

and Wednesday's cost is

$$(25n_3 + 15k_3 + 10q_3) \text{ cents}$$

The eight clean napkins needed for Thursday can be new ones or laundered ones which have returned from Tuesday's dirty shipment to the King's Laundry or Monday's dirty load returned from the Queen's Laundry. This is represented by

$$n_4 + k_2 + q_1 = 8$$

with

$$k_4 + q_4 + d_4 = 8 + d_3$$

and a Thursday cost of

$$(25n_4 + 15k_4 + 10q_4) \text{ cents}$$

We can now proceed to write the remaining equations of our model in a straightforward manner. We assume for discussion purposes that we are interested in the efficient running of only one week of tea parties, and hence, no laundry shipments will be made unless they can be returned to be used on Sunday.

To continue, we have for Friday the equations,

$$n_5 + k_3 + q_2 = 7$$
$$k_5 + d_5 = 7 + d_4$$

with a cost of

$$(25n_5 + 15k_5) \text{ cents;}$$

and for Saturday,

$$n_6 + k_4 + q_3 = 9$$
$$d_6 = 9 + d_5$$

with a cost of

$$(25n_6) \text{ cents;}$$

and for Sunday,

$$n_7 + k_5 + q_4 = 10$$
$$d_7 = 10 + d_6$$

with a cost of

$$25n_7$$

Putting the above equations together, we see that our problem is to find values of n, k, q, and d (these values are either positive or zero, i.e.,

nonnegative) which minimize the cost function $25\,(n_1 + n_2 + n_3 + n_4 + n_5 + n_6 + n_7) + 15\,(k_1 + k_2 + k_3 + k_4 + k_5) + 10\,(q_1 + q_2 + q_3 + q_4)$ and satisfy the linear equations

$$
\begin{aligned}
n_1 &= 5 \\
n_2 &= 6 \\
n_3 + k_1 &= 7 \\
n_4 + k_2 + q_1 &= 8 \\
n_5 + k_3 + q_2 &= 7 \\
n_6 + k_4 + q_3 &= 9 \\
n_7 + k_5 + q_4 &= 10 \\
k_1 + q_1 + d_1 &= 5 \\
k_2 + q_2 + d_2 &= 6 + d_1 \\
k_3 + q_3 + d_3 &= 7 + d_2 \\
k_4 + q_4 + d_4 &= 8 + d_3 \\
k_5 + d_5 &= 7 + d_4 \\
d_6 &= 9 + d_5 \\
d_7 &= 10 + d_6
\end{aligned}
$$

Optimum Solution

Our calculations show that a clean napkin will be available for each guest if the purchases and laundry shipments indicated below are followed. The total cost would be $8.80, with only 21 new napkins purchased to service the 52 guests expected during a week.

Purchases and Shipments

for a Typical Week of Tea Parties

$n_1 = 5$			$d_1 = 0$
$n_2 = 6$			$d_2 = 0$
$n_3 = 7$	$k_1 = 0$		$d_3 = 0$
$n_4 = 3$	$k_2 = 0$	$q_1 = 5$	$d_4 = 0$
$n_5 = 0$	$k_3 = 1$	$q_2 = 6$	$d_5 = 2$
$n_6 = 0$	$k_4 = 3$	$q_3 = 6$	$d_6 = 9$
$n_7 = 0$	$k_5 = 5$	$q_4 = 5$	$d_7 = 10$
Totals: 21	9	22	21

Total cost: 21 x .25 + 9 x .15 + 22 x .10 = $8.80

We feel confident that a mathematical approach to the running of tea parties can bring about great savings. We trust that the above analysis will be implemented and extended throughout the operations of Mad Hatter, Inc.

Historical Footnote

The caterer's problem first appeared in the literature in the guise of a military application. This thinly veiled attempt to hide the true significance and power of this model was quickly surmounted. For completeness, we note the original statement of the problem. Instead of a caterer, a military commander needs to supply aircraft engines (napkins) based on specified requirements (number of guests per day). He has a choice of buying new ones or scheduling the overhaul of repairable engines to make them

available again to meet the requirements. The overhaul can be accomplished in an expedited manner (fast service) or in the normal, less costly fashion (slow service). Of course, a new engine costs more than either of the overhaul procedures.

We cite this military application to demonstrate that the adaptability of many of the linear-programming models is a function of the ingenuity of the analyst. One can, for example, show that the caterer's problem is really a transportation problem in disguise.

8 I'll Take the Middle Slice

The Trim Problem

This was the most unkindest cut of all.

SHAKESPEARE

8.1 HOW TO MAKE AND CUT CELLOPHANE

As part of the course requirements of the graduate school's class in Advanced Topics in Operations Research—OR 41.519/Fall Semester—each student must visit an industrial or operationally oriented organization and develop a case study for some aspect of the company's activities. In the past, ground-breaking term papers submitted included "Mathematical Decision Making as Applied to Garbage Collecting," "Queuing Discipline within a Supermarket," and "Traffic Control within a Hydrous Environment—A Case Study of Traffic Flow in Venice, Italy."

This year's class formed the usual two- or three-person teams; each group was assigned a company and instructed to delimit a problem area. This chapter describes the work of one of these teams.

The company selected for our team was a highly industrialized chemical and chemical-goods manufacturer. Such an organization has the full range of scheduling, inventory, transportation, and related operational problems. After a tour of the various plants, our team decided to study the operations of the company's cellophane-manufacturing activity. It was fascinating to watch how the finished product came to be.

Viscose, which originates from the reaction of wood pulp, sodium hydroxide, and other chemicals, is forced through a long narrow slit into a bath containing sulfuric acid. The viscose is immediately converted into cellulose in sheet form. The film then passes through a series of baths where it is purified, washed, dried over heated rollers, and finally wound—like a carpet—into large rolls of finished cellophane.

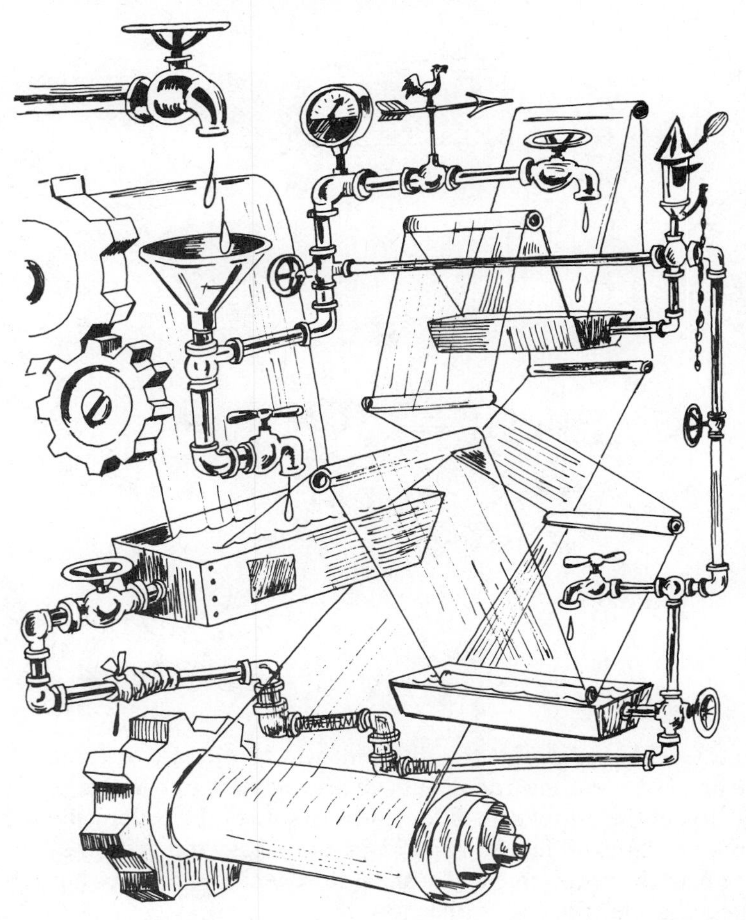

These rolls of cellophane are then stacked in a warehouse waiting to be sliced into rolls of smaller width, depending on the orders to be filled. Each large roll is 5 feet wide. When a roll is to be cut into smaller-sized rolls, it is transported to the cutting room and mounted on the cutting machine. This device unrolls the 5-foot sheet of cellophane, but as it does so, the cellophane is sliced by prepositioned blades, and the smaller widths are then rolled at the other end of the machine. The blades are set to yield a number of smaller rolls for customer orders, and if possible, any leftover roll would have a width which is saleable. For example, a 5-foot roll (60 inches) could be cut into three 15-inch rolls, a 10-inch roll, and a 5-inch roll. But, if no one ever orders a 5-inch roll, then this small roll has to be turned into scrap material or destroyed; it is called the trim loss for that setting of the cutting blades. Each setting of the blades (here we require four blades to cut the 5-foot roll into the five smaller-width rolls) yields rolls which can be marketed or are trim loss.

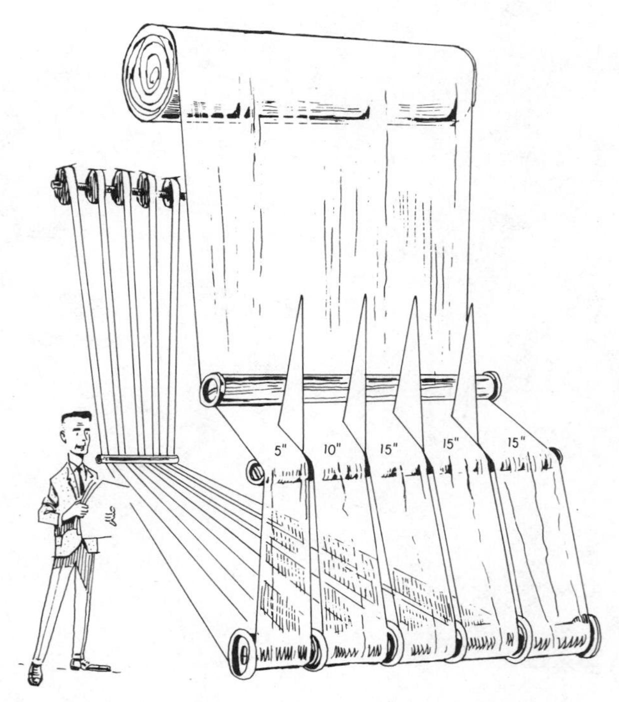

As the week's orders for the different-width rolls are accumulated, the supervisor of the cutting room puts them together into order groups and attempts to cut the 5-foot rolls into widths to satisfy the orders and minimize the trim loss. An experienced supervisor can juggle and combine the orders so as to fill all the requirements for the different sizes and does a pretty good job in keeping the trim loss rather low.

8.2 SAVING TRIM THE MATHEMATICAL WAY

During their tour of the plant the members of the study team noted piles of trim loss being shredded and packed off. Could they cut down on this apparently wasteful and costly part of the operation? They were quick to note that the real decision-making activity was imbedded in the experience of the supervisor, and then attempted to quantify the supervisor's heuristic approach to optimization. Based on work originally done at the Abitibi Power and Paper Company of Canada, the team proposed to try out a linear-programming model which yielded substantial savings for the paper industry. They decided to illustrate the model by comparing the supervisor's decisions with those recommended by the solution of the linear program for the same week's data. In situations like this where the newfangled mathematics is pitted against human experience, it is psychologically and strategically important to obtain the cooperation and interaction of the human element. Although novices at this game, the members of our team were smart enough to make friends of the supervisor and to involve the super in their experiment—or so they thought.

As supplied by the supervisor, the test week's orders for rolls were as follows:

Widths Ordered (inches)	Number of Rolls Ordered
28	30
20	60
15	48

The cutting room was required to cut at least thirty 28-inch-wide rolls, sixty 20-inch-wide rolls, and forty-eight 15-inch-wide rolls from the supply of standard 60-inch-wide rolls. It is assumed that there are enough of the large rolls available to yield a week's orders for the smaller rolls. For the type of cellophane processed that week, any leftover roll less than 15 inches wide was to be considered a trim loss.

Key to the formulation of the corresponding linear model is the definition of the appropriate variables. Each setting of the cutting blades yields a set of smaller rolls, and the problem then becomes the determination of how the blades should be positioned and, for a given setting of the blades, how many large rolls should be cut. For example, to obtain rolls 28 inches wide, two blades can be used to cut the 60-inch-wide roll into three pieces—two rolls 28 inches wide and one roll 4 inches wide. The former rolls can be used to fill the orders, while the latter roll is trim loss. Thus, the first thing to be done is to determine which setting of the blades yields rolls that can be used. Each distinct setting is a decision variable of the problem, and the value of each variable represents how many 60-inch rolls should be cut at the corresponding setting of the blades. We let x_1 be the number of times we cut a 60-inch roll into two 28-inch rolls, with a 4-inch trim loss, and let other appropriate blade settings be variables as noted in the following table:

Widths ordered (inches)	x_1	x_2	x_3	x_4	x_5	x_6	x_7	Number of rolls ordered
28	2	1	1	0	0	0	0	30
20	0	1	0	3	2	1	0	60
15	0	0	2	0	1	2	4	48
Trim loss	4	12	2	0	5	10	0	

Variable x_3 represents the number of times a 60-inch roll is cut into one 28-inch roll and two 15-inch rolls, with a trim loss of 2 inches. The other variables are similarly defined. Each variable is nonnegative (≥ 0)—we cut rolls at that blade setting or we do not. The mathematical model is now rather easy to set down.

In setting up the equations, our team determined that to give greater flexibility or freedom of action in the decision process, the manufacturer allows for more rolls of the desired widths to be cut than there are orders. Any overage can be stored and used to help fill the next week's orders. If this were not the case, then we would have to cut exactly thirty 28-inch rolls, sixty 20-inch rolls, and forty-eight 15-inch rolls. Hence, our constraints allow us to make more than the ordered amounts and are represented by inequalities instead of equalities.

After constructing the above tableau, our team was able to write the three major constraints of the linear-programming model, one constraint for each width. For the 28-inch rolls we have

$$2x_1 + x_2 + x_3 \geq 30$$

for the 20-inch rolls,

$$x_2 + 3x_4 + 2x_5 + x_6 \geq 60$$

and for the 15-inch rolls,

$$2x_3 + x_5 + 2x_6 + 4x_7 \geq 48$$

Also, all the x's must be greater than or equal to zero (≥ 0). Finally, the total trim loss is measured by

$$4x_1 + 12x_2 + 2x_3 + 0x_4 + 5x_5 + 10x_6 + 0x_7$$

The team had to find the set of x's which satisfied the inequalities and minimized the total trim-loss measure.

One solution, not necessarily the minimal solution, should be the one used by the supervisor. Other solutions are readily available. If $x_1 = 15$, $x_4 = 20$, and $x_7 = 12$, and all other x's zero, we see that the constraints are satisfied as equalities with a total trim loss of $4x_1 + 0x_4 + 0x_7 = 4(15) = 60$ inches. This solution states that we should cut 15 large rolls using the blade positions of x_1 (cut two 28-inch rolls and one 4-inch roll), cut 20 large rolls with the cutting pattern of x_4 (cut three 20-inch rolls), and cut 12 large rolls as given by the x_7 (cut four 15-inch rolls)—this yields a loss of only 60 inches. Are there any other solutions with a smaller trim loss?

If $x_3 = 30$ and $x_4 = 20$, we again satisfy the constraints (we have more 15-inch rolls than needed), and we also have a 60-inch trim loss. We leave it to the reader to demonstrate that either of these solutions is the minimal-trim solution.

With their solution in hand, our team reviewed the approach with the supervisor and attempted to compare the super's decisions with the model's results. The supervisor noted that such a comparison was impossible and that the team abstracted the real-life problem to fit their model; the supervisor then proceeded to show them where they went wrong.

The decision process of how to cut a roll could not be made independent of the physical characteristics of an individual roll. Where in the paper industry a roll is usually made without imperfections, the cellophane rolling process in use at this plant rolled the 5-foot roll in an erratic manner. Most rolls looked like a rolled carpet with one end pushed in and the other telescoping out. It was rare that a full 5-foot roll was available. The ends had to be lopped off and the resultant shorter roll used to fill the orders. A good portion of the trim loss being shredded and packed was due to the manufacturing process instead of the cutting process. Also, because of the difficulty in keeping the proper tension on the cellophane as it was rolled from the vats, bad spots appeared in some rolls and had to be cut away like knotholes. This also contributed to the pile of trim loss, as well as shorter rolls.

After finding this out, the team really went to work. They took measurements on how trim loss was generated—bad rolls, bad spots, bad cutting. They found that the supervisor contributed little to the trim loss. In fact, when the supervisor found a bad spot in a roll, adjustments in the cutting positions could be made at that time and the orders regrouped to take advantage of such anomalies.

Our team wrote up their experiences in model building with a recommendation that the cutting process be looked at in detail, and they outlined an approach. It called for instrumenting the cutting machine with electric eyes to detect bad spots and rolls, and tying the measurements to a computer. Then, utilizing linear-programming techniques, the team determined how the blades should be set for an individual roll based on the remaining orders. The blades would be automatically set by the computer. The cost was estimated, and a preliminary cost-effectiveness study was made, comparing the present mode of operation and this proposed technically advanced method. The team concluded with a strong recommendation for keeping the supervisor.

8.3 EPILOGUE

It has been noted that the person trained in dealing with reasonably sized optimization problems can solve both his or her problem and come rather close to the optimum. This was the case with Stigler and the diet problem, and our supervisor. As the problems become more complex, however, the individual tends to lose out as the power of the mathematical model becomes overwhelming. But not always.

9 You're in the Army Now

The Personnel-Assignment Problem

Never volunteer.

ANONYMOUS

9.1 CAMP LP WITH ABLE, BAKER, AND CHARLIE

In the old days of World War II, a draftee was given a variety of aptitude tests to determine the most suitable matching of his skills and the Army's needs. Tests were taken to measure ability in radio, mechanics, and electricity; the scores were analyzed; needs were studied; and the soldier found himself in the infantry.

Today, things are different. Tests are still taken, needs are still analyzed, but there is a good chance that the soldier will end up training for a military career that does have some resemblance to his or her aptitude profile. The problem of matching job requirements to available personnel resources has been a continuing research program of the Army, as well as industrial organizations. Linear programming has been an important research tool in the field of personnel classification. The personnel-assignment problem is central to this research, and it can be stated and solved as a linear-programming model.

Let us look in at a recruitment center when business is slow. Only three soldiers are being processed by the center at Camp LP. For obvious reasons, we shall name them Able, Baker, and Charlie. They have taken a series of tests to determine their aptitude for careers as a radio operator, computer programmer, and clerk. Their scores are shown in the following array:

	Radio	Computer	Clerk
Able	5	4	7
Baker	6	7	3
Charlie	8	11	2

The higher the score, the higher the aptitude of the recruit for the corresponding job. Charlie would probably do really well as a computer programmer, but be a dud as a clerk. The center has been given the quota for three men—our Able, Baker, and Charlie—and the center must assign them to three available jobs: one each in radio, computer, and clerical schools. The problem faced by the center is how should the assignments be made in order to maximize the utility of the new recruits' service to the Army.

The linear-programming and psychological-testing approach to the solution of this problem is to assume that the scores on the tests enable us to measure the worth of a person assigned to the corresponding job and that the total worth of all the assignments is the sum of the scores. For example, if Able is assigned the first job, Baker the second, and Charlie the third, as shown by the assignment table

	Radio	Computer	Clerk
Able	1	0	0
Baker	0	1	0
Charlie	0	0	1

then the total score is $5 + 7 + 2 = 14$. We may argue the validity of measuring the total worth in this manner—we are assuming a linear relationship between the worth of the individual assignments.

The above table is called an assignment table as, rightly enough, it assigns one person to only one of the jobs—each person to a job and each job to a person. Thus, an assignment table is a square array with exactly one 1 in each column and row. Another assignment for the three recruits is given by

	Radio	Computer	Clerk
Able	0	1	0
Baker	1	0	0
Charlie	0	0	1

This assignment has a value of $6 + 4 + 2 = 12$. For our 3×3 problem, there are only $3 \times 2 \times 1 = 6$ possible assignments. It is easy to exhaust all the possibilities and show that the assignment

	Radio	Computer	Clerk
Able	0	0	1
Baker	1	0	0
Charlie	0	1	0

yields the maximum score of $6 + 11 + 7 = 24$. When the number of people to be assigned becomes rather large, such an exhaustive treatment becomes impractical. However, a linear-programming model for this application can be developed in a rather straightforward manner.

9.2 THE ASSIGNMENT MODEL

Here it would help to consider the problem formulation in the light of our knowledge of transportation problems. In essence, we have a number of origins called Able, Baker, and Charlie; each such origin has one unit of material—a recruit—to be shipped to some destination. The destinations—radio, computer, and clerk—each require one unit of the available three units. Unlike the transportation problem, we wish to determine the shipping pattern which *maximizes* the total shipping cost—in reality, the sum of the test scores for the assignments. Arranging the data in a tableau similar to the transportation problem, we have

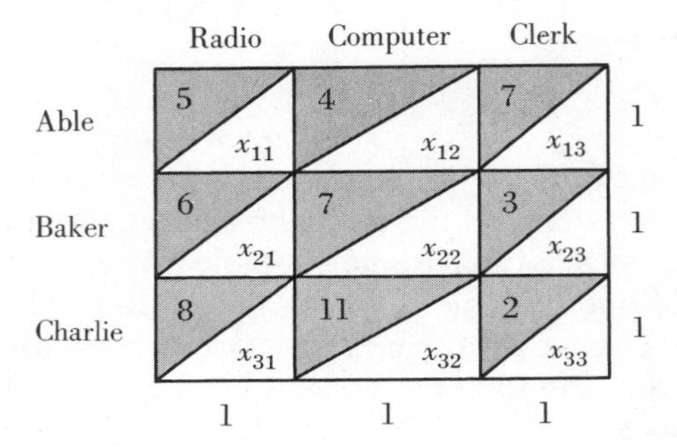

The variables are interpreted as an assignment of the corresponding person to the corresponding job; that is, x_{23} represents the assignment of Baker to the clerical school. Each variable must be positive or zero, and in fact, each variable is restricted to be either a one or a zero. The optimal solution given above states that $x_{21} = 1$, $x_{32} = 1$, $x_{13} = 1$ and all the other x's are zero. Taking our lead from the transportation problem, the linear-programming model of the personnel-assignment problem is to find nonnegative values of the variables, $x_{ij} \geq 0$, which maximizes

$$5x_{11} + 4x_{12} + 7x_{13} + 6x_{21} + 7x_{22} + 3x_{23} + 8x_{31} + 11x_{32} + 2x_{33}$$

subject to the constraints

$$\begin{array}{lllllllllll}
x_{11} + x_{12} + x_{13} & & & & & & & & & & = 1 \\
 & & & x_{21} + x_{22} + x_{23} & & & & & & & = 1 \\
 & & & & & & x_{31} + x_{32} + x_{33} & & & & = 1 \\
x_{11} & & & + x_{21} & & & + x_{31} & & & & = 1 \\
 & x_{12} & & & + x_{22} & & & + x_{32} & & & = 1 \\
 & & x_{13} & & & + x_{23} & & & + x_{33} & & = 1
\end{array}$$

The solution to this problem has an explicit requirement that variables take on integer values—either zero or one. Fortunately, the mathematical structure of the problem is the same as the transportation problem, and we are guaranteed an optimal solution in integers. In the trim problem, we made no mention of integer requirements. The trim model can yield an optimal solution which calls for the application of a blade setting a fraction of a time. For manufacturing problems, there is usually no harm done if the fraction is rounded up—it just costs a little more. When assigning a given number of people, however, we must have a solution in integers.

Some assignment problems might require an objective function which is to be minimized instead of maximized. If we were assigning workers to different types of jobs and the "test values" represented the time it took a worker to perform the corresponding job, we would want to determine the assignment which minimized the total time. In any event, the same computational procedure of linear programming can solve either type of optimization problem.[1]

An interesting variant of the assignment problem—and one which has been used to demonstrate the sociological importance of linear programming—is the marriage problem. Here, for example, we have 100 men and 100 women, and we wish to pair them off so that the total "happiness" of the pairings is maximized. We need to know a numerical score which measures the happiness of a coupling for each of the possible hundred pairings of a man with a woman. If we denote by x_{ij} the fraction of the time that the ith man spends with the jth woman and if our objective is to maximize overall happiness, then the mathematical model is just a large assignment model. As noted earlier, optimal solutions to such models are in terms of integers—here either 0 or 1. Thus, although each man is given a chance to fractionalize his time between a number of women, society's happiness is better served if he doesn't philander. As was pointed out by a reporter to George B. Dantzig, the formulator of this problem and the founder of linear programming, we could be working with the wrong kind of models.

[1] In Chapter 20, we describe a special algorithm for solving the assignment problem.

10 The Simple Furniture Company

The Activity-Analysis Problem

Have you heard of the wonderful one-hoss shay,
That was built in such a logical way
It ran a hundred years to a day?
OLIVER WENDELL HOLMES

10.1 THE MEETING OF THE BOARD

A modern way of looking at the manufacturing operations of an organization is to describe each item produced by the organization in terms of the amount of resources required to manufacture one unit of the item under consideration. The making of a unit—a unit of production activity—represents the utilization of some of the available resources. The problem of the production manager is to determine how many items of each unit to produce—the levels of activity—which enable the company to maximize profits subject to the restrictions imposed by available resources. This rather simplistic approach—the viewing of complex operations in terms of basic interrelated activities—is a powerful method for resolving and understanding a wide range of industrial processes. Inherent in this approach is the linear-programming activity-analysis model. For a discussion of this model, let us look into the monthly board of directors meeting of the Simple Furniture Company—our motto: *Proba mers facile emptorem reperit.*[1] President Simon is just starting to describe the new look for his company's manufacturing technology:

"Gentlemen and, of course, Mother Simon, a recent study of our manufacturing organization conducted by Super Management Consultants has brought new insights into how we can improve the operations of our company. In order to convey an understanding of this new approach in the few minutes allotted to this part of the program, I shall take some liberties and greatly simplify the discussion. I have much of what I am going to relate pictured on these flip-charts.

"On the first chart I have shown the four pieces of furniture that we make out of solid mahogany: chairs, tables, desks, bookcases. You will note that

[1] "Good merchandise finds a ready buyer"—Plautus.

besides the main ingredient of mahogany wood, some of the pieces use leather; we use sliding glass doors for the bookcases; all use glue, plus a few minor items like screws, which we shall ignore. When I mention the word chair, we all picture one of our lovely designs as it appears on the showroom floor. From a manufacturing point of view, we come up with a different picture—like the one I have on the next chart.

"You see, one chair is really equal to 5 board-feet of mahogany, plus 10 hours of labor, 3 ounces of glue, and 4 square feet of leather. We just put these resources together so they look like a chair. A table is equivalent to 20 board-feet, 15 labor-hours, and 8 ounces of glue. Our pride and joy—the Simon-designed desk—is really 15 board-feet, 25 labor-hours, 15 ounces of glue, and 20 square feet of leather, and it is a beauty. The companion bookcase requires 22 board-feet, 20 labor-hours, 10 ounces of glue, and 20 square feet of glass.

"Let us take an average production week. When we start the week off, our v-p for production, Sid Simon, gets a listing of the resources available, does some fast figuring, and tells the production crew how many units of each piece of furniture to make. Now, my purpose here is not to throw Sidney out of a job, but to give him a decision-making aid which will enable him to increase profits. On this chart I have listed the total available resources for the average week:

20,000 board-feet of mahogany
4,000 labor-hours
2,000 ounces of glue
3,000 square feet of leather
500 square feet of glass

"What Sid has to do is to determine how he puts these resources together to make the most money. To do this he gets from our v-p for finance, Harry Simon, the latest profit figures for each unit, which I have listed here:

\$ 45 per chair
\$ 80 per table
\$110 per desk
\$ 55 per bookcase

How many units should we make? How can we get the most profit? Is Sid doing a good job? Let me show you how our new linear-programming model will answer these and other questions.

"I have to get a little mathematical, but I will try and use as little as possible. We've all been out of school a long time. I want to represent the number of units to make of a piece by the first letter of its name, c for chair, t for table, and so on. If $c = 10$, I build 10 chairs. For each chair I build, I use these many resources

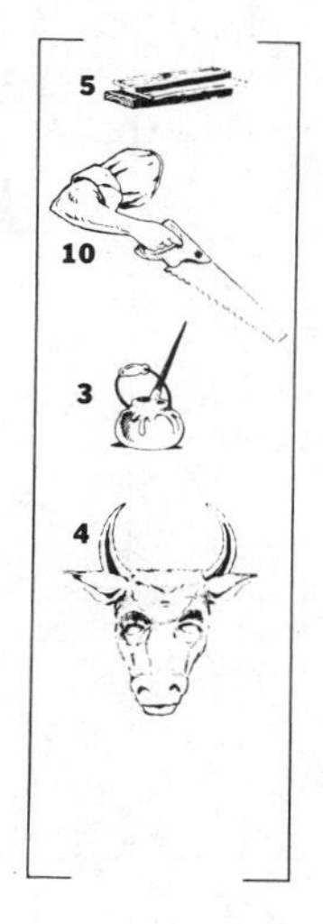

If I build c chairs, I use up c times these resources or

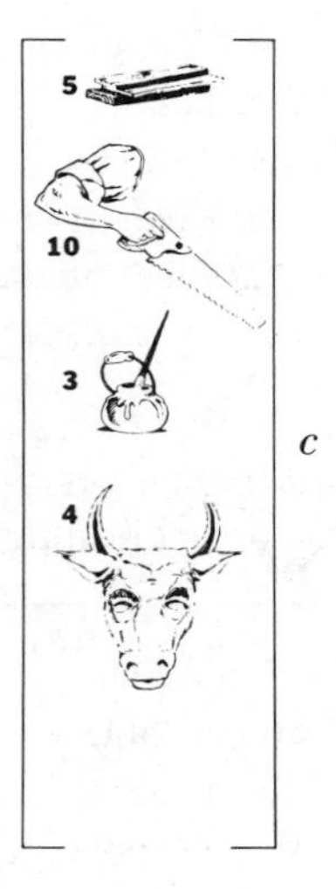

The same thing for the tables, desks, and bookcases. I can let the total amount used up in producing c chairs, t tables, d desks, and b bookcases be represented by

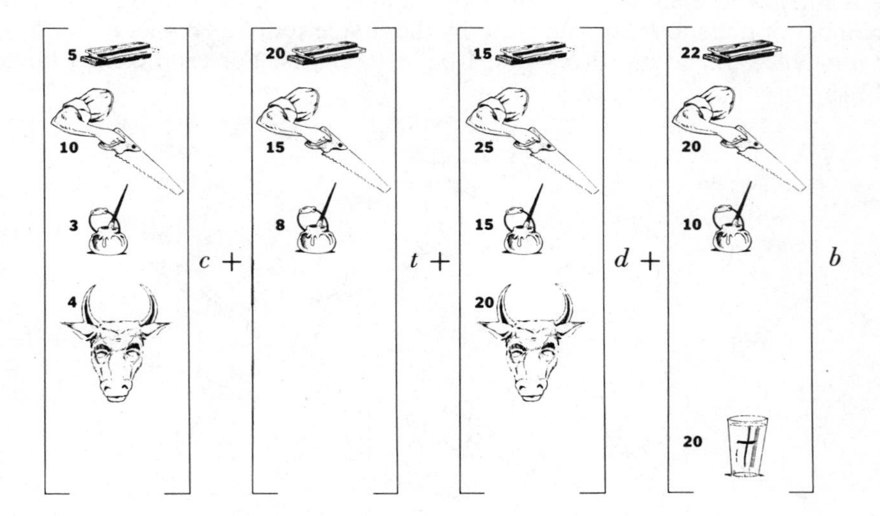

This total amount used must be less than or equal to the total resources

available, so I must have

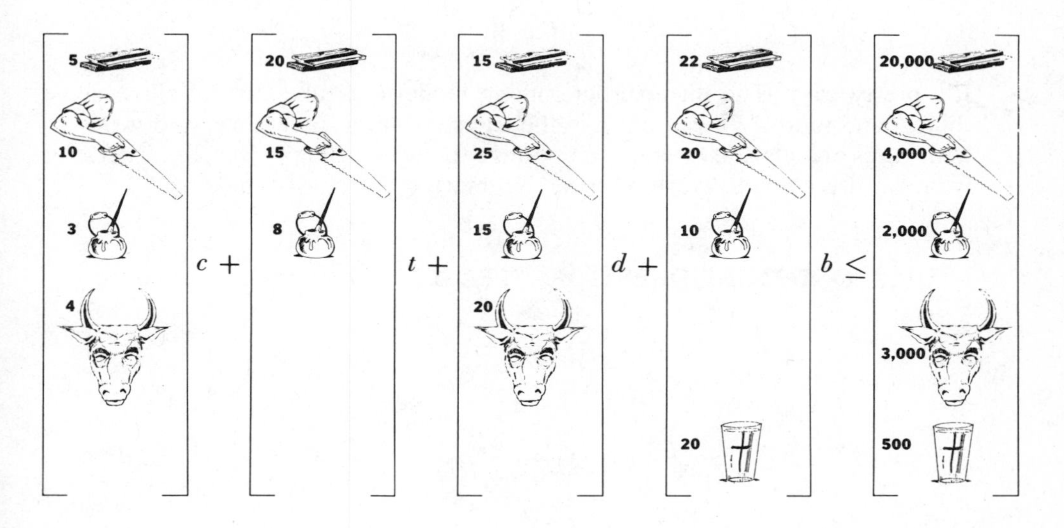

The profit to be made is represented by the sum

$$\$45c + \$80t + \$110d + \$55b$$

All we need to do now is to find the right values of c, t, d, and b which makes that sum as large as possible. This is what Sid tries to do. Linear programming guarantees giving us the maximum profit. Any questions? Sid.''

''You sure left out a lot of things. I have to make sure we meet the sales forecast. The people in the field call in with their orders and they want their orders right away. The way you glibly stated the problem I have no guarantee that I will make any chairs, for example. Also, I might end up making too many chairs, or only chairs. How can you take care of these things?''

''Simple, Sid. Let me write the inequalities without the pictures like this:

$$\begin{aligned}
5c + 20t + 15d + 22b &\leq 20{,}000 \\
10c + 15t + 25d + 20b &\leq 4{,}000 \\
3c + 8t + 15d + 10b &\leq 2{,}000 \\
4c \phantom{{}+ 20t} + 20d \phantom{{}+ 10b} &\leq 3{,}000 \\
20\,b &\leq 500
\end{aligned}$$

If you need at least 50 chairs all you add to the model is the inequality

$$c \geq 50$$

If you don't want more than 30 desks all you add is

$$d \leq 30$$

It's pretty easy. The linear-programming model can take care of all these things, and more. We can bring in the transportation, inventory, and warehousing problems and attempt to optimize the whole Simple System. That's what we are going to work on next. No more questions—time's up.''

10.2 ONE MORE QUESTION, PLEASE?

11 Part II Discussion, Extensions, and Exercises

11.1. The standard form of the linear-programming model can be stated as follows:

$$\text{Minimize } x_0 = c_1x_1 + c_2x_2 + \cdots + c_nx_n$$

subject to

$$\begin{aligned}
a_{11}x_1 &+ a_{12}x_2 + \cdots + a_{1n}x_n = b_1 \\
&\vdots \\
a_{i1}x_1 &+ a_{i2}x_2 + \cdots + a_{in}x_n = b_i \quad (i = 1, \ldots, m) \\
&\vdots \\
a_{m1}x_1 &+ a_{m2}x_2 + \cdots + a_{mn}x_n = b_m \\
& x_j \geq 0 \quad (j = 1, 2, \ldots, n)
\end{aligned}$$

where the c_j are termed *cost coefficients,* the a_{ij} are *technological coefficients,* the b_i are the *right-hand side elements* (we can assume the $b_i \geq 0$), the nonnegative x_j are the decision variables, and x_0 is the unknown value of the *objective function* (measure of effectiveness). A linear-programming problem can also be stated as a maximizing problem and contain a mixture of greater than or equal, less than or equal, and equality relationships. Such problems can always be converted to the above standard form. You will learn how this can be done. The equality model requires that $m < n$ (less equations than variables), otherwise there is no decision problem. Why?

11.2. You should be able to formulate the mathematical model and solve the following linear-programming problem (your first model exercise!) without much difficulty.

The recently elected treasurer of the Junior Class must manage the class's treasury of $2000 so as to have the most money for next year's Senior Class dance. Eight-hundred dollars must be placed in the class's interest-bearing checking account to take care of expected expenditures for the coming year, leaving $1200 for investment. The campus bank offers three types of investment possibilities: a savings passbook account that pays 8% interest, a 6-month certificate that yields 12% interest, and a 12-month, 14% certificate. There are interest penalities if the certificates are cashed in early, but there is no penalty if money is taken out of the passbook account. The bank will allow school organizations to open certificate accounts with no minimum-balance requirement. Any funds invested in a 6-month certificate would be reinvested for the next 6 months at the 12% rate, unless the money was required to pay unexpected expenses.

The treasurer recognizes that not all of the money should be placed into certificates as this would require a penalty payment if the $800 in the checking account were not enough to pay for the current year's expenses. In the hope of avoiding penalties and maintaining a ''liquid'' cash position, the treasurer has decided to impose the following prudent investment condition: some of the $1200 should go into the passbook account and it should open with an amount equal to $2.00 for every dollar in the 6-month certificate account and $4.00 for every dollar in the 12-month certificate account. Assuming that things go well and no certificates have to be turned in early, how should the treasurer invest the $1200?

Letting x_1 be the amount of money to be placed in the passbook account, x_2 the amount to be placed in the 6-month certificate account, and x_3 the amount to be placed in the 12-month certificate account, show that the optimal solution to the Junior Class's treasurer's problem is $x_1 = 800$ and $x_2 = 400$. What is the corresponding value of the objective function?

Reformulate and solve the problem with the following prudent investment conditions for the funds placed into the passbook account:

(a) The passbook account should open with an amount equal to $1.00 for every dollar in the 6-month certificate account and $2.00 for every dollar in the 12-month certificate.

(b) Let the dollar values in item 1 above be $1.00 and $3.00, respectively.

(c) The ratio of the money placed in the passbook account to the sum of the monies placed in both certificate accounts must be 4 to 1.

(*Hint:* To solve these problems you can substitute for x_1 the expression for x_1 obtained from the prudent investment condition. The substitution is allowed as we know that x_1 will be nonnegative in the final solution. We then have a single equation in x_2 and x_3, with the objective function in only x_2 and x_3.)

11.3. The original nut-mix problem is due to Charnes et al. and is as follows. A manufacturer wishes to determine an optimal program for mixing three grades of nuts consisting of cashews, hazels, and peanuts according to the specifications and selling prices given below (the prices are those of 1953!). The program is to determine how much of each grade should be mixed during a day's production period.

Grade mixture	Specifications	Selling price (cents/pound)
Grade A	Not less than 50% cashews. Not more than 25% peanuts.	$0.50
Grade B	Not less than 25% cashews. Not more than 50% peanuts.	$0.35
Grade C	No restrictions	$0.25

Note that hazels may be mixed into a grade in any quantity. The manufacturer has a daily capacity limit on the amount of each nut that can be processed and used as input. These capacities and the costs of the resources are as follows:

Resources	Capacities (pounds/day)	Costs (cents/pound)
Cashews	100	$0.65
Hazels	60	$0.35
Peanuts	100	$0.25

In analyzing such problems, a first task is to determine what are the decision variables. As we go about defining the decision variables and other elements of a problem, we must take great pains to use mathematical notation that is clear, as simple as possible, and aids in the development of the mathematical structure. Sometimes we find it convenient to use letters, names, or mnemonic devices as we did in the diet, caterer, and activity-analysis problems. Other times we need to use standard, straightforward variable notation as in the transportation problem and the trim problem. For the nut-mix problem it appears as if we need to get a handle on the number of pounds of each grade to be mixed during a day. We could use the letters A, B, C to represent these decision variables, but we shall find it convenient to let

X = the number of pounds of Grade A to be produced

Y = the number of pounds of Grade B to be produced

Z = the number of pounds of Grade C to be produced.

Each mixture is a blend of the three nuts, so we can express the X, Y, Z by the simple accounting relations

$$\begin{aligned} x_1 + x_2 + x_3 \qquad\qquad\qquad\qquad\qquad\qquad &= X \\ y_1 + y_2 + y_3 \qquad\qquad\quad &= Y \\ z_1 + z_2 + z_3 &= Z \end{aligned}$$

where

x_1 = the number of pounds of cashews in mixture X

x_2 = the number of pounds of hazels in mixture X

x_3 = the number of pounds of peanuts in mixture X

Similar definitions apply to (y_1,y_2,y_3) and (z_1,z_2,z_3). You should write out these definitions. We see that the decision variables X, Y, and Z are really defined by the corresponding decision variables expressed as x's, y's, and z's.

The capacity restrictions are readily taken care of by the sums

$$\begin{aligned} x_1 + y_1 + z_1 \qquad\qquad\qquad\qquad\qquad\qquad &\leq 100 \\ x_2 + y_2 + z_2 \qquad\qquad\quad &\leq 60 \\ x_3 + y_3 + z_3 &\leq 100 \end{aligned}$$

These constraints just state that the amounts of cashews, hazels, and peanuts that can be processed have to be less than or equal to the stated daily capacities.

More difficult to express are the conditions implied by the specifications for Grades A and B. Grade A must consist of at least 50% cashews and not more than 25% peanuts. In terms of X and x_1 (cashews) and x_3 (peanuts) these conditions state that

$$x_1 \geq \tfrac{1}{2}X$$

and

$$x_3 \leq \tfrac{1}{4}X$$

Grade B must have at least 25% cashews and not more than 50% peanuts;

we then have

$$y_1 \geq \tfrac{1}{4}Y$$

$$y_3 \leq \tfrac{1}{2}Y$$

The only other condition we need to take care of is the one that requires any use of a resource to be either at a positive or zero level, that is, all the decision variables must be nonnegative as expressed by $(x_1, x_2, x_3, y_1, y_2, y_3, z_1, x_2, z_3) \geq 0$. Variables X, Y, and Z are thus forced to be nonnegative. Why?

To complete the model we need to express the measure of effectiveness as a function of the decision variables. We are concerned with net revenues, that is, total revenues from sales minus total cost of resources. Using our notation and the given prices, and letting x_0 equal net revenues, we have as our model (after rearranging a few terms):

$$\begin{aligned} \text{Maximize } x_0 = {} & 0.50X + 0.35Y + 0.25Z - 0.65(x_1 + y_1 + z_1) \\ & - 0.35(x_2 + y_2 + z_2) - 0.25(x_3 + y_3 + z_3) \end{aligned}$$

subject to

$$\begin{array}{llllllllllllcr}
-X & + x_1 & + x_2 & + x_3 & & & & & & & & & = & 0 \\
 & & & & -Y & + y_1 & + y_2 & + y_3 & & & & & = & 0 \\
 & & & & & & & & -Z & + z_1 & + z_2 & + z_3 & = & 0 \\
 & x_1 & & & & + y_1 & & & & + z_1 & & & \leq & 100 \\
 & & x_2 & & & & + y_2 & & & & + z_2 & & \leq & 60 \\
 & & & x_3 & & & & + y_3 & & & & + z_3 & \leq & 100 \\
-\frac{1}{2}X & + x_1 & & & & & & & & & & & \geq & 0 \\
-\frac{1}{4}X & + & & + x_3 & & & & & & & & & \leq & 0 \\
 & & & & -\frac{1}{4}Y & + y_1 & & & & & & & \geq & 0 \\
 & & & & -\frac{1}{2}Y & & & + y_3 & & & & & \leq & 0 \\
\end{array}$$

$$(X, Y, Z, x_1, x_2, x_3, y_1, y_2, y_3, z_1, z_2, z_3) \geq 0$$

The optimal solution to the nut-mix problem is to produce only Grade A at a rate of $X = 200$ pounds per day that consists of $x_1 = 100$ pounds of cashews, $x_2 = 50$ pounds of hazels and $x_3 = 50$ pounds of peanuts. The net revenue $x_0 = \$5.00$ per day. By analyzing the selling prices and the costs of the resources, you should be able to show why it does not pay to make any Grade B or Grade C. (*Hint:* The selling prices are rather low in comparison to the costs of the resources that would go into any mixture. In fact, they hardly cover the resource costs, let alone the costs of manufacturing, storage, shipping, and so on. Why and how this manufacturer stays in business is a good question.)

11.4. Explain why the decision variables X, Y, and Z are really not needed in the mathematical formulation as given in Section 11.3. Rewrite the model in Section 11.3 without these variables. For hand calculations it pays to get rid of equations that are redundant. But, if we were solving the problem on a computer, and since we need to know what X, Y, and Z are, there is no harm in keeping these equations and letting the computer do the calculations for us.

11.5. The diet problem as described earlier (Chapter 6) does not take into consideration the need for variety, taste, and personal biases. To make the diet model more realistic, describe a breakfast-diet problem that includes, besides the two cereals K and C, eggs, milk, toast, bacon, and ham. You will need to collect nutrient data (thiamine, niacin, and calories) for these foods and their costs. You will probably have difficulty in interpreting the answer for it might call for a fraction of an egg; you would then have to scramble it and take the correct measure. How would you make sure that your diet called for at least one egg and a piece of toast?

11.6. Discuss how you would extend the diet problem to include breakfast, lunch, and dinner? How would you separate those foods that you want to eat at a particular time? Many institutions (hospitals, jails) use a variation of the diet model called menu planning. Here the selections are not individual foods, but a menu designed for nutrient content, taste, color, variety, and so on. A menu might consist of an entree, two vegetables, bread, salad, appetizer, and dessert. The problem then is to select a menu plan for a time horizon (say a month) that meets the nutrient requirements, minimizes cost, and does not repeat a menu too often. You might try your skill at formulating a model that yields a menu plan.

11.7. The U.S. Department of Agriculture suggests the following as recommended daily allowances (RDA's) for the important nutrients:

Protein	65 grams
Vitamin A	5000 International Units
Vitamin C	60 milligrams
Thiamin	1.5 milligrams
Riboflavin	1.7 milligrams
Niacin	20 milligrams
Calcium	1.0 milligrams
Iron	18 milligrams

The following table lists the calories and the percentages of the RDA's recommended for a male and female between the ages of 19–22 years:

		Percentages of the RDA's							
	Calories	Protein	A	C	Thiamin	Riboflavin	Niacin	Calcium	Iron
Male	3000	85	100	75	100	110	60	80	60
Female	2100	75	80	75	75	85	35	80	100

Using these figures, construct a diet problem (daily, weekly, or monthly) that involves at least 15 foods, using actual dietary data for each food. You should attempt to use constraints that reflect your preferences for certain foods, for example, steak or vegetables being forced into the solution. How do these preferences impact on the cost of your diet? Describe your approach to modeling the diet problem in terms of the decision framework of Part I. What are your alternative solutions?

11.8. For the trim problem, show that $x_1 = 3$, $x_3 = 24$, and $x_4 = 20$ is also an optimal solution. We now have three solutions with the same trim loss. Explain why you would choose a particular one. (*Hint:* Some solutions yield extra rolls to be stored, while others require less blade-setting changes with reduced setup costs and idle time.)

11.9. Reformulate the caterer's problem if the King's laundry takes one day and the Queen's laundry takes four days. Also, consider the problem if a third laundry, the Prince's laundry, starts operating. Formulate the caterer's problem if the King's laundry takes one day, the Queen's laundry takes two days, and the Prince's laundry takes three days. The cost of laundering a napkin by the Prince is 5 cents. The caterer's problem can be transformed into a transportation problem. Try to do this with the original problem. [*Hint:* Let x_{ij} be the number of dirty napkins sent to a laundry (slow or fast) on day i to be returned as clean napkins on day j. We are allowed to launder napkins in advance and do not store dirty napkins unless they will not be laundered at all.]

11.10. A furniture manufacturing company wishes to determine how many tables, chairs, desks, or bookcases it should make to optimize the use of its available resources. These products utilize two different types of lumber, and the company has on hand 1500 board-feet of the first type and 1000 board-feet of the second. There are 800 labor-hours available for the total job. Sales forecast plus back orders require the company to make at least 40 tables, 130 chairs, 30 desks, and no more than 10 bookcases. Each table, chair, desk, and bookcase requires 5,1,9,12 board-feet, respectively, of the first type of lumber and 2,3,4 and 1 board-feet of the second type. A table requires 3 labor-hours to make, a chair 2 hours, a desk 5 hours, and a bookcase 10 hours. The company makes a total of $12 profit on a table, $5 on a chair, $15 on a desk, and $10 on a bookcase. Write out the linear-programming model of this problem in terms of maximizing profit. How can you cause four chairs to be made for every table?

11.11. The following problem can be handled as an activity-analysis problem. Formulate the linear-programming model.

A bakery starts the day with a certain supply of flour, shortening, eggs, sugar, milk, and yeast. It specializes in making bread, cakes, English muffins, and cookies. The recipes are given in the following table (we ignore such plentiful supplies as salt, water, etc.).

	(1) Bread	(1) Cake	(6) Muffins	(60) Cookies	Available resources
Flour (cup)	12	3	$4\frac{1}{2}$	$1\frac{1}{2}$	500
Shortening (tbsp)	2	12	3	4	350
Eggs	0	3	1	1	120
Sugar (cup)	$\frac{1}{4}$	$\frac{3}{2}$	$\frac{1}{8}$	1	300
Milk (cup)	2	$\frac{3}{4}$	1	0	250
Yeast (cake)	1	0	1	0	100
Profit	\$0.15	\$0.45	\$0.33	\$0.42	

11.12. This problem will test your ability to generalize the formulation of Section 11.11 and combine it with the transportation formulation. The statement of the problem is not very specific and you should attempt to restrict the problem's scope and state assumptions so you will be able to develop a meaningful and useful model.

Assume the products of Section 11.11 can be made at two different oven locations (bakeries) for shipment to the firm's five retail stores and to 10 supermarkets. We want to develop a production planning model that maximizes net profit on a daily basis. You will need to be concerned with the recipes of each product, the costs and availability of the recipe materials (e.g., flour, eggs, etc.), oven production capacities and costs at each bakery (e.g., available oven time and the amount of each product that can be baked), the rates at which the bakers can handle each product and labor costs, the daily demand and sales prices for each product, and the transportation costs from each oven to each outlet. You might want to start this problem by working with a single product, single oven and single sales outlet, and then generalize your formulation. Try and visit a bakery to determine what actually goes on and whether your final model could be of any use to the firm.

11.13. The range of problem situations that can be handled and solved by the linear-programming model is most extensive. The previous discussions illustrate just a few. It is not our purpose in this text to make you an expert in formulating such models. You should, however, recognize the basic forms and be able to adapt and generalize them. We have the diet and blending problems in which raw materials must be mixed to satisfy restrictions, activity-analysis problems that require products to be made out of limited amounts of resources, transportation problems in which items are to be shipped or assigned, and special, but important, structures like the caterer's problem and trim-loss problem. In what follows, we shall describe a few other problem and practice situations you should become familiar with.

At this point we should like to stress that there are no set procedures and techniques you can apply to determine and test the linear-programming formulation of a given problem. The casting of a new and complex problem into a linear-programming form appears to be a process of evolution. Your first impulse is to define what appears to be the variables of the problem and to determine the interrelationships between these variables, the resulting constraints, and appropriate objective function. Problems based on either actual or test data are then solved, using the initial model, and the solutions compared with expected results. These studies will in all probability suggest that new variables must be defined and introduced into the system and that previously defined variables can be eliminated; changes in the objective function and constraints, a reevaluation of the test data, and so on, may also result. Additional problems are then solved, and this process for evolving a correct model continues until you are satisfied that the resulting model approximates the real situation to an acceptable degree (or, possibly, are satisfied that the problem cannot be handled by linear-programming techniques). The above is just another way of saying that the decision-framework process described in Chapter 3 applies to linear-programming model formulation and application.

11.14. The telephone company schedules its operators based on estimates of daily demand by 15-minutes time periods. Similar scheduling problems are faced by hospitals in assigning nurses during the day, police departments in scheduling officers for shifts, and by many other service organizations. Let's look at the police problem as given in Wagner's text.

A police department has the following minimal daily requirements for police officers during its six shift periods:

Time of Day	Period	Minimal Number Required
2 a.m.–6 a.m.	1	22
6 a.m.–10 a.m.	2	55
10 a.m.–2 p.m.	3	88
2 p.m.–6 p.m.	4	110
6 p.m.–10 p.m.	5	44
10 p.m.–2 a.m.	6	33

An officer must start at the beginning of a 4-hour shift and stay on duty for two consecutive shifts (an 8-hour tour). Anyone starting during period 6 stays on duty during period 1 of the next day. The objective of the police department is to always have on duty the minimal number required in a period but to do so with the least number of officers. A source of difficulty is that the required numbers in the successive shift periods fluctuate quite a bit. The decision variables are how many officers to schedule at the beginning of each shift period; we denote these variables by x_1, x_2, x_3, x_4, x_5, and x_6, respectively. The measure of effectiveness, the linear-programming-model objective function, is just to minimize

$$x_0 = x_1 + x_2 + x_3 + x_4 + x_5 + x_6$$

The rest of the formulation proceeds in a straightforward manner. Assuming that the set of minimal number of officers required is the same from day to day, we have that the first-period officers plus those that started in period 6 of the preceding day must be greater than or equal to 22, that is,

$$x_1 + x_6 \geq 22$$

Then the second-period's 55 requirement must be satisfied by x_1 and x_2, that is,

$$x_1 + x_2 \geq 55$$

And in a similar manner, we have

$$\begin{aligned} x_2 + x_3 &\geq 88 \\ x_3 + x_4 &\geq 110 \\ x_4 + x_5 &\geq 44 \\ x_5 + x_6 &\geq 33 \end{aligned}$$

(of course all $x_j \geq 0$ for $j = 1,2,3,4,5,6$).

Putting it all together, the linear-programming model is to minimize

$$x_0 = x_1 + x_2 + x_3 + x_4 + x_5 + x_6$$

subject to

$$\begin{array}{llllllr} x_1 & & & & & +x_6 & \geq 22 \\ x_1 & + x_2 & & & & & \geq 55 \\ & x_2 & + x_3 & & & & \geq 88 \\ & & x_3 & + x_4 & & & \geq 110 \\ & & & x_4 & + x_5 & & \geq 44 \\ & & & & x_5 & + x_6 & \geq 33 \\ & & & & & (x_1, x_2, x_3, x_4, x_5, x_6) & \geq 0 \end{array}$$

You might try your skill at finding a minimal solution. For example, let $x_6 = 11$, $x_5 = 22$, $x_4 = 22$, and so on. Do you think a solution exists if the minimal requirements were made to be exact, that is, if the $\geq$'s were changed to $=$'s? What would you do if the optimal solution involved a fraction of an officer? How would the formulation change if you wanted to minimize costs and the cost of an officer depended on the time of the starting period? Do you think it would matter if the number of officers per period was a bit lower than the requirement? How about for nurses or for telephone operators? (Can you think of a way to modify the model so that the measure of effectiveness is to get as close to the requirements, but you can go above or below?)

11.15. Develop the data for the following trim problems and formulate the corresponding linear-programming models.

(a) A paper company analyzed by Paull uses a 215-inch standard roll to cut orders received from its customers. For the next production period, it must produce these widths and the corresponding numbers of rolls ordered:

Widths Ordered	Number of Rolls Ordered
64	782
60	624
48	142
45	118
33	144
32	826
16	188

You will soon find out that the tedious aspect of this problem is determining the many different combinations of ordered widths that can be cut from the 215-inch roll. (Try not to give up!) Remember that each combination represents a decision variable, for example, from the 215-inch roll the manufacturer can cut three 64-inch orders, one 16-inch order, and have a trim loss of 7 inches. When this type of linear-programming problem was first solved on a computer, the computer was used to calculate the many possible combinations. Now advanced linear-programming techniques are used that do not have to have the explicit statement of all combinations.

The answer to this problem is to use the following combinations and cut them the indicated number of times:

Roll width	Combination						
64	3				1	2	2
60		1		3			
48							1
45		2	1			1	
33			3				
32		2	2	1	4	1	1
16	1				1		
Trim loss (inches)	7	1	7	3	7	10	7
Number cut (solution)	109	12	48	204	79	46	142

Show that the orders are satisfied exactly, except for the 32-inch roll, and calculate the total trim loss.

(b) A generalization of the trim problem uses more than one standard roll. Such a problem from Taha has two standard rolls that are 10-feet and 20-feet wide. The manufacturer must cut 100 rolls that are 5-feet wide, 300 rolls that are 7-feet wide, and 200 rolls that are 9-feet wide. Formulate this problem as a linear program.

11.16. For Section 11.15, problem (b), assume that the ordered rolls came from two customers and were being cut at two factories. Discuss how you would introduce shipping costs and constraints into the problem.

11.17. The diet menu-planning model formulation is illustrated by the following problem due to Balintfy. Here we want to select one entree and one dessert that meets the dietary requirements for protein and vitamin C. All of the necessary information is contained in the tableau:

		Menu items					
		Entrees (serving)			Desserts (serving)		
		Stew x_1	Ham x_2	Veal x_3	Gelatin x_4	Cobbler x_5	Cake x_6
Protein (grams)	38	16	20	25	15	21	25
Vitamin C (milligrams)	20	20	15	12	10	8	5
Cost (cents)		23	30	38	20	25	35

The needs of this problem cause certain complexities that change the formulation from a standard linear-programming model to that of an *integer-programming model.* Here we require that exactly one entree and exactly one dessert be served. If an entree or dessert is served, it must be a complete portion and not a fractional amount. Up to now we have not been too concerned with formulations that gave answers that were not whole numbers. We could live with $2\frac{2}{9}$ ounces of a food as it could be measured out, or $3\frac{1}{3}$ chairs as we could round up or down without causing too much strain to the real-world problem situation. But when assigning people to jobs or police officers to shifts we like to deal with whole numbers. We did note that the transportation and assignment problems yield integer values due to the mathematical structure of the resulting model. This is also true for the police officer problem (11.14); it can be converted into a transportation model. (Try to find the formulation trick.) The mathematical structures of most models do not guarantee that the resulting optimal solutions will be in integers. The menu-planning problem is one of these. For this problem we have to add on two types of constraints. The first requires that all the variables are either zero or one, with a zero value meaning that the corresponding entree is not selected and a one meaning that it is. Also, since we only want one entree and dessert, the variables that represent the selection (serving) for the entrees must sum up to exactly one, and similarly for the variables for the desserts. The resulting model is to minimize

$$x_0 = 23x_1 + 30x_2 + 38x_3 + 20x_4 + 25x_5 + 35x_6$$

subject to

$$\begin{aligned}
16x_1 + 20x_2 + 25x_3 + 15x_4 + 21x_5 + 25x_6 &\geq 38 \\
20x_1 + 15x_2 + 12x_3 + 10x_4 + 8x_5 + 5x_6 &\geq 20 \\
x_1 + x_2 + x_3 \qquad\qquad\qquad\qquad &= 1 \\
x_4 + x_5 + x_6 &= 1 \\
x_j = 0 \quad \text{or} \quad 1 \qquad \text{for } j = 1, 2, \ldots, 6&
\end{aligned}$$

The *nonlinear integer restrictions,* that is, the $x_j = 0$ or 1, introduce a bit of computational complexity. We can solve just about any linear-programming problem using today's computers, but we cannot be sure of solving an integer-programming problem. Linear-programming problems of immense size (5000 constraints and 50,000 variables) are solved readily using the simplex algorithm[1] (not without computer cost of course). For integer problems with more than 100 variables the existing algorithms tend to break down.

So far we have not been too interested in solving our problems. Even though exact, optimizing methods might not exist, the formulation of a problem in mathematical terms can yield both insights into the decision problem and offer a framework for solving it. At worst, you might try solving the integer-programming problem as a linear-programming problem and hope for the best, that is, relaxing the integer constraints to just be $x_j \geq 0$, and for the menu problem restricting $x_j \leq 1$; or possibly work around the noninteger solution by judicious rounding or substitution. Watch out, however. The above menu-planning problem has the optimal integer solution of $x_2 = 1$ and $x_5 = 1$, with a cost of 55 cents. (You should be able to show that this is optimal.) The optimal continuous linear-programming solution is $x_1 = \frac{8}{9}$, $x_3 = \frac{1}{9}$, and $x_5 = 1$, with a cost of 49.67 cents. (Why would you expect the continuous solution to have a lower cost?) The corresponding rounded-integer answer of $x_1 = 1$ and $x_5 = 1$ does not form a feasible solution!

11.18. Discuss the following problem and "mathematical proof" that monogamy is the best of all forms of marriage. A pioneering colony of seven bachelors entices seven prospective brides to visit them. (You might have seen the movie *Seven Brides for Seven Brothers*.) After a suitable courting

[1] The simplex algorithm for solving linear-programming problems is described in Chapter 16.

period, they decide to have a mass wedding. Each bride is given a list of the seven names on which she writes her preferences on a scale of one to ten (remember the movie *10*?) with a ten representing her first choice, a nine her second choice, and so on. She may also cross out names unacceptable to her. We assume that for any assignment of brides to bachelors that the sum of the corresponding preference numbers is a valid measure of the anticipated happiness of the colony. Develop the model of this problem in terms of an assignment problem. Here the objective function is to maximize potential happiness. If there are feasible solutions, there will be an optimal solution that assigns one bride to one bachelor, that is, the optimal solution represents a monogamous assignment. What would happen if all the brides decide to cross out the name of the same bachelor? Do you think that this model can be used by computer-matching companies?

11.19. Try solving the following assignment problem. Five applicants have been interviewed for five jobs and their ratings for each job are as follows:

		Jobs				
		1	2	3	4	5
Applicants	1	2	6	1	2	3
	2	3	9	4	5	3
	3	4	3	5	4	3
	4	1	6	2	3	2
	5	6	5	3	2	3

We want to place the applicants so that the sum of the assigned ratings is maximized. In developing a heuristic algorithm to solve this problem, you might consider finding a good feasible solution to start with, and then systematically trying to make changes that increase the value of the objective function while maintaining a solution that is feasible. A good starting solution might be obtained by first assigning the applicant with the highest rating to the corresponding job and then moving to the remaining applicants and jobs and repeating the process.This turns out to be the right approach for this numerical problem. You might not be able to prove it unless you calculate all the possible $5 \times 4 \times 3 \times 2 \times 1 = 120$ solutions. The value of the objective function for the optimal solution is 26. Can you determine the correct assignments?

This type of algorithm is called a *greedy algorithm* in that you do the best at each step without worrying about future consequences. There are some optimization problems for which the greedy algorithm works. We will discuss one of them in a later chapter. Show that the greedy approach does not work for the following assignment problem:

		Jobs		
		1	2	3
Applicants	1	6	7	11
	2	6	10	9
	3	8	11	12

11.20. The assignment model can also be used when the objective is to be minimized. The ratings must then be interpreted properly, for example, they can be the costs of assigning a piece of work to be processed by a machine, where each machine can do the job. To illustrate, three workers can be assigned to run three different machines. Each employee can run each machine, but due to experience and salary levels the costs of completing the tasks that are processed by an employee on a machine are different. We might have the following cost table:

		Machines		
		1	2	3
Employees	1	11	7	10
	2	8	5	7
	3	7	4	8

You want to find the solution that minimizes the sum of the costs of the assignments. Try the greedy algorithm, and then show it does not find the optimal solution by calculating the possible $3 \times 2 \times 1 = 6$ assignments.

11.21. The following quotation is from the book *Fire Department Deployment Analysis*. It describes a set of interrelated problems faced by big city fire departments:

> A very large fire is raging. Many fire companies, both engine and ladder, are hard at work controlling and extinguishing the blaze. They will be occupied by these tasks for many hours; meanwhile, their firehouses will be empty and the region in which they ordinarily respond to alarms will be left unprotected.
>
> When this happens in a large city, the fire department usually temporarily relocates (moves) some fire companies from their firehouses in parts of the

> city that are still adequately protected to some of the empty houses. In smaller cities it may be necessary to achieve the same effect by borrowing companies temporarily from neighboring communities via a mutual assistance agreement. The purpose of such temporary relocations is clear—to spread out the still available firefighting resources in order to reduce and balance the risks and consequences that would result if other fires occur. The dilemma of relocation is equally clear: How much can be borrowed? When is the borrowing (or lending) justified? Precisely which companies should be relocated? Which of the empty houses should be filled?"

We are not able to analyze all these problems in this text, but our knowledge of the assignment model can be used to determine which of the available fire companies should be sent to the empty houses.

The procedures in the quoted book first determine that there is a need for relocation, then the empty houses to be filled, then the available fire companies that can be relocated, and, finally, the assignment of the available companies to the empty houses. What is required is a suitable measure of effectiveness. The measure used is to minimize total relocation distance, that is, the elements in the assignment "cost" table represent the distances a company to be moved must travel to get to each of the empty houses. This objective tends to keep companies within areas they are familiar with. For New York City, the analysts that wrote the fire department book found that it was rare that they had to move more than five companies to five empty houses. They could then solve this small assignment problem rapidly on a computer by just calculating all of the $5 \times 4 \times 3 \times 2 \times 1 = 120$ possible solutions. If you have a programmable hand calculator, you might want to program it to solve 5×5 assignment problems by finding all 120 possible assignments. Do you think your calculator could solve a 10×10 problem?

You might check with your fire department if relocating companies is a problem and if yes, how assignments are determined. Are you satisfied with the measure of effectiveness? What about minimizing the time for relocating?

11.22. There is an important class of integer problems called *set-covering problems* which you should be familiar with. The police officer scheduling problem (Section 11.14) is of this type. We shall illustrate the problem structure and model formulation by considering the problem faced by space-agency scientists in designing an experimental pay load for an orbiting laboratory. The laboratory will perform three experiments. Packages of instruments can be obtained that can carry out each one of the three experiments; or multipurpose packages can be purchased that can perform more then one experiment. Specifically, six packages have been proposed: packages 1, 2, and 3 can perform only the corresponding experiments; package 4 can do experiments 1 and 3; package 5 can perform experiments 1 and 2; and package 6 can carry out experiments 2 and 3. Each of the six packages have a corresponding weight (or cost). The problem is to select the package(s) to be placed in the laboratory so that each experiment can be accomplished (covered) and the total weight of the package(s) is minimized.

For each package we define an integer variable x_j equal to 1 if package j is chosen for the pay load and equal to 0 if it is not selected ($j = 1, 2, \ldots, 6$). Let the corresponding weight be denoted by c_j.

We arrange the data of the problem in the following tableau:

	x_1	x_2	x_3	x_4	x_5	x_6	Experiments	
	1	0	0	1	1	0	1	Experiment 1
	0	1	0	0	1	1	1	Experiment 2
	0	0	1	1	0	1	1	Experiment 3
Weight	c_1	c_2	c_3	c_4	c_5	c_6		

The one's in the last column represent the statement that each experiment must be accomplished. Thus, the three constraints of the system that correspond to the rows of the tableau can be written

$$
\begin{aligned}
x_1 \qquad\qquad + x_4 + x_5 \qquad &\geq 1 \\
x_2 \qquad\qquad + x_5 + x_6 &\geq 1 \\
x_3 + x_4 \qquad\quad + x_6 &\geq 1 \\
x_j &= 0 \quad \text{or} \quad 1
\end{aligned}
$$

The objective function is to minimize the total weight, that is,

$$
\text{Minimize } x_0 = c_1x_1 + c_2x_2 + c_3x_3 + c_4x_4 + c_5x_5 + c_6x_6
$$

From the above formulation we see that by letting $x_1 = x_2 = x_3 = 1$ and $x_4 = x_5 = x_6 = 0$ that all the experiments are covered exactly with a weight of $x_0 = c_1 + c_2 + c_3$. But, there is nothing in the statement of the problem that rules out the feasible solution of $x_1 = x_2 = x_3 = x_4 = x_5 = x_6 = 1$ with a weight of $x_0 = c_1 + c_2 + c_3 + c_4 + c_5 + c_6$. This solution provides maximum redundancy (coverage or backup) in the ability of the laboratory to perform all the experiments, but this solution weighs the most. Other solutions also provide some backup, that is, only $x_5 = 1$ and $x_6 = 1$. Covering problems are integer-programming problems with greater than or equal to constraints. If the constraints are equal only, they are called *set-partitioning problems*.

11.23. Formulate the following problems as set-covering or set-partitioning problems:

(a) A delivery firm has three large boxes to deliver and six different trucks that can carry loads of varying sizes. Truck 1 can carry only order 1, truck 2 must carry orders 1 and 3 together, truck 3 must carry orders 1 and 2 together, truck 4 must carry all the orders at the same time, and truck 6 can carry only order 3. Each truck can make one trip a day and all deliveries must be made on the same day. The cost of each truckload combination is c_j.

(b) We wish to target multiwarhead missiles so as to hit (cover) four targets with the minimum number of warheads. Missile system 1 can knock out targets 1 and 3 (i.e., has two independently targeted warheads); missile system 2 can take targets 1, 3, and 4; missile system 4 can hit targets 2, 3, and 4; and missile system 5 can knock out targets 1, 2, and 5. Will a target get hit more than once? How does the formulation change if we wish to minimize the number of missiles?

(c) Interpret the data of problem (b) above in terms of an airline-crew-scheduling problem in which the targets are flight legs and a missile represents a possible routing of a crew. What would be an appropriate objective function?

11.24. We are now in a position to develop a mathematical model for part of the museum director's decision problem. Recall that the director's decision hinged on the cost of the guards required by each of the two plans. Plan One, the standard museum design, can be analyzed by using an integer-programming model. This plan, reproduced below, has nine rooms.

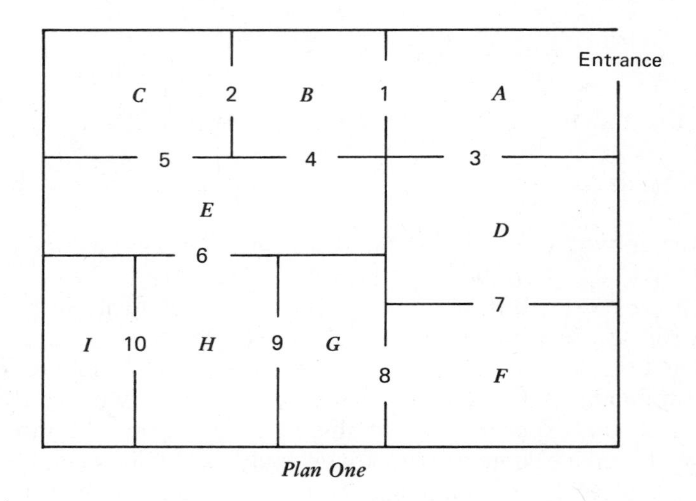

Plan One

A room can be considered to be guarded if a guard is situated in the opening between two rooms. We number the openings and letter the rooms as shown in the plan. Thus, for example, a guard at opening 1 can watch rooms A and B. We want to make sure that each room is guarded (covered) by at least one guard. Denote $x_j = 1$ if a guard is assigned to opening j and $x_j = 0$ if a guard is not assigned to opening j. We next determine the coverage for each room as a linear constraint as follows:

Opening											
1	2	3	4	5	6	7	8	9	10		Room
x_1		$+ x_3$								≥ 1	A
x_1	$+ x_2$		$+ x_4$							≥ 1	B
	x_2			$+ x_5$						≥ 1	C
		x_3				$+ x_7$				≥ 1	D
			x_4	$+ x_5$	$+ x_6$					≥ 1	E
						x_7	$+ x_8$			≥ 1	F
							x_8	$+ x_9$		≥ 1	G
					x_6			$+ x_9$	$+ x_{10}$	≥ 1	H
									x_{10}	≥ 1	I

The objective function is to minimize the number of guards, that is,

$$x_0 = \sum_{j=1}^{10} x_j$$

This is just a standard set-covering problem. Can you find an optimal set of guard positions? (*Hint:* x_{10} must equal one.) There is a covering that requires five guards. Can you do it with less? Determine if a guard can be saved if the walls between rooms B and C were eliminated. For the nine-room plan, can you add any wall space by dividing one of the rooms into two without increasing the number of guards? Once you have the basic model structure, you should be able to respond to such questions raised by the museum director or the architects.

11.25. We next complete our study of the museum director's decision problem by analyzing the guard situation for the zig-zag museum—Plan Two.

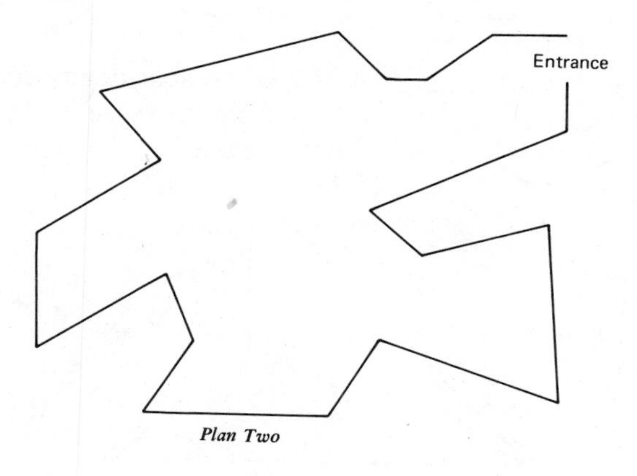

Plan Two

The assumptions of the problem are that the guards are to remain at fixed posts, but they are able to turn around on the spot. Also, the walls must be straight, but they can take on any zig-zag shape. How would you go about solving this problem?

From our decision-framework point of view, it is unclear as to exactly what are the possible alternative solutions. Guards can be located at any interior position; there is a multitude of solutions! We could post a guard in front of each of the 19 walls. That is certainly an inefficient solution. We can reduce that number by putting a guard at every other corner so that a guard is responsible for two intersecting walls. Thus, if n is the number of walls, then this approach requires $\frac{1}{2}n$ guards, if n is even; and $\frac{1}{2}(n + 1)$, if n is odd, with one guard responsible for only one wall. Can we do better? It would be nice if we could develop a formulation and an algorithm that are functions of the number of walls. Then we could indicate to the museum director the results of removing or adding walls to the zig-zag plan. The integer-programming model of the regular museum plan depends on the number of rooms.

Instead of attacking Plan Two directly, let us first look at zig-zag plans that we can solve without any difficulty. Hopefully, we can then generalize our insights to more complex situations.

For example, what if the museum consisted of three intersecting walls and an entrance in one of the walls, as follows:

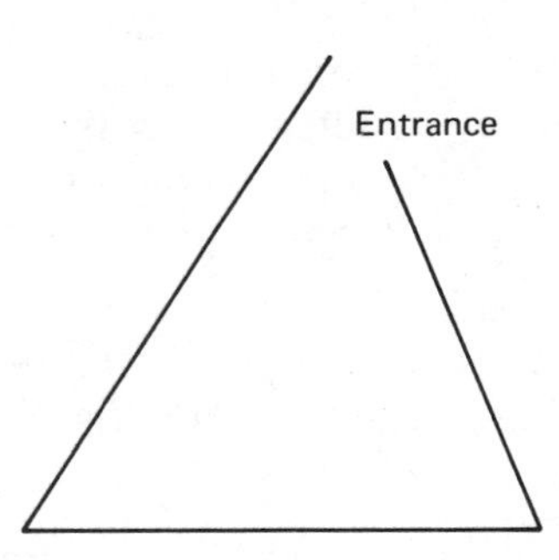

The zig-zag shape is just a triangle; one guard placed at any corner (including the entrance corner) or any interior spot can observe all the walls. For a four-wall museum, we can have something like

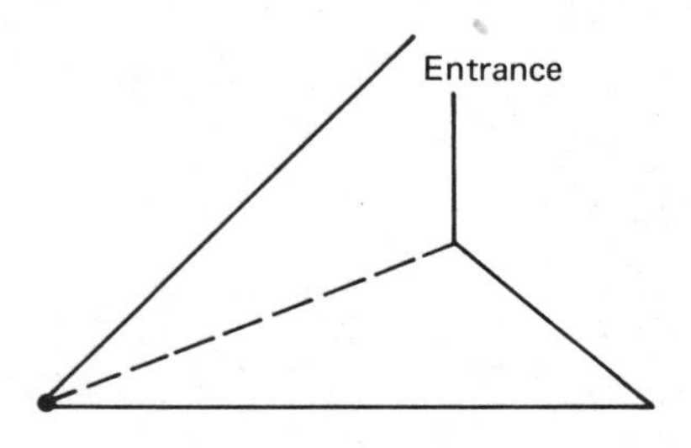

or

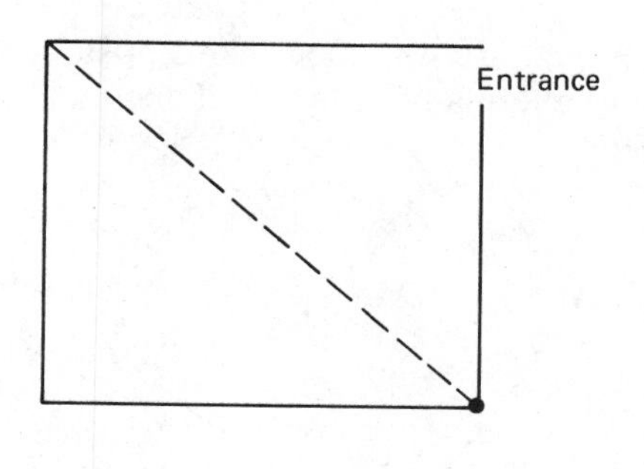

or

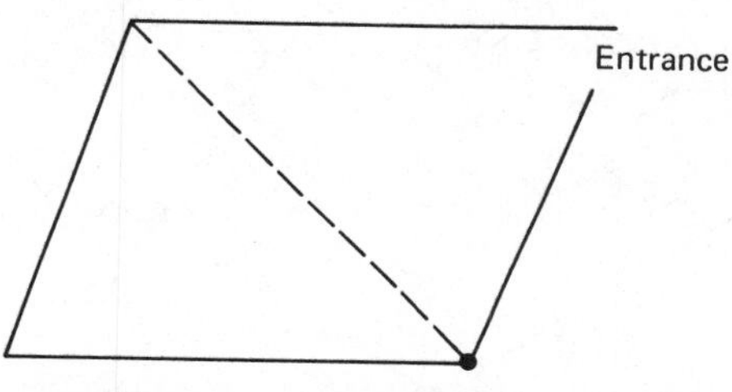

If we divide the four-sided figures into triangles using nonintersecting diagonals (as shown by the dotted lines), we see that the triangles come together at a vertex forming a fanlike figure. A guard (indicated by the dot) posted at such a vertex can observe all the walls that are part of the fan.

Similarly, for a five-sided figure we can have a museum that looks like

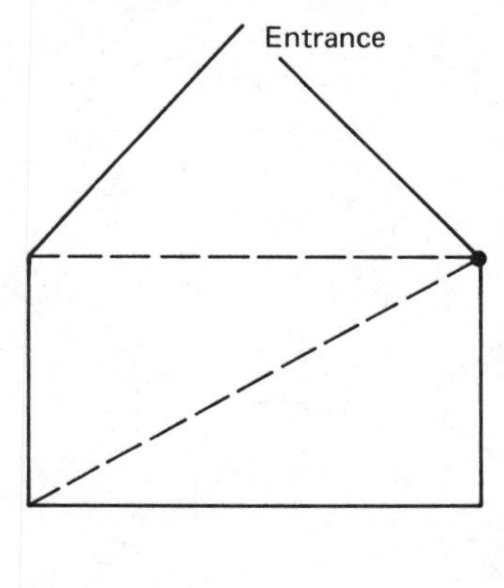

or

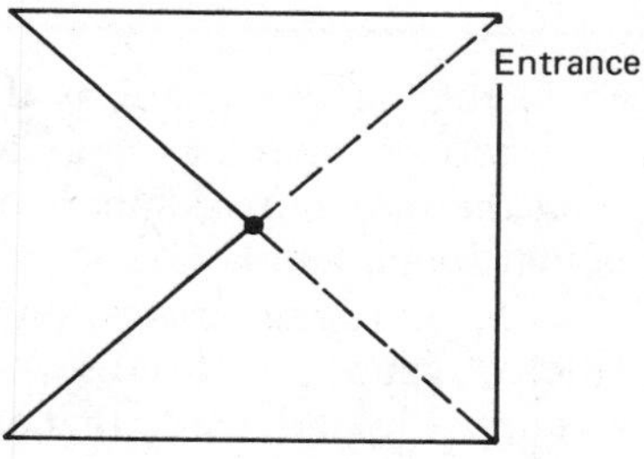

Thus, for $n = 3, 4$, or 5 we need only one guard. Note that in the last figure there are two vertices that are not good guard positions.

For $n = 6$, we can have many different shapes including

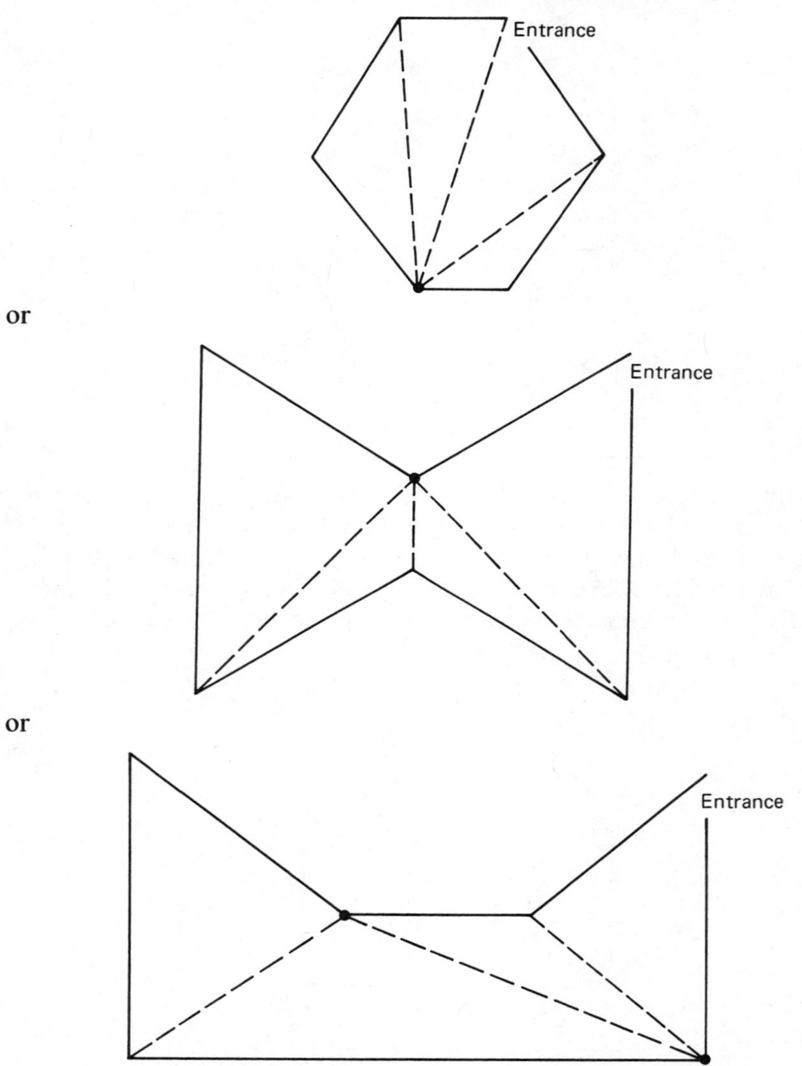

The first two figures would require one guard as they have only one fan; the third figure would require two guards as you cannot divide it into less than two fans. We see that the minimum numbers of guards for an n-sided figure depends on the actual shape and is not a function of n alone. Based on the study of $n = 3, 4, 5, 6$ zig-zag rooms can you hypothesize some relationship on the number of guards required for an n-sided room? What function of n looks like an upper bound? Note that a particular triangulation does not have a unique fan decomposition.

The mathematician Vašek Chvátal has proved that any triangulation of an n-gon can be divided into m fans with $m \leq [\frac{1}{3}n]$, where $[\frac{1}{3}n]$ represents the integer part of $\frac{1}{3}n$, for example, $[\frac{1}{3}n] = 2$ for $n = 7$. For an n-walled museum, the director would never require more than m guards. For $n = 3$, 4, 5, we have $m = 1$; for $n = 6$, we have $m = 2$. For Plan Two we have $n = 19$ and $m \leq [19/3] = 6$. This result is not too satisfying; for a particular triangulation we do not know the exact minimum number of guards and where they should be posted. Trial-and-error fan divisions of a triangulation will give you some feel for whether you have the minimum. We know it cannot be greater than m.

You should divide Plan Two into fans and try to figure out the minimal number of guards and their posts. We can do it with three guards, where one guard is watching 15 walls and the other guards are watching two walls apiece. As this allocation of work is not very fair, some of the 15 walls can be assigned to the two not-so-busy guards, that is, some walls can be watched (covered) by two guards. Can you figure it out?

Chvátal's fan-construction approach assumes that the guard positions are restricted to the corners of the n-gon. (This is not quite correct; a guard stationed at a vertex of a fan can move to many points inside the fan and still see all the walls that are a part of the fan. These interior points are guard positions that are equivalent to the fan vertex.) With this assumption, can you formulate this problem as an integer-programming set-covering problem? Will this enable you to find the minimal number of guards?

A guard placed at a vertex can cover all the walls that form the fan that originates at that vertex. Here the possible guard positions—the decision variables—are just the 19 vertices of Plan Two (including the entrance vertex). Let us identify each vertex by a number and each of the 19 walls by a letter as follows:

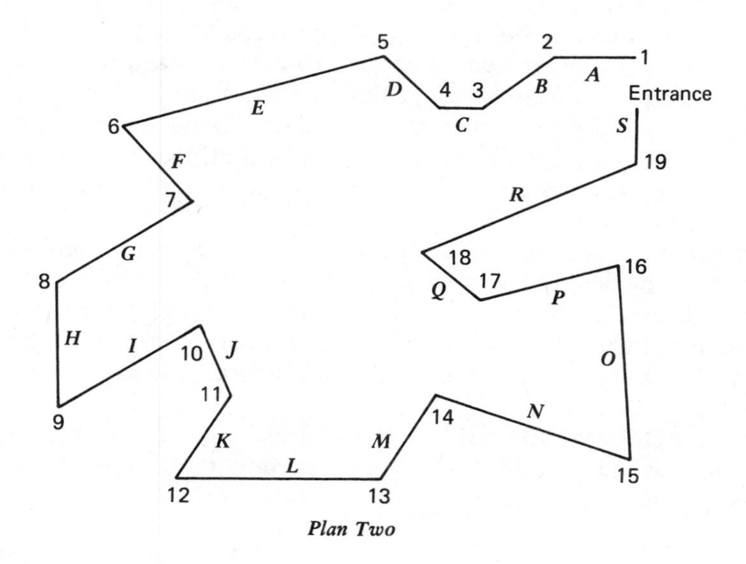

Plan Two

Let x_j ($j = 1, \ldots, 19$) be the binary decision variable corresponding to the placing of a guard at vertex j. If $x_j = 1$, then a guard is placed at vertex j; if $x_j = 0$, then a guard is not placed at vertex j.

For each vertex, we need to construct its largest, most inclusive fan and determine the walls that are covered by a guard placed at that vertex. This fan will be formed by nonintersecting diagonals that originate from the vertex. The fan just tells us what walls can be covered (seen) by a guard from a specified vertex. We illustrate this by constructing the fan for vertex 19.

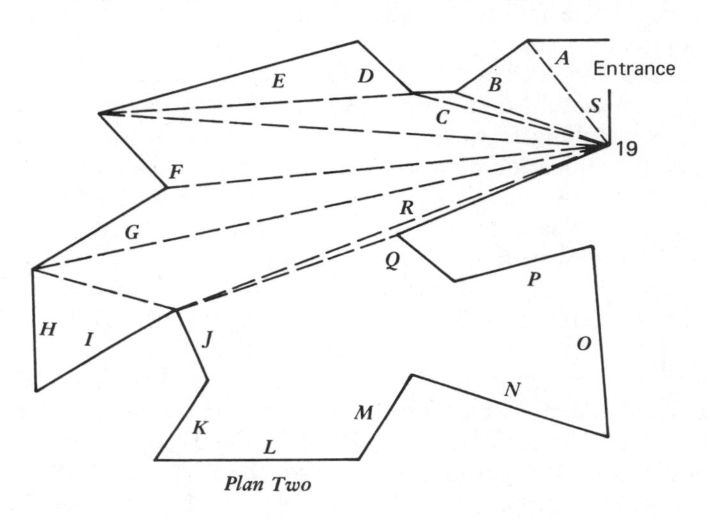

Plan Two

A guard at vertex 19 can observe walls *A, B, C, F, G, R,* and *S*. Note that walls *D, E, H, I,* and *J* are not covered by vertex 19. Once we determined the walls that can be observed from all vertices, we can structure an integer-programming formulation similar to the Plan One museum set-covering model. Each column of the zig-zag model corresponds to a vertex; each row corresponds to a wall. If a guard at vertex j can see wall i ($i = A, \ldots, S$) then the coefficient of the (i, j)th position is 1; if wall i cannot be seen from vertex j, then the (i, j)th coefficient is 0. The right-hand-side elements are all 1, with each row a greater-than-equal-to constraint, that is, each wall must be covered at least once. The objective function is to minimize $x_0 = \sum_{j=1}^{19} x_j$, with $x_j = 0$ or 1.

You should determine the complete fans for each vertex and construct the set-covering model. Use it to pick out—"eyeball"—a minimal solution. Can you determine the minimal three guard positions? Your solution will yield multiple guards for some of the walls, but you should be able to reduce it to one that is equivalent to the solution using three guards with no multiple coverage.

We illustrate what you need to do by displaying the set-covering model for the following 6-gon figure. Check to see how the complete vertex fans and coverings were determined.

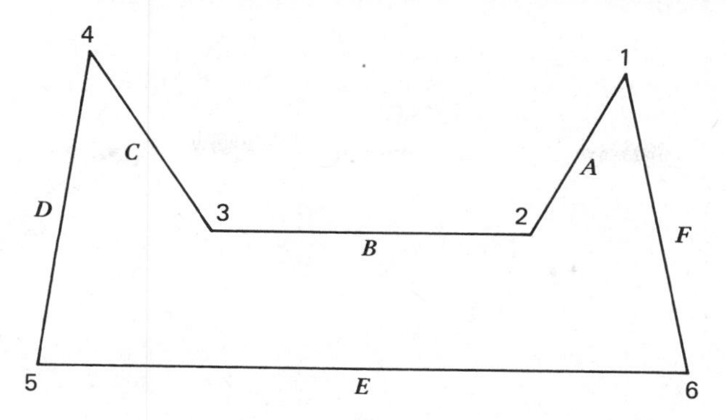

Vertex 1	2	3	4	5	6		Wall
x_1	$+ x_2$				$+ x_6$	≥ 1	A
	x_2	$+ x_3$		$+ x_5$	$+ x_6$	≥ 1	B
		x_3	$+ x_4$	$+ x_5$		≥ 1	C
		x_3	$+ x_4$	$+ x_5$		≥ 1	D
	x_2	$+ x_3$		$+ x_5$	$+ x_6$	≥ 1	E
x_1	$+ x_2$				$+ x_6$	≥ 1	F

The objective function is to minimize $x_0 = \sum_{j=1}^{6} x_j$ with $x_j = 0$ or 1.

All the walls are covered if we put guards at vertices 2 and 4, or at vertices 4 and 6. These solutions cover each wall exactly once. Can you find other such solutions. How about two guard positions that cover some walls more than once? Since wall E is relatively long, we might want to use guard positions 5 and 6. Chvátal's upper bound is $m = [6/3] = 2$.

The covering constraints of this 6-gon have a certain inherent structure. Note that the covering inequality of wall A is the same as that for wall F and similarly for walls B and E and walls C and D. Thus, any solution to the three inequalities that cover walls A, B, and C, will also cover walls D, E, and F. If we look at the reduced problem formed by the first three constraints, it is easy to see that it requires exactly two guards and which vertices they should be located at. Can you show that the linear-programming model of this reduced system will have an integer optimal solution? Is that true for all n-gon covering constraints?

11.26. Even though we have not introduced an algorithm for solving the integer-programming model, organizing the data to form a set-covering problem enables us to select a good, if not optimal, feasible solution. Computer procedures are available for solving more complex integer problems. One is discussed in Chapter 21. You should be convinced that we can go very far in our understanding and solution of a problem just by investigating the form of its corresponding mathematical model. To illustrate, let us take a closer look at set-covering problems.

The general set-covering problem can be stated as follows:

$$\text{Minimize } x_0 = \sum_{j=1}^{n} c_j x_j$$

subject to

$$\sum_{j=1}^{n} a_{ij} x_j \geq 1 \qquad i = 1, \ldots, m$$

$$x_j = 0 \quad \text{or} \quad 1$$

where each a_{ij} is a 0 or 1 and $c_j > 0$. For this formulation, can you think of a greedy algorithm that solves it without much difficulty? (*Hint:* For each column j, let $A_j = \sum_i a_{ij}$, that is, the total number of ones in each column. Calculate the ratios A_j/c_j. Based on these ratios, an x_j should be brought into the solution and set equal to one. Which x_j would you pick and why? What do you do next?) What is the rationale for your greedy algorithm? What if all $c_j = 1$? Does your resultant algorithm seem like a reasonable one to use? Try it out on the museum set-covering problems of Section 11.25, and the problems of Sections 11.22 and 11.23 (you will have to assign values to the c_j).

11.27. We formulated the n-gon museum problem as a covering problem assuming that the guard positions were restricted to the vertices of the n-gon. Can you construct an n-gon all of whose walls can be covered by only one guard, but for which a covering-problem formulation requires at least two guards? (*Hint:* Recall that Chvátal's theorem gives an upper bound $m \leq [n/3]$ on the minimum number of guards. The actual minimum could be less than m. Try constructing an n-gon whose walls can be watched by a guard stationed at the center point, but for which any fan triangulation requires at least two guards.)

11.28. The n-gon museum-guard problem was originally posed in the following terms by the mathematician Victor Klee: "Find the smallest number $f(n)$ such that every set bounded by a single closed n-gon is dominated by a set of $f(n)$ points." [A subset A of a set of points S is said to dominate S if for each point x in S there is a point y in A such that the entire line segment (xy) lies within S.] In his paper, Chvátal restated the problem as finding the minimum number of guards and, using induction, proved that $f(n) \leq [n/3]$. A few years later, S. Fisk developed a short proof using the fact that the vertices of any triangulated n-gon can be colored with three colors, a, b, and c such that no two adjacent vertices have the same color. Let T_a be the set of vertices colored a, T_b be the set colored b, and T_c be the set colored c. Denote the number of vertices in each set by $|T_a|$, $|T_b|$, and $|T_c|$, respectively, and assume $|T_a| \leq |T_b| \leq |T_c|$. It should be clear that each triangle has at most one vertex colored a, and $|T_a| \leq [n/3]$. Guards positioned at the vertices colored a will suffice. Do you see the relationship between both statements of the problem?

11.29. Discuss possible shapes for museums in terms of criteria other than cost or number of guards, for example, reducing the distance covered to see an exhibit. The Guggenheim Museum in New York City, designed by Frank Lloyd Wright, is a six-story descending spiral with the outside wall used for hanging and no inside walls. The Hirshhorn Gallery in Washington, D.C., designed by Gordon Bunshaft, is a four-story circular building with each floor shaped like a torus divided into four-walled rooms. (The Guggenheim can almost fit inside the hole of the Hirshhorn torus!) One advantage of a zig-zag shape is that you can start at the entrance wall and do not have to retrace your steps. What is your final recommendation to the museum director? Should the director pick Plan One or Two, or go back to the drawing board? How would you respond to the suggestion that all or some of the guard positions can be covered by rotating TV cameras? Do you see any problems in using TV cameras to do the job of guarding the pictures? You should study the whole system concerned with protecting the pictures.

11.30. *The Apocryphal Linear Program*

The biblical story of Joseph in Egypt recounts his interpreting Pharaoh's dream of seven fat cows and seven lean cows to mean that Egypt would have seven years of plenty, followed by seven years of famine. Pharaoh placed Joseph in charge of planting, harvesting, and storing food for the future. Joseph's management plan was a great success: "He stored the grain in huge quantities; it was like the sand of the sea, so much that he stopped measuring: it was beyond all measure." Some biblical scholars suggest that Joseph was a brilliant decision maker and manager who must have invented and used advanced planning techniques such as "lean-year" programming, (see Genesis 41).

Part II References

Anonymous: *Nutrition Labeling*, Bulletin No. 382, U.S. Department of Agriculture, April 1975.

Balintfy, J. L.: Menu Planning by Computers, *The Communications of the ACM,* Vol. 7 April 1964.

Balintfy, J. L.: Linear Programming Models for Menu Planning, chapter in *Hospital Industrial Engineering*, H. E. Smalley and J R. Freeman (Eds.), Reinhold, New York, 1966.

Bland, R. G.: The Allocation of Resources by Linear Programming, *Scientific American*, June 1981.

Charnes, A., W. W. Cooper, and A. Henderson: *An Introduction to Linear Programming*, Wiley, New York, 1953.

Chvátal, V.: A Combinatorial Theorem in Plane Geometry, *Journal of Combinatorial Theory*, Vol. 18, no. 1, February 1975.

Chvátal, V.: A Greedy Heuristic for the Set-Covering Problem, *Mathematics of Operations Research*, Vol. 4, no. 3, 1979.

Dantzig, G. B.: *Linear Programming and Extensions*, Princeton University Press, Princeton, N.J., 1963.

Fisk, S.: A Short Proof of Chvátal's Watchman Theorem, *Journal of Combinatorial Theory*, Series B24, 374, 1978.

Gass, S. I.: *An Illustrated Guide to Linear Programming*, McGraw-Hill, New York, 1970.

Gass, S. I.: *Linear Programming: Methods and Applications*, 5th Edition, McGraw-Hill, New York, 1985.

Honsberger, R.: *Mathematical Gems II*, The Mathematical Association of America, Washington, D.C., 1976.

Jewell, W. S.: A Classroom Example of Linear Programming, Lesson No. 2, *Operations Research*, Vol. 8, no. 4, July–August 1960.

Koopmans, T. C. (Eds.): *Activity Analysis of Production and Allocation*, Cowles Commission Monograph 13, Wiley, New York, 1951.

Paull, A. E.: Linear Programming: A Key to Optimum Newsprint Production, *Pulp and Paper Magazine of Canada*, Vol. 57, no. 1, January 1956.

Smith, D.: *Linear Programming Models in Business*, Polytech Publishers, England, 1973.

Stigler, G. J.: The Cost of Subsistence, *The Journal of Farm Economics*, Vol. 27, no. 2, 1945.

Taha, H. A.: *Operations Research*, Macmillan Co., New York, 1971.

Vajda, S.: *Problems in Linear and Nonlinear Programming*, Hafner Press, New York, 1975.

Wagner, H.: *Principles of Management Science*, 2nd Edition, Prentice-Hall, Englewood Cliffs, New Jersey. 1975.

Walker, W. E., J. M. Chaiken, and E. J. Ignall (Eds.): *Fire Department Deployment analysis*, The Rand Fire Project, North-Holland, New York, 1979.

PART III Solving Linear-Programming Problems

The Model and Its Algorithm

Chapter 12	Single-Variable Problems (For the True Beginner) 139
Chapter 13	Two-Variable Problems 142
Chapter 14	A Manufacturing Problem 147
Chapter 15	The Diet Problem (Again) 156
Chapter 16	The Simplex Algorithm 159
Chapter 17	Part III Discussion, Extensions, and Exercises 174

We can solve just about any linear-programming problem. There are computational procedures that enable us to solve small-scale problems by hand. Even for the big (and not so big) problems, the same procedures, when coupled with an electronic computer, have been forged into a major weapon of the problem solver. The first thing most computers "learn" after they have been imbued with the basic rules of arithmetic is how to solve linear-programming problems. This phenomenon attests to the pervasiveness and power of the linear-programming model.

You can understand a great deal about linear programming without knowing how to solve a particular problem. As many organizations, including most colleges, have computers with automated algorithms for solving linear programs, you can go through life without worring about solution procedures. (This is true today for almost any quantitative problem-solving activity.) By doing so, however, we feel that you restrict your abilities as an analyst, decision maker, and thinking person. If you do not know how a problem was solved, you have only limited ability to check the correctness or validity of an answer. It is important to know how certain problem conditions were represented and treated throughout the computation: Were solutions rounded to integers?; Are there solutions other than the optimal that may be better due to conditions not included in the formulation? Your ability to ask such questions enables you to treat the algorithm and computer (human or hardware) as other than a mysterious black box that produces answers you must accept. It is axiomatic that a basic and insightful comprehension of any quantitative field can be gained only by a careful blending of the theoretical, computational, and applied aspects. This part, then, introduces you to the basic ideas used to solve linear-programming problems, plus a little of the theory.

We first discuss the geometry of linear-programming problems and how the geometrical properties are used to solve very simple problems.[1] These insights are then carried over to the development and discussion of the basic algebraic method used to solve linear-programming problems—*the simplex algorithm*.

[1] Most of the material in Chapters. 12–15 is from S. I. Gass, *An Illustrated Guide to Linear Programming*, McGraw-Hill Book Co., New York, N.Y., 1970. Reproduced with permission.

12 Single-Variable Problems (For the True Beginner)

The distance is nothing; it is only the first step which counts.

MADAME DU DEFFAND

12.1 HOW SIMPLE CAN WE GET?

The simplest of all optimization problems is to find the maximum value of the single variable x_1, where x_1 is not subject to any restrictions. Here, the numerical value of x_1 can become as large as we want, and we say that there is no maximum value. Geometrically, we are looking for the largest value of x_1 along the infinite number line or axis. As we have no restrictions on x_1, we can march along this line as far to the right as we care to without

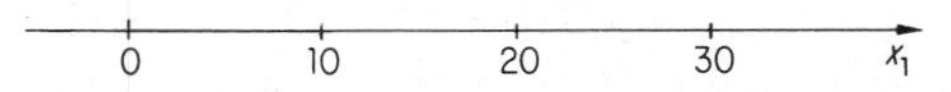

encountering any barriers. The problem becomes more interesting, but still easy to solve by restricting x_1. We shall deal with variables whose values are always restricted by the basic linear-programming requirement of non-negativity—thus, $x_1 \geq 0$.

The problem of maximizing x_1 subject to the inequality constraint $x_1 \leq 20$ allows us to move along the x_1 axis from the point $x_1 = 0$ until we encounter the barrier—a stop sign—at the point $x_1 = 20$. Any point on the line connecting $x_1 = 0$ and $x_1 = 20$ represents a potential solution. This set of points is called the solution set, or solution space. As we are maximizing, the optimal point in the solution space—the point which makes the objective function, maximize x_1, as large as possible—is $x_1 = 20$. If we were looking

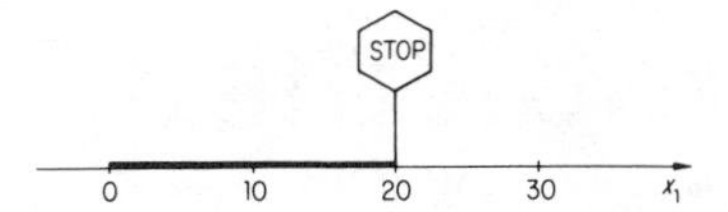

for the minimum value of x_1 subject to $x_1 \leq 20$, the optimal answer—remembering the nonnegativity restriction—would be $x_1 = 0$.

Changing the inequality restriction of the maximizing problem to $2x_1 \leq 20$ restricts our freedom of movement along the x_1 axis to the points between $x_1 = 0$ and $x_1 = 10$, with the optimal answer being $x_1 = 10$. Single-

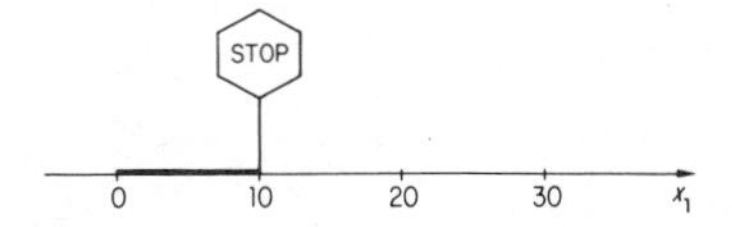

variable, single-restriction problems can readily be solved by such a simple analysis. If the single restriction is an equation like $x_1 = 20$ or $2x_1 = 20$, then the solution space is just a single point—here, $x_1 = 20$ and $x_1 = 10$, respectively. Linear-programming problems can involve equality as well as inequality restrictions, but the geometric basis of such problems is made sharper if we stress the inequality nature of our formulations.

12.2 AN EXAMPLE: SOLVED BY INSPECTION

A typical example of a single-variable linear-programming problem with many restrictions is to maximize x_1 subject to

$$
\begin{aligned}
5x_1 &\leq 75 \\
6x_1 &\leq 30 \\
x_1 &\leq 10
\end{aligned}
$$

with, of course, $x_1 \geq 0$. We need to find the largest value of x_1 which satisfies the three inequalities simultaneously. We first determine the solution space associated with the individual inequalities and then find the solution space which represents the joint solution of all the inequalities.

For $5x_1 \leq 75$ we have that any value of x_1 between 0 and 15 satisfies the constraint:

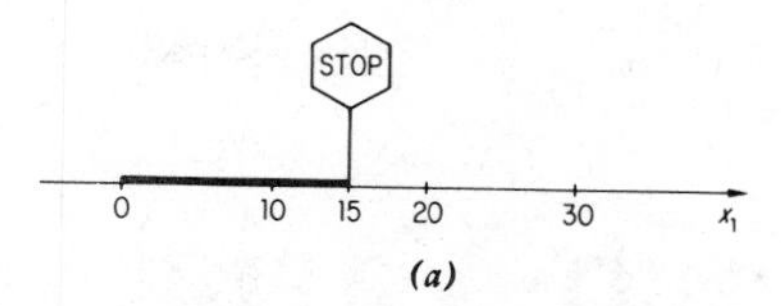

(a)

For $6x_1 \leq 30$, the solution space is between 0 and 5:

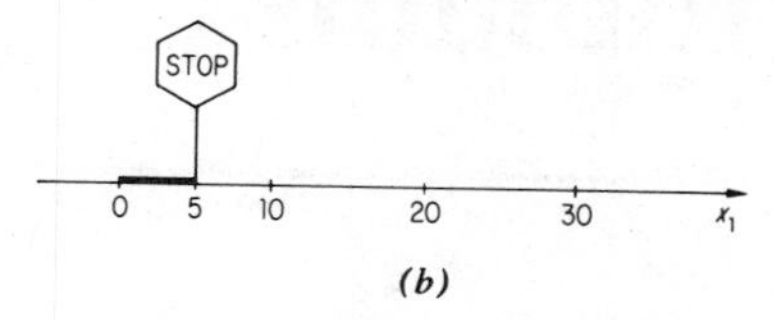

(b)

And finally, for $x_1 \leq 10$ the solution set is from 0 to 10:

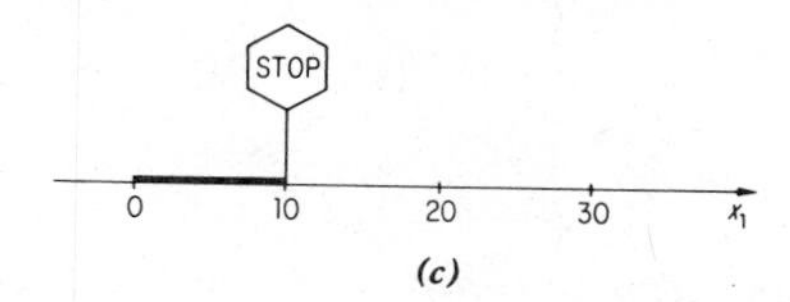

(c)

If we combine all the stop signs onto one graph we have

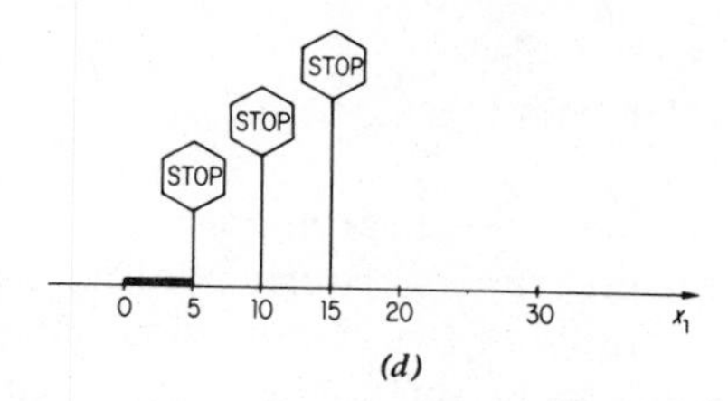

(d)

Thus, our march from the point $x_1 = 0$ to the right cannot get by the point $x_1 = 5$. Any value between 0 and 5 will satisfy the three inequalities simultaneously—this set of points is the solution set of the problem. As we are looking for the largest point in the solution set, the optimal answer to the original three-inequalities problem is $x_1 = 5$. For any problem in one variable we can determine the optimal solution in this way.

13 Two-Variable Problems

Every step forward in the world was formerly made at the cost of mental and physical torture.
NIETZSCHE

13.1 FINDING OUR WAY IN 2-SPACE

We next enter the two-dimensional problem arena, and it is here that the challenge to solve linear programs becomes nontrivial and more instructive. What follows requires graphing two-dimensional constraints—inequalities and equalities—and some understanding of basic high school algebra.

The two variables or dimensions x_1 and x_2 are represented by the right-angled axes

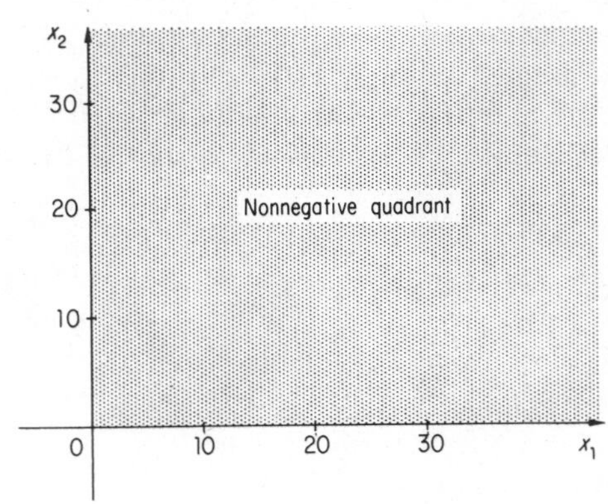

Since we are restricting the range of the variables to the nonnegative set of values, the associated graph is constrained to certain directions. For x_1, we can only move to the right of the point 0, the origin of the axes, and for x_2 we can only move straight up from the origin in a direction perpendicular to the x_1 axis. Mathematically, we say that the solution space is restricted to the nonnegative quadrant.

A value of $x_1 = 20$ sends us 20 units to the right, and a value of $x_2 = 30$ marches us 30 units straight up. Taken together, these directions put us inside the nonnegative quandrant as shown. We use the shorthand notation $(x_1, x_2) = (20,30)$ to represent a point on the two-dimensional graph.

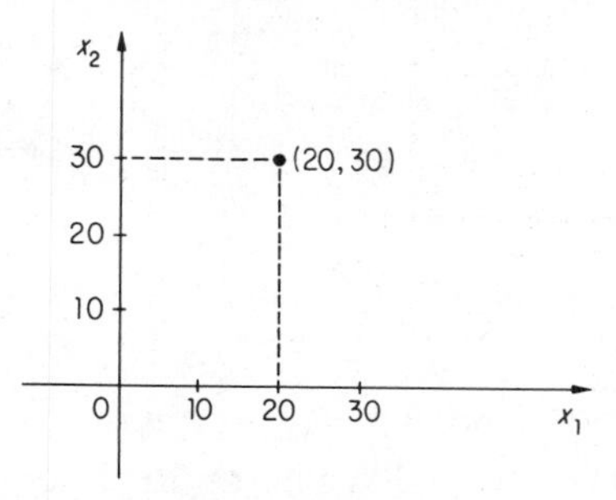

In one dimension the problem of maximizing x_1 subject to the inequality $x_1 \leq 20$ has a solution space which ranges from $x_1 = 0$ to $x_1 = 20$. Any point in this range satisfies the requirement of the inequality—the nonnegative variable x_1 must be less than or equal to 20. Thus, there are an infinite number of solutions contained in this solution space. To list a few, we have 0, 1, 2, 10, $\sqrt{3}$, and $\frac{3}{4}$ as permissible values of x_1. However, the problem was not to find the solution space, but to find a point in the solution space which maximizes x_1. The addition of the objective function enables us to "home in" on a particular point contained in the infinite set of solution points. This same graphical analysis carries over to the two-dimensional problem. We must, in general, first determine the infinite set of solutions to the constraints of the problem and then select a point which optimizes the objective function. To illustrate how we accomplish this, we next solve some rather simple two-dimensional problems.

13.2 EXAMPLES: SOLVED BY ENUMERATION

We wish to maximize x_1 subject to the conditions $x_1 \leq 20$ and $x_2 \leq 30$, with $x_1 \geq 0$ and $x_2 \geq 0$. In the two-dimensional graph we plot the barrier lines $x_1 = 20$ and $x_2 = 30$. The nonnegativity conditions $x_1 \geq 0$ and $x_2 \geq 0$, taken together with the barrier lines, force us to confine our search for solution points to the shaded area and its boundaries—this is the solution space. No matter where we look in the solution space the largest value of x_1 is 20. But here we notice a strange thing—there are many points in the solution space

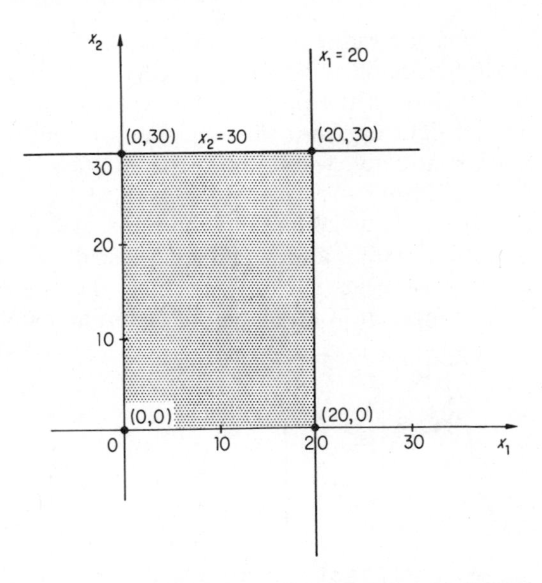

which yield a value of $x_1 = 20$. The points (20,0) and (20,30) are both in the solution space and have $x_1 = 20$. The optimal solution is not unique—a rather common occurrence in linear programming, one which offers us no additional concern. In fact, we really have an infinite number of optimal solutions, as any point on the line segment joining the points (20,0) and (20,30) is a solution with $x_1 = 20$.

An important peculiarity of linear-programming problems is that well-behaved problems, like the one under discussion, have optimal-solution points which include the vertices formed by the intersection of the boundary lines. The present problem has four such vertices—(0,0), (0,30), (20,0), and (20,30)—with the latter two being optimal solutions for the objective function maximize x_1.

If we change the objective function to maximize x_2, we have the optimal vertices (0,30) and (20,30), along with any point on the line segment joining these vertices. If we wanted to minimize x_1 or minimize x_2, we find the former objective function minimized by the vertices (0,0) and (0,30); the latter is minimized by (0,0) and (20,0). Of course, points on the lines connecting the pairs of vertices are also optimal solutions for the respective objective functions.

Let us complicate the problem by changing the objective to finding the values of x_1 and x_2 such that the sum $x_1 + x_2$ is minimized, subject to the same constraints. This objective function takes on its smallest value at the point (0,0)—that is, minimum $x_1 + x_2 = 0$. This optimal solution is unique in that any other point in the solution space has a positive value for at least one of the variables.

What about the solution if the objective function was to maximize $x_1 + x_2$? A slight search into the possibilities reveals the one solution (20,30) for a maximum of $x_1 + x_2 = 50$.

By this time you might have rightly concluded that no matter what the objective function, the optimal solution will occur at one of the four vertices and possibly on some other boundary points. This is true for all well-behaved linear-programming problems—an ill-behaved problem being one in which the objective function is unbounded, as was the case for our first, single-variable problem. At this point it would be correct to inquire what all the fuss is about. If the optimal solution to a linear-programming problem occurs at one of the vertices, all we need do is to determine all the vertices, evaluate the objective function for each vertex, and pick out the one that optimizes the objective function. For example, let the objective function be to maximize $3x_1 - 2x_1$ and set up the table

Vertex	Value of $3x_1 - 2x_2$	Maximum
(0,0)	$3x_1 - 2x_2 = 3(0) - 2(0) = 0$	
(20,0)	$3x_1 - 2x_2 = 3(20) - 2(0) = 60$	60
(0,30)	$3x_1 - 2x_2 = 3(0) - 2(30) = -60$	
(20,30)	$3x_1 - 2x_2 = 3(20) - 2(30) = 0$	

The optimal value is 60, and the unique optimal solution point is (20,0). You can do the same for any linear objective function involving x_1 and x_2, subject to the given constraints. This looks rather easy, and it is for such toy examples. The difficulty in this approach for the general problem is that the number of vertices can be immense, and their enumeration is quite a computational task. The above enumeration approach is acceptable for simple problems. We next refine the process to bring it more in line with the actual algebraic computational process used to solve linear-programming problems, the simplex method. To do this, we must attack more involved problems. But first, we shall introduce some of the basic terminology of linear programming.

13.3 SOME DEFINITIONS

The solution space of the two-variable problem is a *convex set* of points and the vertices are *extreme points* of the convex set. A set of points is convex if the set contains the entire line segment joining any two of its points. The following are convex sets:

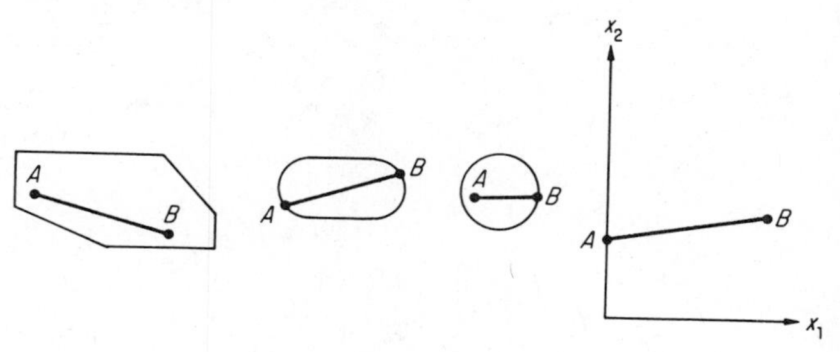

The following are not convex sets:

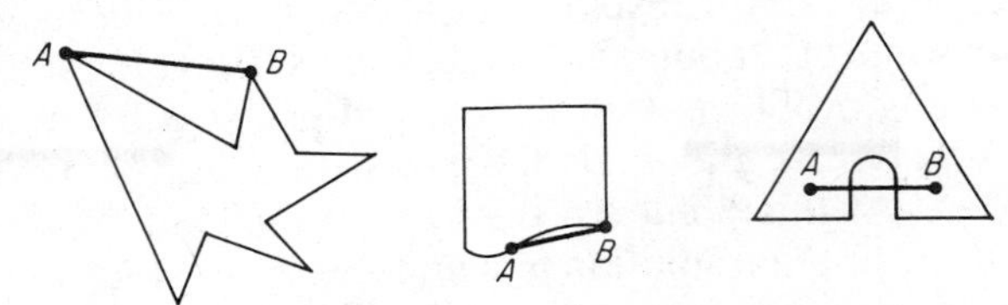

A point of a convex set is an extreme point if it does not lie on a line segment joining two other points of the set. A convex set can have a finite number, infinite number, or no extreme points. We leave it to you to supply examples of each type. The set of solutions to a linear-programming problem is a convex set with a finite number of extreme points. In general, the solution set is called a *convex polyhedron*; if the set is bounded, it is termed a *convex polytope*.

14 A Manufacturing Problem

It is a bad plan that admits no modification.
PUBLILIUS SYRUS

14.1 MAKING CHAIRS AND TABLES: GRAPHICAL TECHNIQUES

We next solve a smaller version of the manufacturing problem encountered by the Simple Furniture Company. To reduce it to a two-variable problem, we shall consider only the making of chairs and tables, subject to the board-foot and labor-hour restrictions, and change the amounts of the available resources to facilitate the discussion of a graphical solution. Our problem then is to maximize the profit function $\$45x_1 + \$80x_2$ subject to

$$5x_1 + 20x_2 \leq 400$$
$$10x_1 + 15x_2 \leq 450$$

Here, x_1 stands for the number of chairs (c) to be manufactured and x_2 for the number of tables (t), with $x_1 \geq 0$ and $x_2 \geq 0$. We have a total of 400 board-feet of mahogany and 450 labor-hours to combine into a manufacturing schedule for chairs and tables.

The first thing to be done is to delineate the solution space. We start out with the solution space restricted to the nonnegative quadrant

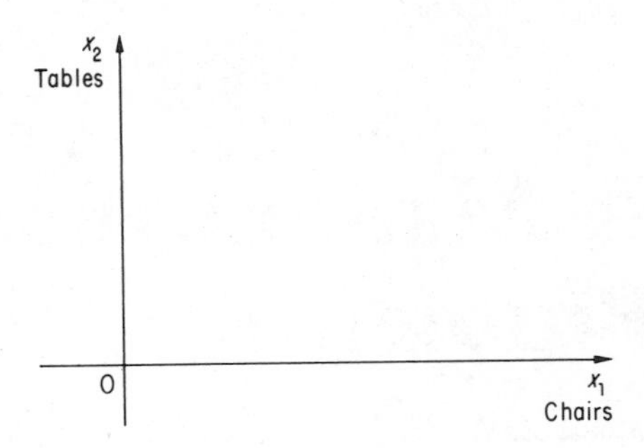

The first inequality, $5x_1 + 20x_2 \leq 400$, requires us to stay in that part of the nonnegative quadrant where the points have coordinates (x_1,x_2) which, when substituted in the expression $5x_1 + 20x_2$, yield a sum less than or equal to 400. The easiest way to determine this region is to plot the barrier line $5x_1 + 20x_2 = 400$. We do this by finding the two points—one on the x_1 axis and one on the x_2 axis—through which the line passes. If $x_1 = 0$, we must have $x_2 = 20$; if $x_2 = 0$, we have $x_1 = 80$. The two points (80,0) and (0,20) lie on the barrier line. Next, we take a convenient test point, say $x_1 = 0$ and $x_2 = 0$, and see if it satisfies the inequality. If it does, then all points that lie on the same side of the barrier line as the test point also satisfy the inequality condition; if not, then all points on the other side satisfy the inequality. (Can you prove that this is so?) The intersection of the inequality solution space and the nonnegative quadrant is the joint solution space as shown in the shaded part of the figure. All points in the shaded area and its boundary satisfy the inequality $5x_1 + 20x_2 \leq 400$, and $x_1 \geq 0$ and $x_2 \geq 0$.

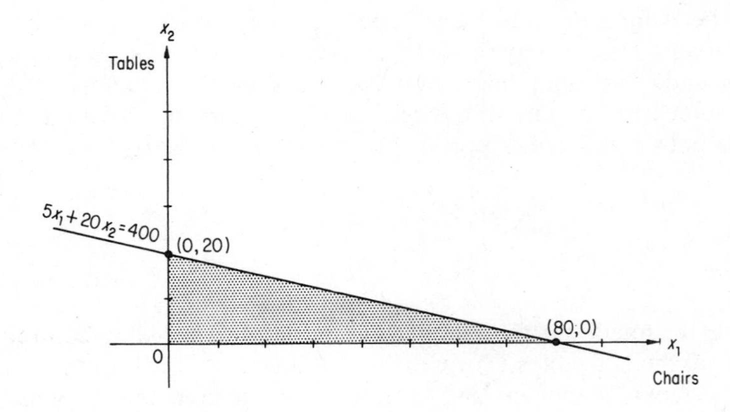

For the second inequality, the barrier line is $10x_1 + 15x_2 = 450$, and the intersection points on the axes are (45,0) and (0,30). This is shown by

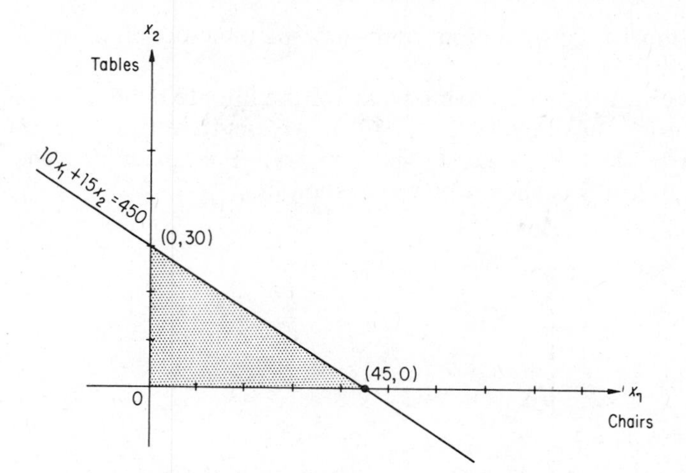

Any point in this convex set satisfies $10x_1 + 15x_2 \leq 450$. The joint solution of the two constraints is found by superimposing both shaded areas onto one graph—the areas which overlap form the required solution space.

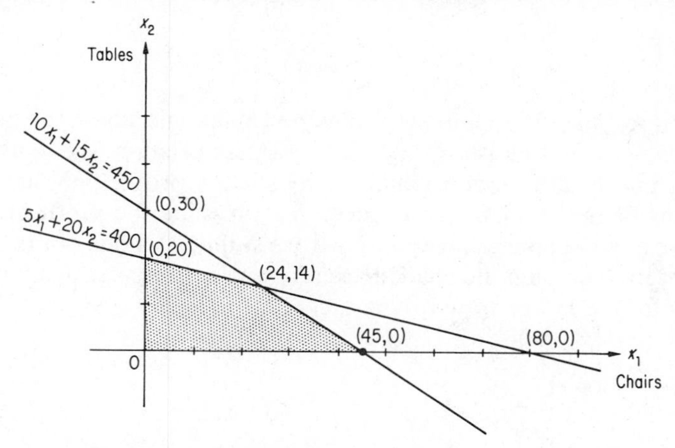

This combined solution space represents the many combinations of chairs and tables which can be manufactured subject to the availability of 400 board-feet and 450 labor-hours. Only points in this region satisfy the constraints of the problem; it is a convex polytope.

Up to now we have identified all the extreme points of the convex set of solutions except the one generated by the intersection of the two barrier lines. This point represents the manufacturing of an amount of chairs and tables which uses up all the available resources. Any point (x_1,x_2) on the line $5x_1 + 20x_2 = 400$, for example, (0,20), uses up exactly 400 board-feet. Similarly, any point on the line $10x_1 + 15x_2 = 450$ requires the full amount of available labor-hours. By a little algebraic juggling we find that the point $x_1 = 24$ chairs and $x_2 = 14$ tables satisfies both constraints exactly. You should show how this answer was obtained by solving the two equations simultaneously. This solution represents the intersection of the corresponding two straight lines.

The problem is to determine which of the infinite number of points in the shaded region maximizes $45x_1 + 80x_2$. We can treat this expression like an equation with the variables x_1 and x_2. For example, the equation $45x_1 + 80x_2 = 0$ is shown by the dashed line

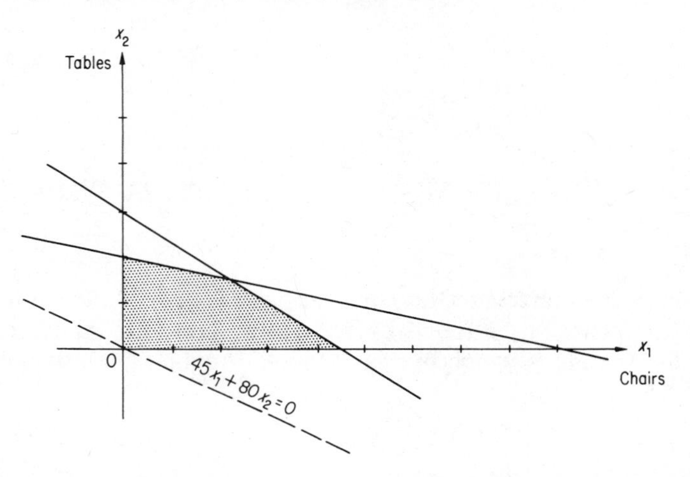

If the Simple Furniture Company does not make any chairs or tables, that is $x_1 = 0$, $x_2 = 0$, this profit line shows a gain of zero. We want to move the profit line in a direction which increases the profit, but subject to the restrictions of the available resources. We can slide the profit line into the shaded area and keep moving it until we are stopped by one of the barriers—we cannot look outside the shaded area for any solutions. Such a movement of the profit line is illustrated by

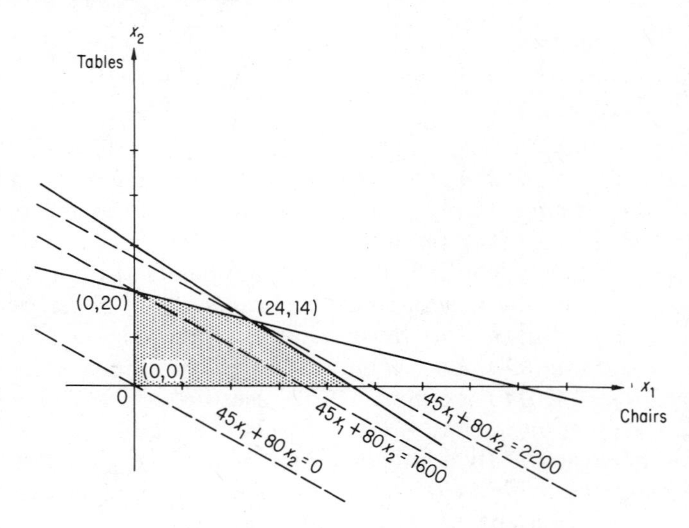

The optimal manufacturing plan is to make $x_1 = 24$ chairs and $x_2 = 14$ tables, which yields a profit of ($45 × 24) + ($80 × 14) = $2200. The sweeping of the objective function through the convex set of solutions is, in effect, what happens when we solve the problem using the algebraic techniques of the simplex method.

The simplex process starts at any extreme point like $x_1 = 0$, $x_2 = 0$; determines that there are more profitable extreme-point solutions; and tries one out for size. The process keeps improving the solution by shifting from one extreme point to a better one—here to $x_1 = 0$, $x_2 = 20$—and continues in this manner until it can demonstrate that the current extreme point cannot be beat. The saving feature of the simplex method is that it is able to pick and choose its path from one extreme point to a better, adjacent, extreme point by trying out a small subset of all the possible extreme points. In this respect, the process is quite efficient, but no one knows exactly why.

The point (24,14) represents an optimal production schedule for a wide range of objective functions. As the profit for a chair or a table changes, the slope of the profit line also changes—it can be visualized as pivoting about the point (24,14). A big enough change in the profit coefficients will twist the profit line so that the optimum is at a different point. For any profit line, the optimum would occur at one of the four extreme points. Multiple optimal solutions arise when the objective function takes up a final position which includes a boundary of the convex set. The objective function $\$50x_1 + \$75x_2$ is optimized by the extreme points (24,14) and (45,0) and all the points on the boundary joining them.

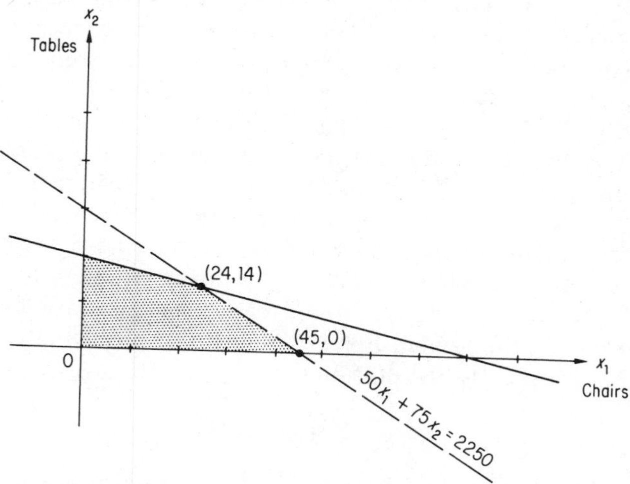

The above approach can be used to solve any two-variable inequality problem. The only points which we are required to calculate are those needed to graph the barrier lines and the extreme point at which the objective function makes its last point of contact with the convex set of solutions.

The same general graphical interpretation applies to problems with higher dimensions. We have a convex polyhedron defined by the constraints, and we wish to move the objective function through the polyhedron in the direction which yields the optimum. You can picture this process by viewing the room you are sitting in as a convex set defined by the constraints of a linear program. One corner of the room is the origin of the three-dimensional world. The floor, walls, and ceiling are barrier constraints, with the convex set being a part of the nonnegative orthant of three-dimensional space. Any objective function, such as $x_1 + x_2 + x_3$, is maximized at one of the eight extreme points. If we remove the ceiling constraint, then the convex set is unbounded. If x_3 is the vertical direction, then the objective function x_3 to be maximized is unbounded. We leave it to your imagination to conjure up the wide variety of situations that can arise in the three-dimensional world. Again, we emphasize that problems with more than two variables are not solved graphically, but many of them can be solved with pencil and paper and the simplex method.

14.2 SHOULD WE ALSO MAKE DESKS?: THE DUAL PROBLEM

You have probably noted that the optimal solution to the chair–table manufacturing problem, make 24 chairs and 14 tables, uses all the available resources. For this solution the mahogany inequality constraint $5x_1 + 20x_2 \leq 400$ is an exact equality, $5(24) + 20(14) = 400$. Similarly, for labor-hours we have $10(24) + 15(14) = 450$. A natural question to ask is what would happen to the Company's profits if it could increase the available supply of a particular resource? Also, for the given resources, if the Simple Company now considers the making of desks, as well as chairs and tables, how can it decide whether or not it is profitable to switch its resources to desk production? These questions and their answers are basic to economic theory and the running of an industrial organization. The first question concerns what is termed *marginal analysis,* while the second deals with *opportunity costs*. For linear-programming models like the chair–table problem—in fact, for any linear-programming problem—these questions can be answered by solving a related linear-programming problem called the *dual problem*; the original problem is the *primal.* The dual problem is formed using the data from the primal. For the chair–table problem it is as follows: Minimize the resource valuation function $w_0 = 400w_1 + 450w_2$ subject to

$$
\begin{aligned}
5w_1 + 10w_2 &\geq 45 \\
20w_1 + 15w_2 &\geq 80 \\
w_1 \qquad\quad &\geq 0 \\
w_2 &\geq 0
\end{aligned}
$$

Here w_1 is the unknown value (accounting price) of the mahogany resource, and w_2 is the unknown value of the labor resource. (You should note the switch between the objective-function coefficients and the right-hand sides of these problems; also note that the rows of the primal coefficients are the column coefficients of the dual.) As both these resources are bottlenecks in the optimal solution—that is, they are scarce and limit the total production—we would expect these dual values to be positive in the optimal solution to the dual. The optimal solution is $w_1 = 1$ and $w_2 = 4$, with a value of the objective function of $w_0 = \$2200$, assuming that the physical units of the w's are measured in dollars per unit of resource. We interpret this answer as follows.

The marginal-profit contribution of an additional unit of mahogany is 1; for labor it is 4. If one of the resources was not scarce, for example an excess of mahogany, then an increase in its availability would not cause a change in profits, and its marginal value would be zero. The main theorem of linear programming (*the duality theorem*) states that if the primal has a finite optimal solution, so does the dual, and the optimal values of the objective functions are equal. Thus, we note that the minimum valuation of the resources is equal to the total profit of \$2200.

For the primal problem, since the optimal solution requires us to produce both chairs and tables, their opportunity costs are zero. This is reflected in the optimal solution to the dual by the fact that $w_1 = 1$ and $w_2 = 4$ make both dual constraints equalities. If, for example, the optimal solution required the production of only tables, then the opportunity cost for chairs, given by the difference between the left- and right-hand sides of the first inequality, $5w_1 + 10w_2 - 45$, would be positive; that is, the value of the resources that make a chair are worth more than the profit from a chair and resources required to make a chair are more valuable when applied to the making of tables. A similar analysis can be made to determine if it would be more profitable to make a third product like desks. If the profit of a desk is \$110 and a desk uses 15 units of mahogany and 25 units of labor, should the Simple Furniture Company make desks? We would need to consider the new dual constraint $15w_1 + 25w_2 \geq 110$. We see that the current dual solution $w_1 = 1$ and $w_2 = 4$ does satisfy this new constraint, that is, $15(1) + 25(4) - 110 = \$115 - \$110 = \$5$. Thus, the profit from a desk (110) is smaller than the accounting value (115) of the resources that go into the making of a desk and we would not make any desks. What if the profit of a desk was greater that \$115? Show what happens by solving the dual graphically and letting the profit of a desk vary from \$110 to \$120.

One of the important contributions of linear programming is its ability to answer questions dealing with marginal values and opportunity costs. To answer such questions we do not need to solve the dual problem explicitly. Using the simplex algorithm, the answer to the dual problem can be found directly from the solution to the primal. The dual answer is usually part of the output of a computer solution to a linear-programming problem. The form of the dual problem as a minimization problem with its inequalities is not a haphazard choice. The duality theorem applies to primal and dual problems that are related in the stated manner (with some variations allowed).

As the concepts of the dual problem and marginal values are important and are somewhat difficult to understand, we offer the following discussion to clarify the above explanation. The key to interpreting the dual of any problem is the definition of the dual variables. We noted that the dual variables, w_1 and w_2, are measured in dollars per unit of resource. This has to be the case for the dual constraints to make any sense in terms of the dimensions and quantities implied by the inequalities

$$5w_1 + 10w_2 \geq 45$$
$$20w_1 + 15w_2 \geq 80$$

The righthand sides of the inequalities are in terms of dollars per unit produced; thus, the left-hand sides must also be in dollars per unit produced. The numerical coefficients of the w's are in units of resource per unit of product produced. The only way the dimensions on the left-hand sides can be equivalent to the dimensions on the right-hand sides (dollar per unit produced) is to define the dual variables in terms of dollars per unit of resource. You should check this out by writing the dual constraints with the dimensions and seeing that things cancel and combine correctly.

As the resources have been bought and paid for by the manufacturer, you might be wondering why the value of the resources are not just their costs or open-market values. For example, the manufacturer has a work force that can contribute up to 450 hours to the making of the chairs and tables. The cost of this resource is reflected in the net profits of $45 for a chair (that uses 10 labor-hours) and $80 for a table (that uses 15 labor-hours). The actual cost of a labor-hour is, let us say, $10. This is what the manufacturer has to pay. Then, what is the meaning of the dual variable that represents the marginal value of an hour of labor, $w_2 = 4$?

The optimal solution uses all of the available labor-hours, that is, labor-hours is a scarce resource for the manufacturer. It is reasonable for the Simple Furniture Company management to ask what if it had an extra hour of labor—451 hours instead of 450—how much extra profit can it make? The $w_2 = 4$ says that if Simple had the hour, and enough additional resources to make full use of it, profits would increase by $4.00. Thus, in this case, if Simple had to pay more than $4.00 for that extra hour, say the $10, then the Company might make an additional chair or table, but it would lose money on the transaction. The Company would consider paying up to $4.00 for that extra hour. A similar discussion applies to an extra board-foot of lumber. Lumber is also a scarce resource and has a marginal value of $w_1 = 1$. If a resource was not used up by the manufacturing process, that is, the Company had a surplus of this resource, then the Company would not be willing to pay anything for an extra unit. Also, the Company's competitors would not be willing to help it out by buying any surplus resources. The marginal value of surplus resources is zero.

We can consider the Simple Furniture Company to be a self-contained economic entity in which the Company has the ability to combine its economic resources to make a profit. Thus, the opportunities open to the Company to mix its resources to make products that maximize its net profits represent the Company's economy. The value of the Company's resources has to be considered within this framework. The optimal dual variables represent artificial (Company internal) accounting values. These values are determined so that the accounting value of the resources used to make the optimal set of products is minimized.

A board-foot of lumber probably has a value on the open market of more than $1.00. The Simple Furniture Company could probably sell this resource to a competitor for the higher value and not make any products. It could also fire its labor force. But that is not the economic situation being analyzed. Our analysis determines the optimal mix of the given resources to maximize profits and a valuation of the resources that minimizes the value of the resources used. The optimal solutions to the primal and dual problems state that the Company's economy is in equilibrium, that is, the maximum profit is equal to the minimum value of the resources used to make the profit. (This last statement can also be applied to a set of competing firms and generalized to a "competitive" equilibrium.) If the Company was willing to sell its resources, it would require the prices paid for its resources to be such that the total sales price would be at least equal to the $2200 it could make if it stayed in business. In turn, the buyer would want to keep the prices paid as small as possible. This "bargaining" leads to the marginal values of the resources being the prices and a competitive equilibrium is established.

15 The Diet Problem (Again)

The best doctors in the world are Doctor Diet,
Doctor Quiet, and Doctor Merryman.

JONATHAN SWIFT

15.1 THE RIGHT MIX

To illustrate a slightly different geometric situation, we return to the housewife's breakfast problem. We let x_1 stand for ounces of Krunchies and x_2 for ounces of Crispies. The mathematical formulation of the problem is to find nonnegative values of x_1 and x_2 which minimizes the total cost

$$3.8x_1 + 4.2x_2$$

subject to

$$\begin{aligned} 0.10x_1 + \quad 0.25x_2 &\geq 1 \\ 1.00x_1 + \quad 0.25x_2 &\geq 5 \\ 110.00x_1 + 120.00x_2 &\geq 400 \end{aligned}$$

Plotting the barrier lines as before, we have the shaded solution space shown below. This convex set is a convex polyhedron and is unbounded in the direction of increase for both x_1 and x_2.

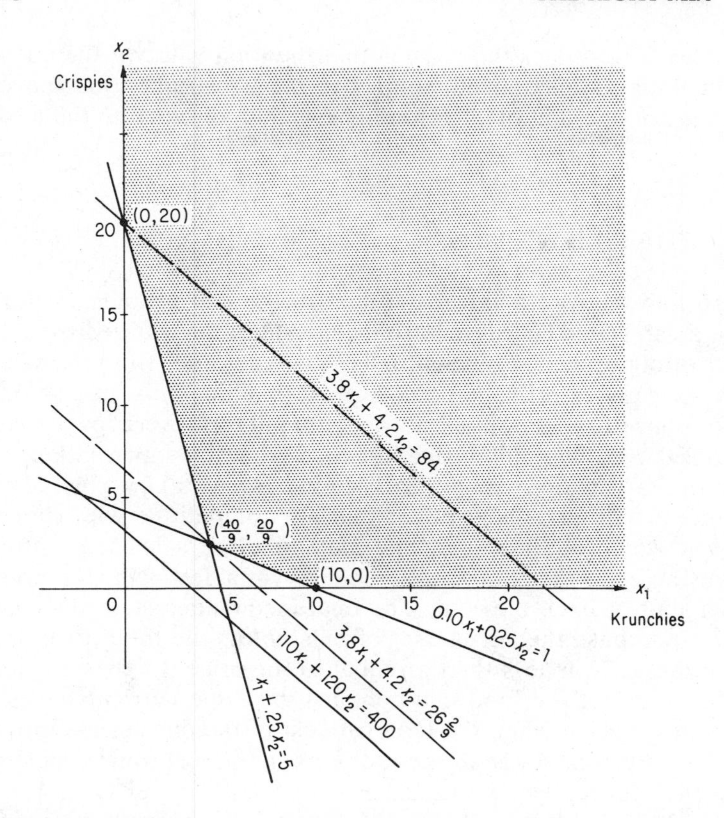

Our graphical analysis has revealed that the constraint on calories will always be satisfied if the other two constraints are satisfied. The barrier line for calories, $110x_1 + 120x_2 = 400$, does not form part of the boundary of the convex set of solutions. It is redundant to the problem and can really be dropped from further consideration. In the figure, we have drawn two traces of the cost line, $3.8x_1 + 4.2x_2 = 84$ cents and the minimal-cost line of $3.8x_1 + 4.2x_2 = 26\frac{2}{9}$ cents. The optimal solution is unique and requires the housewife to purchase $x_1 = \frac{40}{9}$ ounces of Krunchies and $x_2 = \frac{20}{9}$ ounces of Crispies.

The dual to the diet problem is as follows: Find nonnegative values of w_1, w_2, and w_3 which maximize

$$w_0 = w_1 + 5w_2 + 400\,w_3$$

subject to

$$\begin{aligned} 0.10w_1 + 1.00w_2 + 110.00w_3 &\leq 3.8 \\ 0.25w_1 + 0.25w_2 + 120.00w_3 &\leq 4.2 \end{aligned}$$

Here a w_i can be interpreted as the value of one unit of nutrient i. Thus, the dual problem determines the w_i such that the total value of the nutrients in the minimal diet is maximized, subject to the constraints that the value of the nutrients in one unit of each food is less than or equal to the cost of the food.

15.2 THE PILL PEDDLER

An interesting interpretation of the dual to the diet problem is due to the mathematician David Gale. He introduces a salesman who represents a manufacturer of concentrated nutrient pills. You can order from the salesman a pill with so many calories, vitamin A pills, thiamine pills, and so on. If the prices for concentrated units of nutrients in pill form were low enough, you might consider buying pills instead of food from the supermarket. (We are ignoring the satisfaction you get from eating versus taking pills.) The salesman wants to sell his pills at prices that are competitive with supermarket prices, but which at the same time maximize the salesman's profit. The unknowns here are the pill prices—the dual variables—which represent the value per unit of each nutrient. The dual inequalities state that the dieter would not purchase the correct set of pills unless the total price of the nutrients in the set was less than or equal to the price the dieter would have to pay for a unit of the food that contains the same nutrients. We have an equilibrium situation when the minimal cost of the diet (primal problem) is equal to the maximal value of the nutrients in the diet (dual problem). (See Section 17.17 for further discussion.)

You might have noticed that in the Simple Furniture Company problem the primal is maximized and the dual is minimized; the opposite is the case for the diet problem. We can call either the maximum or minimum problem the primal or dual. The one with the more natural (or original) formulation is usually considered to be the primal.

16 The Simplex Algorithm

16.1 INEQUALITIES TO EQUATIONS

To find an optimal solution to a nontrivial linear-programming problem, you need to be able to solve a set of linear equations. For the discussion that follows, we assume your mathematical training has included the basic solution method known as *complete elimination* or *Gauss–Jordan elimination*. (We suggest that such material be reviewed in class, as required.) Except for computational and numerical refinements, our approach to the simplex method is equivalent to the computer procedures used to solve very large problems.

When we want to solve a linear-programming problem, it is more convenient to deal with exact equations than with inequalities. The furniture and diet problems were stated in terms of linear inequalities, while the models of the transportation, caterer, and trim problems were linear equations. (A problem can have a mixture of equations and inequalities.) There is a very simple way to convert any linear inequality into an equation.

For example, the breakfast-diet problem has a greater than or equal inequality for calories that states that the total amount of calories obtained by eating a mixture of the two cereals has to be greater than or equal to 400, that is,

$$110.00K + 120.00C \geq 400$$

There is no reason to expect that the optimal mixture of K and C would make the left-hand side of the constraint equal to exactly 400. In fact, for this problem, the optimal solution of $K = 4\frac{4}{9}$ and $C = 2\frac{2}{9}$ produces $755\frac{5}{9}$ calories, an overage of $355\frac{5}{9}$ calories. No matter what feasible values we choose for K and C, the left-hand side will be either equal to 400 or greater than 400. The difference between the left-hand and right-hand sides of the inequality can be treated as a variable that measures the overage, or the surplus as it is termed. Here the surplus is positive and equal to $355\frac{5}{9}$. The measure of the surplus, called the *surplus variable,* is either positive or zero. It will never be negative as we are dealing with the nonnegative quantity

$$(110.00K + 120.00C - 400) \geq 0$$

We define the expression in parenthesis as the nonnegative variable s_3 and write

$$s_3 = 110.00K + 120.00C - 400$$
$$s_3 \geq 0$$

(We use the subscript three as we are working with the third inequality of the problem.) We can then rewrite the original inequality as the equation

$$110.00K + 120.00C - s_3 = 400$$

Introducing different surplus variables, one for each of the three diet problem constraints, we have the following equation form:

$$\text{Minimize } 3.8K + 4.2C$$

subject to

$$\begin{aligned}
0.10K + 0.25C - s_1 \qquad\qquad &= 1 \\
1.00K + 0.25C \qquad - s_2 \qquad &= 5 \\
110.00K + 120.00C \qquad\qquad - s_3 &= 400 \\
(K, C, s_1, s_2, s_3) &\geq 0
\end{aligned}$$

Note that we now have a standard linear-programming equation model; all the variables are nonnegative. Any solution will make the surplus variables positive or zero. The optimal diet produces the exact required levels of thiamine and niacin (constraints one and two) and thus, $s_1 = 0$ and $s_2 = 0$ in the optimal solution. This solution corresponds to the intersection (the simultaneous solution) of the linear equations

$$\begin{aligned}
0.10K + 0.25C &= 1 \\
1.00K + 0.25C &= 5
\end{aligned}$$

It is the optimal extreme point of the convex set of solutions shown in the problem's graphical solution. In general, the surplus variables do not appear in the objective function; they are given cost coefficients of zero. However, if there is a penalty or cost in having a surplus amount and this can be measured properly, you can introduce the appropriate linear terms in the objective function.

We transform a less than or equal to inequality in a similar fashion. The original Simple Furniture Company problem has five such inequalities. The first one considers the number of board-feet of mahogany available as a resource. It is given by

$$5c + 20t + 15d + 22b \leq 20{,}000$$

Any feasible solution to the problem will make the left-hand side less than or equal to 20,000. We define the difference between the right-hand and left-hand sides as a new nonnegative variable that measures the amount of resource not used, here termed the *slack variable* and expressed by

$$0 \leq 20{,}000 - (5c + 20t + 15d + 22b)$$

We denote the first slack variable by s_1. It is defined by

$$s_1 = 200{,}000 - (5c + 20t + 15d + 22b)$$

or

$$5c + 20t + 15d + 22b + s_1 = 20{,}000$$

Any solution that does not use all of the available 20,000 board-feet will have $s_1 > 0$. We introduce similar slack variables, one for each constraint, and obtain the equation form of the Simple Furniture Company production problem:

$$\text{Maximize } 45c + 80t + 110d + 55b$$

subject to

$$\begin{array}{rcl}
5c + 20t + 15d + 22b + s_1 & = & 20{,}000 \\
10c + 15t + 25d + 20b \qquad + s_2 & = & 4000 \\
3c + 8t + 15d + 10b \qquad\qquad + s_3 & = & 2000 \\
4c \qquad + 20d \qquad\qquad\qquad\qquad + s_4 & = & 3000 \\
20b \qquad\qquad\qquad\qquad + s_5 & = & 500 \\
(c, t, d, b, s_1, s_2, s_3, s_4, s_5) & \geq & 0
\end{array}$$

The problem now has five equations and nine variables. You might think that we have complicated the problem by adding more variables. But, if we tried to solve the inequality form, we would have to deal with the slack or surplus variables in an implicit manner anyway.

16.2 SOLUTIONS AND EXTREME POINTS

We stated earlier that a finite optimal solution to a linear-programming problem occurs at an extreme point (and possibly boundary points) of the convex set of solutions. You should be able to convince yourself that this is true by studying the geometry of the situation. Related to this is the not-so-obvious fact that each extreme point corresponds to a solution of the model's equations in which a number of variables will be zero, with this number being equal to the difference between the total number of variables and the total number of equations. This is a bit difficult to illustrate geometrically as we are dealing with many dimensions (variables). If you look at the graph of the breakfast-diet problem, you see that the optimal extreme point has $s_1 = 0$ and $s_2 = 0$. The equation form of this problem has five variables and three equations. If you solved the three-equation system with $s_1 = 0$ and $s_2 = 0$, you would determine the optimal solution $K = 4\frac{4}{9}$, $C = 2\frac{2}{9}$, and $s_3 = 355\frac{5}{9}$, with the objective function equal to $26\frac{2}{9}$ cents.

The values of the other extreme points also can be determined by solving square sets of equations, that is, sets that have the same number of equations as variables. The extreme point $K = 10$, $C = 0$ is associated with the square set of equations obtained by letting $C = 0$ and $s_1 = 0$. You should show that $K = 10$, and $s_2 = 5$ and $s_3 = 700$. If we were able to plot and interpret the five dimensions of the problem, we would see that the point $K = 10$, $s_2 = 5$, and $s_3 = 700$ was an extreme point of the five-dimensional convex set of solutions.

You might have concluded from this discussion that all we need to do is to set the right number of variables equal to zero and solve the resulting square set of equations. You would be right up to a point. Which variables do you set equal to zero? For the 3×5 equation diet problem there are 10 choices. For the 5×9 equation furniture problem there are 126 choices. (Try to figure out the formula for determining the number of choices. It concerns finding the number of combinations of n things taken m at a time. Your hand calculator may have the appropriate function key.) The number of choices can grow very quickly, that is, exponentially, as we vary the number of variables and equations. A 50-variable problem with five equations has 2,118,760 choices. But a 50-variable problem with 48 constraints has only 1225 choices.

For most realistic problems, it would be quite tedious and impractical, if not impossible, to enumerate all combinations. But we know how to find the solution to a square set of equations. It is rather easy to do. What we need is a procedure that searches through the large number of combinations that corresponds to the possible extreme points. As a square set of equations might not have a solution in terms of nonnegative variables, we have to be able to filter out these infeasible combinations. This is where the simplex algorithm comes in.

16.3 GEORGE B. DANTZIG AND HIS SIMPLEX ALGORITHM

The mathematician George B. Dantzig was the first one to develop the theory and general formulation of the linear-programming model. He also invented the simplex algorithm that is used throughout the world. The simplex algorithm, as programmed for computers, has solved problems of immense size—tens of thousands of equations and millions of variables. It is difficult to measure the economic and other benefits of our ability to solve linear-programming problems on a regular, routinized basis.

The simplex algorithm is a numerical procedure that systematically finds extreme-point solutions to a linear-programming problem. It also, and most importantly, provides a test for determining if any such solution is optimal. If the problem was to maximize an objective function, a solution that passes the simplex optimality test is guaranteed to have a value of the objective function that is greater than or equal to all other feasible solutions to the problem, extreme point or otherwise. The algorithm also applies to a minimization problem, with the optimality test providing information on whether an extreme-point solution yields the smallest value of the objective function for all possible solutions.

Some linear-programming problems have solutions such that the value of the objective function increases (or decreases) without bound, that is, the value of the objective function can be made as large (small) as you want. Real-life problems that have unbounded values are rare, if not impossible. It would be quite strange for a manufacturer, with finite amounts of resources, to be able to make a set of products that would cause the company's profits to get as large as possible. (We have been looking for that type of business for years!)

16.4 APPLYING THE SIMPLEX ALGORITHM

We next develop the major elements of the simplex algorithm by a straightforward attack on the smaller version of the Simple Furniture Company's problem. Then, as it is always nice to do for an algorithm, we shall routinize the computational procedure for easier application by hand, hand calculator, or computer.

The problem is to maximize

$$45x_1 + 80x_2$$

subject to

$$\begin{aligned} 5x_1 + 20x_2 &\le 400 \quad \text{(board-feet of mahogany)} \\ 10x_1 + 15x_2 &\le 450 \quad \text{(labor-hours)} \\ x_1 &\ge 0 \\ x_2 &\ge 0 \end{aligned}$$

We next introduce the nonnegative slack variables x_3 for the first resource constraint (board-feet of mahogany) and x_4 for the second resource constraint (labor-hours). It is also convenient to define the variable $x_0 = 45x_1 + 80x_2$. Here x_0 is just the value of the objective function for any feasible values of x_1 and x_2. For this problem x_0 will always be nonnegative, but that would not be the case for problems with objective functions that have negative coefficients. We now have the following equivalent problem formulation:

$$\text{Maximize } x_0$$

subject to

$$\begin{aligned} x_0 - 45x_1 - 80x_2 \qquad\qquad &= 0 \\ 5x_1 + 20x_2 + x_3 \qquad &= 400 \\ 10x_1 + 15x_2 \qquad + x_4 &= 450 \\ (x_1, x_2, x_3, x_4) &\ge 0 \end{aligned}$$

Note that the value of x_0 is not restricted to be nonnegative.

To initiate the simplex algorithm, we need a starting extreme-point solution. Extreme points to a set of m equations in n variables are found by setting $n - m$ of the variables equal to zero and solving the resulting square set of equations for the values of the remaining variables. If these values are nonnegative, we have an extreme-point solution. This type of solution is also called a *basic feasible solution*. The variables in the square set of equations are termed *basic variables*; the $n - m$ variables that were set equal to zero are called *nonbasic variables*. There is a direct correspondence between a basic feasible solution and an extreme point. Every basic feasible solution is an extreme point; every extreme point corresponds to a basic feasible solution. The Simple Furniture Company's problem has two equations in four variables (we do not include the equation that defines the objective function). Thus, we need to set $n - m = 4 - 2 = 2$ variables equal to zero. Which ones?

Computational life can be made easier depending on our choice. If we set $x_3 = x_4 = 0$, then we have to solve the square set of equations

$$\begin{aligned} 5x_1 + 20x_2 &= 400 \\ 10x_1 + 15x_2 &= 450 \end{aligned}$$

This requires a bit of work and we also do not know if the answers will be such that $x_1 \geq 0$ and $x_2 \geq 0$. The easy thing to do (you probably have guessed it already), is to set $x_1 = x_2 = 0$ and solve the square set of equations

$$\begin{aligned} x_3 \quad\quad &= 400 \\ x_4 &= 450 \end{aligned}$$

But, there's the solution! No work to be done. How good is the solution in terms of the objective function? Well, not very good as $x_0 = 0$. We really want a solution that makes x_0 as large as possible. Hopefully, it will involve x_1 and/or x_2 as they are the only variables in the objective function with positive coeffficients. It would be nice to start out with basic feasible solutions that include x_1 and/or x_2; but, as we noted, they require some computation and we are not sure we would be able to use the answers because of the nonnegativity restrictions. So, we compromise and use the "sure but poor" basic feasible solution of $x_3 = 400$ and $x_4 = 450$ to start us off.

For the purpose of exposition, it is convenient to rewrite the equations as

$$\begin{aligned} x_0 \quad\quad\quad &= 0 + 45x_1 + 80x_2 \\ x_3 \quad &= 400 - 5x_1 - 20x_2 \\ x_4 &= 450 - 10x_1 - 15x_2 \end{aligned}$$

For $x_1 = 0$, $x_2 = 0$, we have the first basic feasible solution of $x_3 = 400$, $x_4 = 450$, and $x_0 = 0$. We see from the equation that defines the value of the objective function, x_0, that if we can find a feasible solution that makes x_1 positive, the objective function would increase by $45x_1$. Similarly, if we can make $x_2 > 0$, then x_0 will increase by $80x_2$.

It is easier to work with only one nonbasic variable at a time to see if it should or can be increased. A good rule is to select the one that has the largest positive coefficient in the right-hand side of the objective function; this choice will tend to (but not always) give us the largest increase in the objective function. A nonbasic variable that has a negative coefficient would not be of interest because if the variable was made positive, it would decrease the value of the objective function. Let's work on x_2.

We would like to find a basic feasible solution that involves x_2 at a positive level, but do not want to search around without some assurance of success. We still keep $x_1 = 0$ and investigate the equations

$$\begin{aligned} x_3 \quad\quad &= 400 - 20x_2 \\ x_4 &= 450 - 15x_2 \end{aligned}$$

These equations define the relationships between x_2, x_3, and x_4. As we increase x_2, we must remember that we can do so only if we keep

$$x_3 = 400 - 20x_2 \geq 0$$

and

$$x_4 = 450 - 15x_2 \geq 0$$

These inequalities tell us how big we can make x_2, that is,

$$\begin{aligned} 400 - 20x_2 &\geq 0 \\ 450 - 15x_2 &\geq 0 \end{aligned}$$

or

$$\begin{aligned} 400 &\geq 20x_2 \\ 450 &\geq 15x_2 \end{aligned}$$

or

$$\begin{aligned} \tfrac{400}{20} &= 20 \geq x_2 \\ \tfrac{450}{15} &= 30 \geq x_2 \end{aligned}$$

The last inequalities bound x_2 (the bounds are the ratios) and must be satisfied simultaneously. Hence, the largest value of x_2 is 20. Note that this value corresponds to the minimum ratio. If we let $x_2 = 20$, then

$$\begin{aligned} x_3 &= 400 - 20(20) = 0 \\ x_4 &= 450 - 15(20) = 150 \end{aligned}$$

By letting x_2 assume this largest value, we obtain, surprisingly, another extreme-point solution; now $x_3 = 0$ with $x_1 = 0$, and the basic variables are $x_2 = 20$ and $x_4 = 150$. The value of the objective function for this basic feasible solution is

$$x_0 = 0 + 45(0) + 80(20) = 1600$$

Let's use this information to obtain a better solution if one exists.

The original equations were

$$\begin{aligned} x_0 - 45x_1 - 80x_2 \qquad\qquad &= 0 \\ 5x_1 + 20x_2 + x_3 \qquad &= 400 \\ 10x_1 + 15x_2 \qquad + x_4 &= 450 \end{aligned}$$

As we normally write all the variables on left-hand side of the equation sign, we need to interpret the above discussion for this arrangement of the equations. Here, for a maximization problem, we would select x_2 as it has the smallest (algebraically negative) coefficient. The largest value of x_2 is the minimum of the ratios 400/20 and 450/15. To obtain the basic feasible solution that involves x_2 and x_4 we have to perform one step of the complete elimination procedure. We have to eliminate variable x_2 from all equations (including the objective-function equation) except from the row in which the minimum ratio occurred. This row is called the *pivot row* and the coefficient of x_2 in that row, the 20, is the *pivot element*. The column containing x_2 is the *pivot column*. To illustrate this, we first work with the resource equations; the pivot element has been circled:

$$\begin{aligned} 5x_1 + \textcircled{20}\, x_2 + x_3 \quad &= 400 \\ 10x_1 + 15x_2 \quad + x_4 &= 450 \end{aligned}$$

Dividing the pivot row by 20, we obtain

$$\begin{aligned} \tfrac{5}{20} x_1 + x_2 + \tfrac{1}{20} x_3 \quad &= 20 \\ 10x_1 + 15x_2 \quad + x_4 &= 450 \end{aligned}$$

To eliminate x_2 from the last equation, we multiply the transformed pivot row by 15 and subtract it from the last equation to obtain

$$\begin{aligned} \tfrac{5}{20} x_1 + x_2 + \tfrac{1}{20} x_3 \quad &= 20 \\ \tfrac{125}{20} x_1 \quad - \tfrac{15}{20} x_3 + x_4 &= 150 \end{aligned}$$

This is a new basic feasible solution with $x_2 = 20$, $x_4 = 150$, and $x_1 = x_4 = 0$. The question that comes up now is can we find a better solution? To answer it, we need to look at the form of the objective function for this basic feasible solution.

We first rewrite the above equations as

$$\begin{aligned} x_2 &= 20 - \tfrac{5}{20} x_1 - \tfrac{1}{20} x_3 \\ x_4 &= 150 - \tfrac{125}{20} x_1 + \tfrac{15}{20} x_3 \end{aligned}$$

We have expressed the new basic variables in terms of the current nonbasic variables. Just as we did with the starting basic feasible solution of x_3 and x_4, we need to determine for the new basic feasible solution of x_2 and x_4 whether or not we can increase the objective function by making either of the nonbasic variables x_1 or x_3 greater than zero. To do this, we take the above expressions of x_2 and x_4 and substitute them for x_2 and x_4 in the original objective function

$$x_0 = 0 + 45x_1 + 80x_2$$

We only need to be concerned with

$$x_2 = 20 - \tfrac{5}{20}x_1 - \tfrac{1}{20}x_3$$

as x_4 has a zero coefficient in the objective function. Making this substitution, we obtain

$$x_0 = 0 + 45x_1 + 80(20 - \tfrac{5}{20}x_1 - \tfrac{1}{20}x_3)$$

or

$$x_0 = 1600 + 45x_1 - 20x_1 - 4x_3$$

or

$$x_0 = 1600 + 25x_1 - 4x_3$$

This equivalent form of the objective function, given in terms of the nonbasic variables, tells us that it would pay to increase x_1 as any $x_1 > 0$ increases the objective function by $25x_1$; it would not pay to increase x_3 as any $x_3 > 0$ would decrease the objective function by $4x_3$. Note that with $x_1 = 0$ and $x_3 = 0$, then $x_0 = 1600$, the value we had calculated before.

We can systematize this transformation of the objective function by just eliminating x_2 from the original form of the objective function. We had the system

$$\begin{aligned} x_0 - 45x_1 - 80x_2 \qquad\qquad &= 0 \\ 5x_1 + \textcircled{20}\, x_2 + x_3 \qquad &= 400 \\ 10x_1 + 15x_2 \qquad + x_4 &= 450 \end{aligned}$$

Pivoting on the pivot element 20 and then eliminating x_2 from the objective-function equation, as well as the other equation, yields the sequence of equations:

$$\begin{aligned} x_0 - 45x_1 - 80x_2 \qquad\qquad &= 0 \\ \tfrac{5}{20}x_1 + x_2 + \tfrac{1}{20}x_3 \qquad &= 20 \\ 10x_1 + 15x_2 \qquad + x_4 &= 450 \end{aligned}$$

$$\begin{aligned} x_0 - 45x_1 - 80x_2 \qquad\qquad &= 0 \\ \tfrac{5}{20}x_1 + x_2 + \tfrac{1}{20}x_3 \qquad &= 20 \\ \tfrac{125}{20}x_1 \qquad - \tfrac{15}{20}x_3 + x_4 &= 150 \end{aligned}$$

$$\begin{aligned} x_0 - 25x_1 \qquad + 4x_3 \qquad &= 1600 \\ \tfrac{5}{20}x_1 + x_2 + \tfrac{1}{20}x_3 \qquad &= 20 \\ \tfrac{125}{20}x_1 \qquad - \tfrac{15}{20}x_3 + x_4 &= 150 \end{aligned}$$

The last set is the form of the equations that expresses the basic feasible solution of $x_2 = 20$, $x_4 = 150$, and $x_0 = 1600$, with $x_1 = x_3 = 0$.

We have now carried out one *iteration* of the simplex method and have moved to another extreme point of the convex set of solutions. If you look at the graph of this problem discussed in Chapter 14, you can see that we were originally at the origin extreme point ($x_1 = 0$, $x_2 = 0$, and $x_0 = 0$).

We have moved to the *adjacent extreme point* ($x_1 = 0$, $x_2 = 20$, and $x_0 = 1600$). Extreme points are adjacent if they have a boundary of the convex set in common. The simplex algorithm moves from one adjacent extreme point to another, always trying to improve the value of the objective function. From the graph, you can see that the selection criterion we used for choosing a nonbasic variable to be made basic will not always give us the best improvement in the objective function. If we had selected x_1 instead of x_2 (even though x_1 had an objective-function coefficient of -45 versus x_2's coefficient of -80), we would have moved from the origin to the extreme point of $x_1 = 45$, $x_2 = 0$, and $x_0 = 2025$.

We now work with the new extreme-point solution that has x_2 and x_4 as basic variables, that is, the transformed equations

$$\begin{aligned} x_0 - 25x_1 \qquad + 4x_3 \qquad &= 1600 \\ \tfrac{5}{20}x_1 + x_2 + \tfrac{1}{20}x_3 \qquad &= 20 \\ \tfrac{125}{20}x_1 \qquad - \tfrac{15}{20}x_3 + x_4 &= 150 \end{aligned}$$

The objective function rewritten as

$$x_0 = 1600 + 25x_1 - 4x_3$$

tells us that we should try and make $x_1 > 0$ and not to bother with x_3. We then investigate the following relationships to determine how big x_1 can be made:

$$\begin{aligned} x_2 &= 20 - \tfrac{5}{20}x_1 \geq 0 \\ x_4 &= 150 - \tfrac{125}{20}x_1 \geq 0. \end{aligned}$$

The nonbasic x_1 can be increased as long as

$$\begin{aligned} 20 - \tfrac{5}{20}x_1 &\geq 0 \\ 150 - \tfrac{125}{20}x_1 &\geq 0 \end{aligned}$$

or

$$\begin{aligned} 20 &\geq \tfrac{5}{20}x_1 \\ 150 &\geq \tfrac{125}{20}x_1 \end{aligned}$$

or

$$\begin{aligned} \frac{20}{\frac{5}{20}} &= 80 \geq x_1 \\ \frac{150}{\frac{125}{20}} &= 24 \geq x_1. \end{aligned}$$

The left-hand-side ratios are bounds on x_1 and indicate that if $x_1 = 24$, then a new basic feasible solution can be obtained with $x_4 = 0$ and $x_2 = 14$. This can be seen to be the case as

$$\begin{aligned} x_2 &= 20 - \tfrac{5}{20} x_1 = 20 - \tfrac{5}{20}(24) = 14 \\ x_4 &= 150 - \tfrac{125}{20} x_1 = 150 - \tfrac{125}{20}(24) = 0 \end{aligned}$$

The value of the objective function would then be

$$x_0 = 1600 + 25x_1 - 4x_3 = 1600 + 25(24) = 2200$$

This solution is obtained by introducing x_1 into the basic feasible solution. In terms of the elimination process, that means we eliminate x_1 from all the equations but the last, using the coefficient $\frac{125}{20}$ as the pivot element. This is shown by the following sets of equations, with the first set being a repeat of the transformed equations that show explicitly the values of x_2 and x_4:

$$\begin{array}{rrrrrl} x_0 & -\ 25x_1 & & +\ 4x_3 & & = 1600 \\ & \tfrac{5}{20}x_1 & +\ x_2 & +\ \tfrac{1}{20}x_3 & & = 20 \\ & \boxed{\tfrac{125}{20}}x_1 & & -\ \tfrac{15}{20}x_3 & +\ x_4 & = 150 \\ \hline x_0 & -\ 25x_1 & & +\ 4x_3 & & = 1600 \\ & \tfrac{5}{20}x_1 & +\ x_2 & +\ \tfrac{1}{20}x_3 & & = 20 \\ & x_1 & & -\ \tfrac{15}{125}x_3 & +\ \tfrac{20}{125}x_4 & = 24 \\ \hline x_0 & & & +\ x_3 & +\ 4x_4 & = 2200 \\ & & x_2 & +\ \tfrac{2}{25}x_3 & -\ \tfrac{1}{25}x_4 & = 14 \\ & x_1 & & -\ \tfrac{15}{125}x_3 & +\ \tfrac{20}{125}x_4 & = 24 \end{array}$$

You should describe the steps taken to go from the second set of equations to the third set. This last set is a new basic feasible solution with $x_1 = 24$, $x_2 = 14$, and $x_0 = 2200$, with $x_3 = x_4 = 0$. The transformed objective function now states in terms of the zero-valued nonbasic variables that

$$x_0 = 2200 - x_3 - 4x_4$$

From this relationship we see that the last basic feasible solution is also an optimal one. Why? Making either $x_3 > 0$ or $x_4 > 0$ would decrease the value of the objective function from its present value of 2200. And if x_3 and x_4 both remain equal to zero, then the solutions must involve x_1 and/or x_2. The only extreme point available to us that includes x_1 and x_2 and that make the slack variables $x_3 = x_4 = 0$ is the one associated with the square set of equations

$$\begin{aligned} 5x_1 + 20x_2 &= 400 \\ 10x_1 + 15x_2 &= 450 \end{aligned}$$

Our last basic feasible solution is just the solution to this set. Just by looking at these equations you could not have concluded that a solution exists with nonnegative values. The systematic approach that we used, moving from one basic feasible solution to another, and improving the value of the objective function, guarantees (with minor exceptions) that we will always determine an optimal solution. Each iteration, that is, a variable exchange, is equivalent to one step of the elimination method.

You have now solved your first linear-programming problem using the simplex algorithm. The algorithm consists of a standard procedure for solving square sets of equations—the complete elimination method; a slight modification of the method to ensure that all basic feasible solutions have nonnegative values—the simplex feasibility ratio test; and the transforming of the objective function to an expression in only the nonbasic variables—the simplex optimality test.

As it took us a long time to get this far, we recapitulate the computational process. We considered the problem:

$$\text{Maximize } x_0$$

subject to

$$\begin{aligned} x_0 - 45x_1 - 80x_2 \qquad\qquad &= 0 \\ 5x_1 + 20x_2 + x_3 \qquad &= 400 \\ 10x_1 + 15x_2 \qquad + x_4 &= 450 \\ x_j &\geq 0 \qquad (j = 1, \ldots, 4) \end{aligned}$$

This problem contains the explicit, basic-feasible solution of $x_3 = 400$, $x_4 = 450$, and $x_0 = 0$. As we are maximizing and are keeping all variables on one side of the equation signs (here, for now, the left-hand side), we see from the objective function that if a nonbasic variable has a negative coefficient, we should be able to increase the value of the objective function. For a maximizing problem, the rule is to choose the nonbasic variable that has the algebraically smallest negative coefficient. For this form of the objective function, if all objective-function coefficients are positive, then the current basic feasible solution is also a maximizing solution. We obtain a new basic feasible solution by introducing the chosen nonbasic variable into the solution by having it replace a current basic variable. The particular variable replaced corresponds to the row in which the minimum ratio occurs. These ratios are used to ensure that a new nonnegative solution will be obtained.

We next organize and routinize the steps of the simplex algorithm by using the following detached-coefficient simplex tableaus. This procedure saves writing information and aids in the analysis and computation of the coefficients of the equations. When using the tableau approach, you should not lose sight of the fact that you are dealing with systems of equations, that is, each row of a tableau corresponds to an equation. The original set of equations is organized in the first tableau below, with the other simplex elimination steps following it. Each tableau corresponds to a basic feasible solution. You should be able to follow the application of the simplex algorithm described above without any difficulty. We have moved the solution side of the equations to the left, as this is the traditional and efficient form used when linear-programming problems are solved on a computer. (Why?) The row corresponding to the objective function is labeled row 0. We consider x_0 as always being in the solution, that is, we never pivot in row 0. Tableau III expresses the optimal solution $x_1 = 24$, $x_2 = 14$, $x_3 = x_4 = 0$, and $x_0 = 2200$. We note that we are maximizing in the row 0 cell of the ratio column.

I.

Row	Basis	Solution	x_0	x_1	x_2	x_3	x_4	Ratio
0	x_0	0	1	−45	−80	0	0	Max
1	x_3	400	0	5	(20)	1	0	$\frac{400}{20}$
2	x_4	450	0	10	15	0	1	$\frac{500}{15}$

II.

Row	Basis	Solution	x_0	x_1	x_2	x_3	x_4	Ratio
0	x_0	1600	1	−25	0	4	0	Max
1	x_2	20	0	$\frac{5}{20}$	1	$\frac{1}{20}$	0	$20/\frac{5}{20}$
2	x_4	150	0	($\frac{125}{20}$)	0	$-\frac{15}{20}$	1	$150/\frac{125}{20}$

III.

Row	Basis	Solution	x_0	x_1	x_2	x_3	x_4	Ratio
0	x_0	2200	1	0	0	1	4	Max
1	x_2	14	0	0	1	$\frac{2}{25}$	$-\frac{1}{25}$	
2	x_1	24	0	1	0	$-\frac{3}{25}$	$\frac{4}{25}$	

You should get use to solving linear-programming problems by the tableau format. The row 0 coefficients for the decision variables not in the basis indicate whether or not a particular variable should be chosen to enter the basis. For a maximization problem, we would select the variable with the most negative row 0 coefficient to enter the basis. If all row 0 coefficients are positive or zero, then the corresponding basic solution is optimal. For a minimization problem, the row 0 coefficients would all have to be negative or zero for a solution to be optimal; a variable would be selected to enter the basis if its row 0 coefficient was the most positive.

17 Part III Discussion, Extensions, and Exercises

17.1. Test your knowledge of the simplex algorithm by solving the following problem:

$$\text{Maximize } x_0 = 6x_1 + 12x_2 + 4x_3$$

subject to

$$\begin{aligned} x_1 + 2x_2 + x_3 &\le 5 \\ 2x_1 - x_2 + 3x_3 &\le 2 \\ x_j &\ge 0 \qquad (j = 1, 2, 3) \end{aligned}$$

The optimal solution is $x_1 = \frac{9}{5}$, $x_2 = \frac{8}{5}$, $x_3 = x_4 = x_5 = 0$, and $x_0 = 30$, (x_4 and x_5 are slack variables). The starting tableau is

Row	Basis	Solution	x_0	x_1	x_2	x_3	x_4	x_5	Ratio
0	x_0	0	1	-6	-12	-4	0	0	Max
1	x_4	5	0	1	2	1	1	0	
2	x_5	2	0	2	-1	3	0	1	

The simplex algorithm selects x_2 to enter the basis as its row 0 coefficient is the smallest, that is, -12. You next have to determine which of the current basic variables will be replaced by the ratio test. The x_2 column has coefficients of 2 and -1 in the first and second rows, respectively. Can you use the -1 as a pivot element? No, because if you tried, it would make the value of $x_2 = \frac{2}{-1} = -2$, which is a "no-no" as all variables of the problem are restricted to be nonnegative. Let us look into this situation a bit deeper. For the basic solution of x_4 and x_5 and the choice of x_2 to enter the basis, we consider the equations

$$\begin{aligned} x_4 &= 5 - 2x_2 \\ x_5 &= 2 + x_2 \end{aligned}$$

We want to make $x_2 > 0$ and as large as possible while keeping

$$\begin{aligned} x_4 &= 5 - 2x_2 \geq 0 \\ x_5 &= 2 + x_2 \geq 0 \end{aligned}$$

or

$$\begin{aligned} 5 - 2x_2 &\geq 0 \\ 2 + x_2 &\geq 0 \end{aligned}$$

or

$$\begin{aligned} 5 &\geq 2x_2 \\ x_2 &\geq -2 \end{aligned}$$

or

$$\begin{aligned} \tfrac{5}{2} &\geq x_2 \\ x_2 &\geq -2 \end{aligned}$$

The second inequality that defined $x_5 = 2 + x_2$ does not restrict how big we can make x_2. It does not enter into the ratio calculations. The same would be true if a coefficient was zero. Thus, a variable selected to enter the basis that has negative or zero coefficients in its column (for rows 1 to m) is not restricted by the corresponding ratios. For this problem, the first simplex iteration would select x_2 to enter the basis and replace x_4, with the pivot element being 2. Continue the computations.

17.2. As you read the discussion of Section 17.1, you probably wondered what do you do if all the coefficients (row 1 to m) of a variable selected to enter the basis were negative or zero. Where do you pivot? You don't, because if that happens you have a problem whose objective function can be made as large as you wish (for a maximization problem). To demonstrate this, we change the coefficient 2 in column x_2 to -2. We then have the tableau:

Row	Basis	Solution	x_0	x_1	x_2	x_3	x_4	x_5	Ratio
0	x_0	0	1	-6	-12	-4	0	0	Max
1	x_4	5	0	1	-2	1	1	0	
2	x_5	2	0	2	-1	3	0	1	

We select x_2 to enter the solution and note that the value of x_2 is restricted by

$$\begin{aligned} x_4 &= 5 + 2x_2 \geq 0 \\ x_5 &= 2 + x_2 \geq 0 \end{aligned}$$

Any $x_2 > 0$ will keep both $x_4 \geq 0$ and $x_5 \geq 0$. What happens to the objective function? It is given by

$$x_0 = 0 + 6x_1 + 12x_2 + 4x_3$$

As $x_1 = x_3 = 0$, and the cost coefficients of the slack variables x_4 and x_5 are zero, the objective function reduces to

$$x_0 = 12x_2$$

We now have a feasible solution (not basic) of

$$\begin{aligned} x_0 &= 0 + 12x_2 \\ x_4 &= 5 + 2x_2 \\ x_5 &= 2 + x_2 \end{aligned}$$

For any $x_2 > 0$ (as large as you want, e.g., 10^9) we will always have nonnegative values of x_4 and x_5 that satisfy the constraints of the problem; x_0 can thus be made very large [e.g., $12(10^9)$ or $12(10^{12})$ if $x_2 = 10^{12}$, etc.]. A strange situation. A simple unbounded example that can be solved by graphical techniques is

$$\text{Maximize } x_0 = x_1 + x_2$$

subject to

$$\begin{aligned} -x_1 + x_2 &\leq 1 \\ x_j &\geq 0 \qquad (j = 1, 2) \end{aligned}$$

From the following graph you can see that the convex set of feasible solutions is unbounded. Also, the optimizing direction of the objective function enables us to move the function through the feasible set so that its value is not restricted by the boundaries of the convex set of solutions.

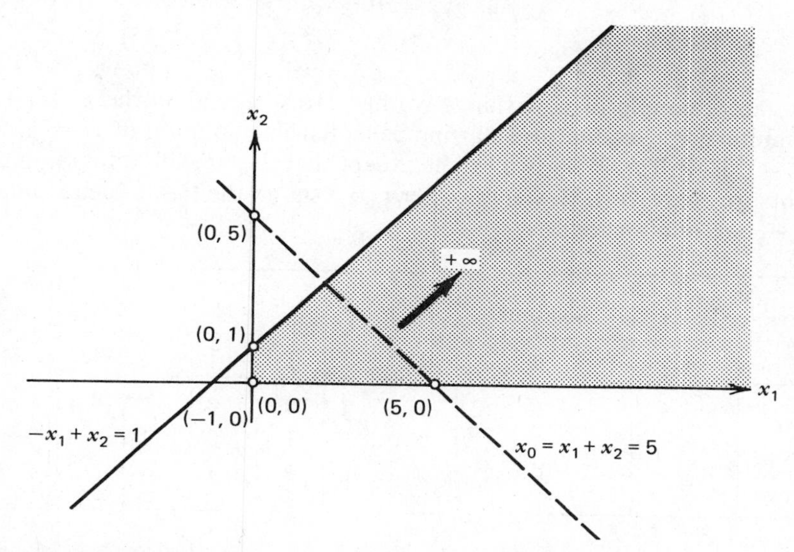

The first simplex tableau (x_3 a slack variable) is

Row	Basis	Solution	x_0	x_1	x_2	x_3	Ratio
0	x_0	0	1	-1	-1	0	Max
1	x_3	1	0	-1	1	1	

This solution corresponds to the origin extreme point $x_1 = 0, x_2 = 0$. We have the choice of selecting x_1 or x_2 to enter the basis as we have a tie for the most negative coefficient in row 0. If you select x_2, you would move to the extreme point $x_1 = 0, x_2 = 1$. Your next choice would force you to select x_1 and the unbounded situation would arise. Try it. But we can see from the above tableau that if we select x_1 first, we have an unbounded solution. Algebraically, it is given by unbounded $x_1 > 0$

$$x_0 = 0 + x_1$$
$$x_3 = 1 + x_1$$

17.3. Solve the following linear-programming problem:

$$\text{Minimize } x_0 = 2x_1 - x_2 + x_3 - 5x_4 + 22x_5$$

subject to

$$\begin{aligned} x_1 \qquad\quad - 2x_4 + x_5 &= 6 \\ x_2 \quad + x_4 - 4x_5 &= 3 \\ x_3 + 3x_4 + 2x_5 &= 10 \\ x_j &\geq 0 \qquad (j = 1, \ldots, 5) \end{aligned}$$

There are a couple of new things we have to worry about here. First, we are minimizing. Second, the starting basic feasible solution of $x_1 = 6$, $x_2 = 3$, $x_3 = 10$, with $x_0 = 19$ is all right except that the objective-function coefficients for these variables are not zero. We set up the first tableau and note that we are minimizing:

Row	Basis	Solution	x_0	x_1	x_2	x_3	x_4	x_5	Ratio
0	x_0	0	1	−2	1	−1	5	−22	Min
1	x_1	6	0	1	0	0	−2	1	
2	x_2	3	0	0	1	0	1	−4	
3	x_3	10	0	0	0	1	3	2	

We need to express the objective function as a function of only the nonbasic variables x_4 and x_5. This can be done by solving for x_1, x_2, and x_3 in terms of x_4 and x_5 using the original equations and performing the substitution directly into the objective function. Or, we can use the tableau and complete the elimination of x_1, x_2, and x_3 from row 0 as follows:

Row	Basis	Solution	x_0	x_1	x_2	x_3	x_4	x_5	Ratio
0	x_0	19	1	0	0	0	3	−14	Min
1	x_1	6	0	1	0	0	−2	1	
2	x_2	3	0	0	1	0	1	−4	
3	x_3	10	0	0	0	1	3	2	

You should check the row 0 coefficients by doing the actual substitution. Since we are minimizing, the row 0 selection criteria is to select the variable to enter the basis that has the largest positive coefficient (why?)—here x_4. Complete the solution to this problem. The optimal solution is $x_1 = 12$, $x_4 = 3$, $x_3 = 1$, and $x_0 = 10$.

17.4. As you solve a linear-programming problem using the simplex algorithm and the tableau format, some computational considerations might occur that need to be clarified. In selecting a nonbasic variable to enter the basis, you can choose anyone that would improve the value of the objective. The rule we used above—select the variable with the most negative row 0 coefficient for a maximizing problem or the most positive row 0 coefficient for a minimizing problem—has proved to be a good one in practice. In using this rule, you might find that more than one nonbasic variable has the most negative (most positive) row 0 coefficient. You may select any one of these tied variables to enter the basis. A systematic rule would be to select the one with the smallest column index.

Ties may also occur in computing the ratio test that determines which variable must leave the basis to preserve the feasibility (nonnegativity) of the basic solution. Here there are ties for the pivot row; any one may be chosen. A systematic rule is to choose the variable with the smallest index to leave the basis.

In calculating the ratio test, you might encounter a ratio that equals zero. This could happen if at least one of the basic feasible variables has a solution value of zero. This is termed a *degenerate basic feasible solution*. If a variable selected to enter the basis has a positive column coefficient in the same row as a zero variable value, then the minimum ratio would be zero. The elimination step continues without any change, but the value of the objective function for the new basic feasible solution would be the same as the preceding solution. In this case we have found two basic feasible solutions that have the same value of the objective function. This situation offers theoretical problems for the simplex algorithm in that there is a chance that you would continue to find solutions that do not change the value of the objective function and you are not at the optimal solution. This situation could lead to the simplex algorithm phenomenon of *cycling*. A cycle occurs when you pass through a succession of basic feasible solutions, all having the same value of the objective function, and you return to the solution that started the cycle. Examples have been constructed that do cycle; one is presented below. You do not need to worry about such problems as they are rare. Computational procedures exist for breaking a cycle if you think you have encountered one.

Many linear-programming problems have multiple optimal basic feasible solutions, that is, there are two or more extreme points at which the objective function is optimal. You can detect this in the final simplex tableau in that a nonbasic variable will have a row 0 coefficient equal to zero. If this variable is introduced into the basis, the value of the objective function will not change. Why?

We illustrate the above considerations by the following tableaus and examples. You should work each problem through to completion.

(a) Ties in selecting a variable to enter the basis and in the ratio test. Arrange the following problem in a simplex tableau and solve:

$$\text{Maximize } x_0$$

subject to

$$
\begin{aligned}
x_0 - 2x_1 + 3x_2 - 2x_3 \qquad\qquad &= 0 \\
x_1 - 2x_2 + 3x_3 + x_4 \qquad &= 3 \\
-2x_1 + 3x_2 + 4x_3 \qquad + x_5 \qquad &= 4 \\
x_1 + 2x_2 + 3x_3 \qquad\qquad + x_6 &= 3 \\
x_j &\geq 0 \qquad (j = 1, \ldots, 6)
\end{aligned}
$$

If you happen to make the right choices, you can find the optimal solution in one step ($x_1 = 3$, $x_4 = 0$, $x_5 = 10$; $x_0 = 6$).

(b) A degenerate basic feasible solution and multiple optimal solution. Find the maximum solution to the following linear-programming problem:

Row	Basis	Solution	x_0	x_1	x_2	x_3	x_4	x_5	Ratio
0	x_0	0	1	−1	0	0	0	0	Max
1	x_3	0	0	1	−1	1	0	0	
2	x_4	2	0	1	2	0	1	0	
3	x_5	1	0	2	0	0	0	1	

The optimal solutions are $x_1 = \frac{1}{2}$, $x_2 = \frac{1}{2}$, $x_4 = \frac{1}{2}$; $x_0 = \frac{1}{2}$ and $x_1 = \frac{1}{2}$, $x_2 = \frac{3}{4}$, $x_3 = \frac{1}{4}$; $x_0 = \frac{1}{2}$.

(c) A problem that cycles.

Show that the following problem cycles after six iterations using the tie-breaking rule that selects the variable with the smallest index to leave the basis:

$$\text{Minimize } x_0 = -\tfrac{3}{4}x_1 + 150x_2 - \tfrac{1}{50}x_3 + 6x_4$$

subject to

$$
\begin{aligned}
\tfrac{1}{4}x_1 - 60x_2 - \tfrac{1}{25}x_3 + 9x_4 + x_5 \qquad\qquad &= 0 \\
\tfrac{1}{2}x_1 - 90x_2 - \tfrac{1}{50}x_3 + 3x_4 \qquad + x_6 \qquad &= 0 \\
x_3 \qquad\qquad\qquad + x_7 &= 1 \\
x_j &\geq 0 \qquad (j = 1, \ldots, 7)
\end{aligned}
$$

The optimal solution is $x_1 = \frac{1}{25}$, $x_3 = 1$, $x_5 = \frac{3}{100}$; $x_0 = -\frac{1}{20}$.

Algorithmic rules (due to R. G. Bland) that will enable you to find the optimum are: (1) if there is more than one variable with a negative (positive) row 0 coefficient, always select the one with the smallest index to enter the basis; *and* (2) if there are ratio ties, always select the variable with the smallest index to leave the basis. Rule 1 ignores the size of the row 0 coefficient and would be a poor one to employ for all problems. Its use would usually require more iterations and slow down the convergence to the optimal solution.

17.5. Up to now, all the linear-programming problems that we had to solve contained an explicit starting (first) basic feasible solution. This may not be the case for many problems. For example, what about the following problem:

$$\text{Minimize } x_0 = 2x_1 + x_2$$

subject to

$$\begin{aligned} 3x_1 + x_2 &\geq 3 \\ 4x_1 + 3x_2 &\geq 6 \\ x_1 + 2x_2 &\geq 2 \\ x_j &\geq 0 \qquad (j = 1, 2) \end{aligned}$$

Its constraint structure is like that of a diet problem; it can be interpreted as one and solved graphically. You should do so. But what if you were asked to solve it using the simplex algorithm?

The simplex algorithm requires the problem to be stated in equation form. We do that by subtracting out a surplus variable from each inequality to obtain:

$$\text{Minimize } x_0$$

subject to

$$\begin{aligned} x_0 - 2x_1 - x_2 \phantom{{}- x_3 - x_4 - x_5} &= 0 \\ 3x_1 + x_2 - x_3 \phantom{{}- x_4 - x_5} &= 3 \\ 4x_1 + 3x_2 \phantom{{}- x_3} - x_4 \phantom{{}- x_5} &= 6 \\ x_1 + 2x_2 \phantom{{}- x_3 - x_4} - x_5 &= 2 \\ x_j &\geq 0 \qquad (j = 1, \ldots, 5) \end{aligned}$$

We have no readily available first basic feasible solution. We cannot let $x_1 = x_2 = 0$ and use x_3, x_4, and x_5 as we would have a negative solution of $x_3 = -3$, $x_4 = -6$, and $x_5 = -2$. What do we do?

Even though we know this problem does have feasible (nonnegative) solutions, it is computationally inefficient to try and find a basic feasible solution by algebraic manipulation. It would be nice if all problems contained an explicit feasible basis. Can we make that always to be the case?

If you encountered this problem in the following form, you would not have any difficulty setting up the simplex tableau and solving the problem:

$$\text{Minimize } x_0$$

subject to

$$\begin{array}{llllllllll}
x_0 - 2x_1 - x_2 & & & & & & & = 0 \\
3x_1 + x_2 - x_3 & & & & + y_1 & & & = 3 \\
4x_1 + 3x_2 & & - x_4 & & & + y_2 & & = 6 \\
x_1 + 2x_2 & & & - x_5 & & & + y_3 & = 2 \\
 & & & & & & (x_j, y_i) & \geq 0 \\
 & & & & & & (j = 1, \ldots, 5; i = 1, 2, 3) &
\end{array}$$

Where did the variables y_1, y_2, and y_3 come from and what do they have to do with the problem? These variables are a computational artifice—an algorithmic crutch—that enables us to start the simplex process in an easy (and lazy) fashion. They have no physical interpretation in terms of the original problem formulation. The modified problem has the obvious basic feasible solution of $y_1 = 3$, $y_2 = 6$, and $y_3 = 2$. If you solved this problem, you might find that the optimal basic feasible solution includes one of the y_i's. As they have no meaning, we do not want this to happen. What do you suggest doing about it? Remember that you have an optimizing algorithm that tries to make x_0 as small as possible by including the "right" variables in an optimal solution.

Recall that the original objective function is to minimize $x_0 = 2x_1 + x_2$. For a minimizing problem, you would expect a nonnegative variable with a very large positive objective-function coefficient not to be in the optimal solution. What do you think would happen if we tried to solve the modified problem with y_1, y_2, and y_3 using the objective function minimize

$$x_0 = 2x_1 + x_2 + (10^{12})y_1 + (10^{12})y_2 + (10^{12})y_3?$$

The variables y_1, y_2, and y_3 have been given large positive cost coefficients. Any positive value of y_1 or y_2 or y_3 would make x_0 rather large. If there are any feasible solutions, we would expect the original variables $(x_1, x_2, x_3, x_4, x_5)$ to eventually appear in a basic solution without any of the y_i's. This intuitive approach for starting a simplex computation when there is no explicit first basic feasible solution is just what is done in practice—and it works!

The added variables are called *artificial variables* and they form an *artificial basis*. The simplex algorithm will try to force these artificial variables out of the basis. Eventually, it will find a basic feasible solution, if one exists, consisting of only the equation problem's original variables. For a problem that has no feasible solutions, then the simplex algorithm, using artificial variables, will terminate with the final tableau containing at least one artificial variable with a positive value. As the row 0 coefficients will all be negative or zero (for a minimizing problem), there exist no other solutions with a smaller value of the objective function.

To carry out the simplex tableau computations, you do not need to give large values to the cost coefficients of the artificial variables. What we really want to do is to first try to solve the problem with artificial variables by using the objective function minimize

$$y_0 = y_1 + y_2 + y_3$$

Since we restrict all $y_i \geq 0$, the minimum value of y_0 will be zero if the original problem has any feasible solutions. Why? If we optimize on y_0 (ignoring x_0 for now), we will have found a basic feasible solution involving only x_j's when all y_i's are out of the basis, that is, $y_0 = 0$. (We might have to worry about a degenerate optimal situation where a basic y_i has a value of zero. Here $y_0 = 0$; an artificial variable is still in the solution, but the simplex algorithm can proceed. Explain why.) We adapt the simplex algorithm as follows.

The problem is now:

$$\text{Minimize } x_0 + y_0$$

subject to

$$\begin{array}{llllllllllll}
y_0 & & & & & & & - y_1 & - y_2 & - y_3 & = 0 \\
& x_0 & - 2x_1 & - x_2 & & & & & & & = 0 \\
& & 3x_1 & + x_2 & - x_3 & & & + y_1 & & & = 3 \\
& & 4x_1 & + 3x_2 & & - x_4 & & & + y_2 & & = 6 \\
& & x_1 & + 2x_2 & & & - x_5 & & & + y_3 & = 2 \\
& & & & & & & & & (x_j, y_i) & \geq 0 \\
\end{array}$$

$$(j = 1, \ldots, 5;\ i = 1, 2, 3)$$

We first find the solution that optimizes y_0; if minimal $y_0 = 0$, then we have found a basic feasible solution involving the original variables (x_1, x_2, x_3, x_4, x_5). We then find the solution that minimizes x_0. This is a *two-phased simplex procedure*, with each phase applying the rules of the simplex algorithm exactly as they have been developed. In phase one, we minimize y_0; in phase two we switch and minimize x_0. We accomplish the switch in a straightforward manner by keeping the equation that defines x_0 in the problem and transforming it under the simplex rules. The tableau for this problem is arranged to include the artificial variables as follows. Note that we separate the two objective functions y_0 and x_0 by adding a row (-1) to the tableau.

Row	Basis	Solution	y_0	x_0	x_1	x_2	x_3	x_4	x_5	y_1	y_2	y_3	Ratio
-1	y_0	0	1	0	0	0	0	0	0	-1	-1	-1	Min
0	x_0	0	0	1	-2	-1	0	0	0	0	0	0	Min
1	y_1	3	0	0	3	1	-1	0	0	1	0	0	
2	y_2	6	0	0	4	3	0	-1	0	0	1	0	
3	y_3	2	0	0	1	2	0	0	-1	0	0	1	

We want to solve the problem by minimizing y_0. The problem is not in the correct form to apply the simplex algorithm as the basic variables y_1, y_2, and y_3 have coefficients that are not zero in their objective-function row (-1). We must eliminate y_1, y_2, and y_3 from this row by applying the elimination transformation to obtain

Row	Basis	Solution	y_0	x_0	x_1	x_2	x_3	x_4	x_5	y_1	y_2	y_3	Ratio
-1	y_0	11	1	0	8	6	-1	-1	-1	0	0	0	Min
0	x_0	0	0	1	-2	-1	0	0	0	0	0	0	Min
1	y_1	3	0	0	③	1	-1	0	0	1	0	0	$\frac{3}{3} = 1$
2	y_1	6	0	0	4	3	0	-1	0	0	1	0	$\frac{6}{4} = 1\frac{1}{2}$
3	y_3	2	0	0	1	2	0	0	-1	0	0	1	$\frac{2}{1} = 2$

This tableau indicates that $y_1 = 3$, $y_2 = 6$, $y_3 = 2$, and $y_0 = 11$. Note that the coefficients of the nonbasic variables in row (-1) are obtained by just summing the coefficients in rows 1, 2, and 3 for these variables, for example, $8 = 3 + 4 + 1$, $6 = 1 + 3 + 2$. $y_0 = 11$ is also obtained in that manner, that is, $11 = 3 + 6 + 2$. Explain why this happens. As long as there are artificial variables in the basis, we use the row (-1) coefficients to select a variable to come into the basis. Since we are minimizing y_0, we select x_1, as its row (-1) coefficient of 8 is the largest. The minimum ratio is the first, $\frac{3}{3} = 1$; x_1 comes into the basis and replaces y_1. The coefficient 3 of row 1 is the pivot element. When an artificial variable is removed from the basis, we will never let it come back in; we drop it from the tableau. After eliminating x_1 from all rows but the first, the next tableau is given by

Row	Basis	Solution	y_0	x_0	x_1	x_2	x_3	x_4	x_5	y_1	y_2	y_3	Ratio
-1	y_0	3	1	0	0	$\frac{10}{3}$	$\frac{5}{3}$	-1	-1		0	0	Min
0	x_0	2	0	1	0	$-\frac{1}{3}$	$-\frac{2}{3}$	0	0		0	0	Min
1	x_1	1	0	0	1	$\frac{1}{3}$	$-\frac{1}{3}$	0	0		0	0	$1/\frac{1}{3} = 3$
2	y_2	2	0	0	0	$\frac{5}{3}$	$\frac{4}{3}$	-1	0		1	0	$2/\frac{5}{3} = 1\frac{1}{5}$
3	y_3	1	0	0	0	($\frac{5}{3}$)	$\frac{1}{3}$	0	-1		0	1	$1/\frac{5}{3} = \frac{3}{5}$

The next simplex iteration introduces x_2 into the basis and removes the artificial variable y_3. The new tableau is given by

Row	Basis	Solution	y_0	x_0	x_1	x_2	x_3	x_4	x_5		y_2		Ratio
−1	y_0	1	1	0	0	0	1	−1	1		0		Min
0	x_0	$\frac{11}{5}$	0	1	0	0	$-\frac{3}{5}$	0	$-\frac{1}{5}$		0		Min
1	x_1	$\frac{4}{5}$	0	0	1	0	$-\frac{2}{5}$	0	$\frac{1}{5}$		0		—
2	y_2	1	0	0	0	0	(1)	−1	1		1		$\frac{1}{1} = 1$
3	x_2	$\frac{3}{5}$	0	0	0	1	$\frac{1}{5}$	0	$-\frac{3}{5}$		0		$\frac{3}{5}/\frac{1}{5} = 3$

We next introduce x_3 and remove the last artificial variable y_2. The new tableau contains a basic feasible solution that involves the problem's original variables. Here $y_0 = 0$, and all transformed cost coefficients in row (-1) are zero.

Row	Basis	Solution	y_0	x_0	x_1	x_2	x_3	x_4	x_5	Ratio
−1	y_0	0	1	0	0	0	0	0	0	—
0	x_0	$\frac{14}{5}$	0	1	0	0	0	$-\frac{3}{5}$	$\frac{2}{5}$	Min
1	x_1	$\frac{6}{5}$	0	0	1	0	0	$-\frac{2}{5}$	$\frac{3}{5}$	$\frac{6}{5}/\frac{3}{5} = 2$
2	x_3	1	0	0	0	0	1	−1	(1)	$\frac{1}{1} = 1$
3	x_2	$\frac{2}{5}$	0	0	0	1	0	$\frac{1}{5}$	$-\frac{4}{5}$	—

This tableau shows a basic feasible solution of $x_1 = \frac{6}{5}$, $x_2 = \frac{2}{5}$, $x_3 = 1$; $x_0 = \frac{14}{5}$. We now switch to row 0 to determine which variable, if any, should come into the basis. Here x_5 is selected and x_3 is removed from the basis. Performing the elimination transformation, we obtain the optimal solution tableau:

Row	Basis	Solution	x_0	x_1	x_2	x_3	x_4	x_5
0	x_0	$\frac{12}{5}$	1	0	0	$-\frac{2}{5}$	$-\frac{1}{5}$	0
1	x_1	$\frac{3}{5}$	0	1	0	$-\frac{3}{5}$	$\frac{1}{5}$	0
2	x_5	1	0	0	0	1	−1	1
3	x_2	$\frac{6}{5}$	0	0	1	$\frac{4}{5}$	$-\frac{3}{5}$	0

The optimal solution is $x_1 = \frac{3}{5}$, $x_2 = \frac{6}{5}$, $x_3 = 0$, $x_4 = 0$, $x_5 = 1$; $x_0 = \frac{12}{5}$.

17.6. You can solve any linear-programming problem by adding a full set of artificial variables to form an artificial starting basic feasible solution. However, if some of the original problem variables appear in only one equation (not including the objective function), they can be used as part of the basic feasible solution, along with as many artificial variables as required. We illustrate this situation by the following example:

$$\text{Maximize } x_0$$

subject to

$$\begin{aligned} x_0 - \quad x_1 - 2x_2 - 3x_3 + x_4 &= 0 \\ x_1 + 2x_2 + 3x_3 \qquad &= 15 \\ 2x_1 + \quad x_2 + 5x_3 \qquad &= 20 \\ x_1 + 2x_2 + \quad x_3 + x_4 &= 10 \\ x_j &\geq 0 \qquad (j = 1, \ldots, 4) \end{aligned}$$

The variable x_4 is a singleton (appears in only one constraint equation) and can be used to form a basic feasible solution. Of course, the right-hand sides must be nonnegative, as is the case. As the coefficient of x_4 is not zero in the objective function, we make it zero. Why? We do so by subtracting the third equation from the objective function to obtain the following transformed problem:

$$\text{Maximize } x_0$$

subject to

$$\begin{aligned} x_0 - 2x_1 - 4x_2 - 4x_3 \qquad &= -10 \\ x_1 + 2x_2 + 3x_3 \qquad &= \quad 15 \\ 2x_1 + \quad x_2 + 5x_3 \qquad &= \quad 20 \\ x_1 + 2x_2 + \quad x_3 + x_4 &= \quad 10 \end{aligned}$$

We can now set up an initial simplex tableau that requires two artificial variables.

Row	Basis	Solution	y_0	x_0	x_1	x_2	x_3	x_4	y_1	y_2	Ratio
−1	y_0	−35	1	0	−3	−3	−8	0	0	0	Max
0	x_0	−10	0	1	−2	−4	−4	0	0	0	Max
1	y_1	15	0	0	1	2	3	0	1	0	
2	y_2	20	0	0	2	1	5	0	0	1	
3	x_4	10	0	0	1	2	1	1	0	0	

Note that we are maximizing, and we have to use Max $y_0 = -y_1 - y_2$ as the artificial objective function. Explain why. You should be able to show how this tableau was developed and continue on to find the optimal solution.

It will take three iterations. The optimal solution is $x_1 = \frac{5}{2}$, $x_2 = \frac{5}{2}$, $x_3 = \frac{5}{2}$; $x_0 = 15$. You should also solve this problem by ignoring x_4 and using three artificial variables as a starting solution.

17.7. For the following problem, find a starting basic feasible solution that does not require any artificial variables:

$$\text{Minimize } x_0 = 2x_1 - 3x_3 - 4x_4$$

subject to

$$\begin{aligned} x_1 - x_2 \qquad + x_4 &\leq -6 \\ 6x_1 \qquad + 2x_3 + 2x_4 &= 5 \\ 3x_1 \qquad\qquad &\leq 3 \\ x_j &\geq 0 \qquad (j = 1, \ldots, 4) \end{aligned}$$

17.8. Solve the following problem using the simplex algorithm:

$$\text{Minimize } x_0 = x_2 - 3x_3 + 2x_5$$

subject to

$$\begin{aligned} x_1 + 3x_2 - x_3 \qquad + 2x_5 \qquad &= 7 \\ -2x_2 + 4x_3 + x_4 \qquad\qquad &= 12 \\ -4x_2 + 3x_3 \qquad + 8x_5 + x_6 &= 10 \\ x_j &\geq 0 \qquad (j = 1, \ldots, 6) \end{aligned}$$

The optimal solution $x_2 = 4$, $x_3 = 5$, $x_6 = 11$; $x_0 = -11$ can be found in two iterations.

17.9. Solve the following problem using the simplex algorithm:

$$\text{Minimize } x_0 = -3x_1 + x_2 + 3x_3 - x_4$$

subject to

$$\begin{aligned} x_1 + 2x_2 - x_3 + x_4 &= 0 \\ 2x_1 - 2x_2 + 3x_3 + 3x_4 &= 9 \\ x_1 - x_2 + 2x_3 - x_4 &= 6 \\ x_j &\geq 0 \qquad (j = 1, \ldots, 4) \end{aligned}$$

The optimal solution is $x_1 = 1$, $x_2 = 1$, $x_3 = 1$, $x_4 = 0$; $x_0 = 7$.

17.10. Solve the following problem using the simplex algorithm:

$$\text{Maximize } x_0 = x_4 - x_5$$

subject to

$$\begin{aligned}
2x_2 - x_3 - x_4 + x_5 &\geq 0 \\
-2x_1 + 2x_3 - x_4 + x_5 &\geq 0 \\
x_1 - 2x_2 - x_4 + x_5 &\geq 0 \\
x_1 + x_2 + x_3 &= 1 \\
x_j &\geq 0 \qquad (j = 1, \ldots, 5)
\end{aligned}$$

The optimal solution is $x_1 = \frac{2}{5}$, $x_2 = \frac{1}{5}$, $x_3 = \frac{2}{5}$, $x_4 = 0$, $x_5 = 0$; $x_0 = 0$.

17.11. Show that the following problem has no feasible solutions:

$$\text{Minimize } x_0 = x_1 - 2x_2 + 3x_3$$

subject to

$$\begin{aligned}
-2x_1 + x_2 + 3x_3 &= 2 \\
2x_1 + 3x_2 + 4x_3 &= 1 \\
x_j &\geq 0 \qquad (j = 1, 2, 3)
\end{aligned}$$

17.12. Solve the following problem by graphical techniques and by the simplex algorithm:

$$\text{Minimize } x_0 = 2x_1 + x_2$$

subject to

$$\begin{aligned}
3x_1 + x_2 &\geq 3 \\
4x_1 + 3x_2 &\geq 6 \\
x_1 + 2x_2 &\geq 2 \\
x_j &\geq 0 \qquad (j = 1, 2)
\end{aligned}$$

17.13. Solve the following problem by graphical techniques and the simplex algorithm:

$$\text{Maximize } x_0 = x_1 + x_2$$

subject to

$$\begin{aligned}
x_1 + x_2 &\geq 1 \\
x_1 - x_2 &\leq 1 \\
-x_1 + x_2 &\leq 1 \\
x_j &\geq 0 \qquad (j = 1, 2)
\end{aligned}$$

17.14. We have not proved mathematically that the optimal solution to a linear-programming problem is an extreme point of the convex set of solutions. Discuss why this is true using a two-variable inequality problem. (*Hint:* Assume the optimal solution is a point in the interior of the convex set, i.e., not on the boundary.)

17.15. When we solve a linear-programming problem using the simplex algorithm we also obtain the optimal solution to the problem's dual. We illustrate this using the optimal simplex tableau for the Simple Furniture Company's problem:

$$\text{Maximize } x_0$$

subject to

$$\begin{aligned} x_0 - 45x_1 - 80x_2 \qquad\qquad &= 0 \\ 5x_1 + 20x_2 + x_3 \qquad &= 400 \\ 10x_1 + 15x_2 \qquad + x_4 &= 450 \\ x_j &\geq 0 \qquad (j = 1, \ldots, 4) \end{aligned}$$

The variables x_3 and x_4 are slack variables for the first resource (board-feet of mahogany) and second resource (labor-hours), respectively.

The optimal simplex tableau is

Row	Basis	Solution	x_0	x_1	x_2	x_3	x_5
0	x_0	2200	1	0	0	1	4
1	x_2	14	0	0	1	$\frac{2}{25}$	$-\frac{1}{25}$
2	x_1	24	0	1	0	$-\frac{3}{25}$	$\frac{4}{25}$

The optimal solution is $x_1 = 24$, $x_2 = 14$, $x_3 = x_4 = 0$; $x_0 = 2200$. The dual problem is given by

$$\text{Minimize } w_0$$

subject to

$$\begin{aligned} w_0 - 400w_1 - 450w_2 &= 0 \\ 5w_1 + 10w_2 &\geq 45 \\ 20w_1 + 15w_2 &\geq 80 \\ w_i &\geq 0 \qquad (i = 1, 2) \end{aligned}$$

The optimal values of the dual variables are $w_1 = 1$ and $w_2 = 4$, with $w_0 = 2200$. You will note that the optimal values of the dual variables w_1 and w_2 are the row 0 numbers for the slack variables x_3 and x_4, respectively. The optimal primal solution uses all the resources, that is, $x_3 = x_4 = 0$. Thus, $w_1 = 1$ is the marginal value of an extra unit of the first resource and $w_2 = 4$ is the marginal value of an extra unit of the second resource.

You should solve the above dual problem using the simplex algorithm to demonstrate that the optimal dual tableau also solves the primal. You will need to use two artificial variables. The optimal values of the primal problem variables x_1 and x_2 will be found in the row 0 cells corresponding to the surplus variables. Be careful in interpreting these values as their signs in row 0 will not be correct. Why?

The solution to a primal's dual will be contained in the optimal primal tableau if the equation form of the primal problem contains a set of m unit columns, for example, a full set of slack or surplus variables. If this is not the case, there are other means of calculating the optimal dual values based on advanced linear-programming techniques.

17.16. Solve the dual to the diet problem of Section 15.1. It is given by

$$\text{Maximize } w_0 = w_1 + 5w_2 + 400w_3$$

subject to

$$\begin{aligned} 0.10w_1 + 1.00w_2 + 110.00w_3 &\leq 3.8 \\ 0.25w_1 + 0.25w_2 + 120.00w_3 &\leq 4.2 \\ w_i &\geq 0 \qquad (i = 1, 2, 3) \end{aligned}$$

You should add two slack variables, w_4 and w_5, to form a first basic feasible solution. The optimal simplex tableau row 0 values of the slack variables will be equal to the optimal values of the primal diet variables x_1 and x_2.

17.17. The U.S. Space Agency, NASA, is trying to reduce the weight of supplies being carried to the manned orbiting space laboratory. As it costs \$1000 to carry a pound of material into space, someone had the bright idea of substituting nutrient pills (iron, vitamin C, calcium, etc.) for foods. The astronauts would receive the proper nutrition by popping pills. (You might recall from science fiction how this method of food intake was part of any futuristic scene.) NASA has called in a chemical manufacturer and has asked the manufacturer to sell them various nutrients to be combined into pills. As this is a new business, the manufacturer needs to determine prices for units of the nutrients. The space agency wants to minimize the cost of the diet (the cost of purchasing the diet in equivalent pills). The manufacturer, of course, wants to maximize profits. For discussion purposes, the manufacturer, knowing something about linear programming, starts to analyze a sample two-food, three-nutrient diet that has the following data:

Food 1	Food 2		
0.15	0.10	Thiamine	(mg/oz)
0.75	1.70	Phosphorus	(mg/oz)
1.30	1.10	Iron	(mg/oz)
0.20	0.167	Cost	(cents/oz)

From this two-food diet, an astronaut must obtain at least 1.0 milligrams of thiamine, 7.5 milligrams of phosphorus, and 10.0 milligrams of iron.

The manufacturer's decision variables are the prices per milligram of each nutrient; denoted by w_1 (thiamine), w_2 (phosphorus), and w_3 (iron). The manufacturer knows that NASA can obtain the nutrients in an ounce of Food 1 for 20 cents; similarly, the nutrient content of Food 2 can be bought for 16.7 cents per ounce. The manufacturer must supply a food's nutrient content for the same or less cost, or the space agency would reject the company's proposal. This means that the following two constraints must be satisfied by the nonnegative prices (w_1, w_2, w_3):

$$
\begin{aligned}
0.15w_1 + 0.75w_2 + 1.30w_3 &\le 0.20 \\
0.10w_1 + 1.70w_2 + 1.10w_3 &\le 0.167
\end{aligned}
$$

The manufacturer wants to set prices to maximize profit. Here profit is a function of the prices of each nutrient and the total amount of each nutrient to be supplied. Thus, the profit function $1.0w_1 + 7.5w_2 + 10.0w_3$ must be maximized.

Of course, this is just the dual of the space-agency's sample diet problem:

$$\text{Minimize } 0.20x_1 + 0.167x_2$$

subject to

$$
\begin{aligned}
0.15x_1 + 0.10x_2 &\ge 1.0 \\
0.75x_1 + 1.70x_2 &\ge 7.5 \\
1.30x_1 + 1.10x_2 &\ge 10.0 \\
x_1 \qquad\quad &\ge 0 \\
x_2 &\ge 0
\end{aligned}
$$

where x_1 and x_2 are the amounts of Foods 1 and 2 to be purchased, respectively. Do you think the astronauts would go along with the pill diet?

You can work out a similar story about the dual to the transportation problem. Here a trucker wants to purchase the goods at the origins for costs

u_i's, and sell them back at the destinations for prices v_j's. The dual takes on the form $-u_i + v_j \leq c_{ij}$ and the objective function $-\sum_i a_i u_i + \sum_j b_j v_j$ (the net profit to the mover). You have to restate the primal transportation problem in terms of less than or equal inequalities, that is, each origin i cannot ship any more than a_i, and each destination j must receive at least b_j. Write out the complete mathematical statements (inequality form) for the transportation problem primal and dual, and interpret the decision variables. For the general linear-programming problem, we can summarize the primal–dual problems as follows:

Primal

$$\text{Maximize } x_0 = \sum_{j=1}^{n} c_j x_j$$

subject to

$$\sum_{j=1}^{n} a_{ij} x_j \leq b_i \qquad i = 1, \ldots, m$$

$$x_j \geq 0$$

Dual

$$\text{Minimize } w_0 = \sum_{i=1}^{m} b_i w_i$$

subject to

$$\sum_{i=1}^{m} a_{ij} w_i \geq c_j \qquad j = 1, \ldots, n$$

$$w_i \geq 0$$

The *duality theorem* of linear programming states that if either the primal or dual problem has a finite optimal solution, then the other problem has a finite optimal solution and maximum x_0 = minimum w_0. Also, if either problem has an unbounded feasible solution, then the other problem has no feasible solutions. You should be able to show that if either problem is given in terms of equations, then the other problem is still in terms of inequalities, but its variables can be positive or negative. Show that this is the case for the equality form of the transportation problem, and that the trucker's dual-problem interpretation holds.

17.18. Solve the Section 11.10 furniture manufacturing company problem using the simplex algorithm. Write out the corresponding dual problem and determine the dual solution from the optimal primal simplex tableau.

Interpret the meaning of the dual variables. The primal optimal solution is to produce 130 tables, 130 chairs and 30 desks, with a profit of $2660.

17.19. Solve the following problem by graphical techniques and by the simplex algorithm:

$$\text{Maximize } x_0 = x_1 + 2x_2$$

subject to

$$\begin{aligned} -x_1 + 3x_2 &\leq 10 \\ x_1 + x_2 &\leq 6 \\ x_1 - x_2 &\leq 2 \\ x_1 + 3x_2 &\geq 6 \\ 2x_1 + x_2 &\geq 4 \\ x_j &\geq 0 \qquad (j = 1, 2) \end{aligned}$$

The optimal solution is $x_1 = 2$, $x_2 = 4$; $x_0 = 10$.

17.20. Describe in words and/or by a computer programming flowchart diagram how to solve a linear-programming problem using the simplex algorithm.

17.21. When we solve a linear-programming problem by hand computation, we need to state it in equation form with all the right-hand sides nonnegative. When you use a computer mathematical-programming system (MPS) to solve a linear-programming problem, you do not need to convert the problem to equations. The computer receives instructions to add or subtract the appropriate slack or surplus variables depending on the form of each constraint. The MPS usually starts with a full set of artificial variables unless specified otherwise.

17.22. We introduced the concept of *sensitivity analysis* when we discussed the automobile trade-in problem in Section 5.17. Whenever we solve a decision problem, we need to be concerned with the accuracy of the data and how inaccuracies may cause the wrong solution to be chosen. Sensitivity analysis is the term that describes the process used to investigate changes in a model's solution as a function of changes in selected data elements. Any analysis, especially one that arises from a decision problem, is incomplete unless it includes a sensitivity study.

Some model structures can be manipulated easily to perform a sensitivity analysis. For others it might be rather difficult and/or involve a great deal of additional calculations, for example, in the automobile problem we had to perform the complete computation every time we changed one of the cost figures. Sensitivity studies of a linear-programming model are an outcome of the simplex algorithm and can be accomplished, in general, with little extra effort. We shall illustrate how it can be done for the activity-analysis model. Advanced linear-programming procedures generalize this approach so it can be applied to any linear-programming problem.

Sensitivity analysis of a linear-programming problem involves three sets of data: the objective-function cost coefficients, the constraint right-hand sides, and technological coefficients. For the activity-analysis problem, the objective-function coefficients represent the profit made per unit of production. Under what conditions does the optimal solution stay the same if the profit margin of a product is in error? What happens if the manufacturer decides to increase or decrease a profit margin?

The right-hand sides represent resource levels available to the manufacturer, for example, labor-hours, tons of material, equipment time. Some of these data may not be accurate. Who knows exactly how many square feet of lumber are in the yard? Other data may be estimates; for example, labor-hours may represent an average level of hours based on past experience with sick leave, vacations, hiring, and firing. For what ranges of the resources will the optimal production activities remain the same?

The technological coefficients represent the amount of a resource that is required to produce one unit of the corresponding product, for example, labor-hours required to make one chair or table. These numbers are usually average amounts and are measured with reasonable accuracy by the production staff. What if they are in error or what if a new process or machine is introduced that changes one of these coefficients. Will the optimal solution change?

You should spend a few moments thinking about how you would answer these questions before reading ahead. As we shall illustrate sensitivity analysis using the Simple Furniture Company problem, you should study its graphical solution given in Chapter 14. It will be very instructive if you carry out the geometrical implications of a sensitivity analysis in parallel with the algebraic discussion given below.

The Simple Furniture Company problem is:

$$\text{Maximize } x_0 = 45x_1 + 80x_2$$

subject to

$$\begin{aligned} 5x_1 + 20x_2 + x_3 &= 400 \\ 10x_1 + 15x_2 + x_4 &= 450 \\ x_j &\geq 0 \qquad (j = 1, \ldots, 4) \end{aligned}$$

Here x_1 is the number of chairs to be manufactured, x_2 is the number of tables to be manufactured, x_3 is the slack variable for the 400 board-feet of mahogany, and x_4 is the slack variable for the 450 labor-hours. A profit of \$45 is made for every chair produced, and a profit of \$80 is made for every table produced. We first investigate what happens if we make a change in the profit margin of chairs. To do this, we represent the profit of a chair by the expression $45 + p_1$, where p_1 can be a positive or negative profit change to the original \$45. The new objective function is

$$x_0 = (45 + p_1)x_1 - 80x_2$$

or

$$x_0 - (45 + p_1)x_1 - 80x_2 = 0$$

We treat p_1 as a parameter and carry it along as we apply the simplex algorithm. The initial tableau is

Row	Basis	Solution	x_0	x_1	x_2	x_3	x_4	Ratio
0	x_0	0	1	$-45 - p_1$	-80	0	0	Max
1	x_3	400	0	5	20	1	0	
2	x_4	450	0	10	15	0	1	

We perform the simplex calculations ignoring the p_1 (conceptually letting it equal zero) and determine the solution as before in two iterations (you should fill in the missing steps). The optimal tableau is

Row	Basis	Solution	x_0	x_1	x_2	x_3	x_4
0	x_0	$2200 + 24p_1$	1	0	0	$1 - \frac{3}{25}p_1$	$4 + \frac{4}{25}p_1$
1	x_2	14	0	0	1	$\frac{2}{25}$	$-\frac{1}{25}$
2	x_1	24	0	1	0	$-\frac{3}{25}$	$\frac{4}{25}$

This tableau is like the optimal one we had calculated previously, except that some of the row 0 coefficients are now functions of p_1. For $p_1 = 0$ (the original problem), the tableaus are the same.

With p_1 active, this solution ($x_1 = 24$ and $x_2 = 14$) will be optimal if all the row 0 coefficients for the nonbasic variables are nonnegative. This implies that any numerical value of p_1 must satisfy the inequalities

$$1 - \tfrac{3}{25}p_1 \geq 0 \quad \text{and} \quad 4 + \tfrac{4}{25}p_1 \geq 0$$

These inequalities are consistent, that is, have a solution, in that at least $p_1 = 0$ satisfies them. Solving for p_1, we have, respectively,

$$1 \geq \tfrac{3}{25}p_1$$

or

$$\tfrac{25}{3} \geq p_1$$

and

$$\tfrac{4}{25}p_1 \geq -4$$

or

$$p_1 \geq -25$$

Together, they restrict p_1 to the range

$$-25 \leq p_1 \leq \tfrac{25}{3}$$

As long as p_1 is in that range, the basis of x_1 and x_2 will be optimal for the chair profit of $45 + p_1$. The value of the objective function will change to $x_0 = 2200 + 24p_1$ as given in the tableau. We would, of course, expect this to be the new value of x_0 as 24 chairs are produced and the profit change per chair is p_1.

Note that if p_1 equals -25, then the row 0 coefficient of x_4 will equal zero. This means that the surplus variable x_4 could enter the basis and form a multiple optimal solution of x_2 and x_4. What has happened is that the profit of x_1 is now $45 - 25 = 20$ and an optimal solution now exists that produces only tables. Similarly, if $p_1 = \tfrac{25}{3}$, then the profit of a chair would be $45 + \tfrac{25}{3} = \tfrac{160}{3}$ and a multiple optimal solution exists that produces only chairs. You should calculate these multiple solutions.

You can use the graphical solution to the Simple Furniture Company problem to show how the objective functions shifts from the original extreme-point solution to these other extreme points for $p_1 = -25$ and $p_1 = \tfrac{25}{3}$. What happens when p_1 is between these values? Note that the profit coefficient of x_1 can vary between $45 + \tfrac{25}{3} = \tfrac{160}{3}$ and $45 - 25 = 20$.

We performed sensitivity analysis on the profit coefficient c_1 the hard way. You probably noticed that the row 0 coefficients for x_3 and x_4 can be found directly from the original optimal simplex tableau. They are sums formed by the original row 0 coefficients and the products of p_1 and the optimal tableau numbers that are in the row position corresponding to the basic variable x_1 (row 2) and the columns of the nonbasic variables x_3 and x_4. This is the case for the activity-analysis problem as it contains an explicit starting basis (the slack variables). You can show mathematically that the required sensitivity-analysis data will appear in the starting basis positions of the optimal tableau.

The sensitivity-analysis study on the table profit coefficient $c_2 + p_2$ is easily done as follows. The change p_2 must satisfy the constraints

$$1 + \tfrac{2}{25}p_2 \geq 0 \quad \text{and} \quad 4 - \tfrac{1}{25}p_2 \geq 0$$

The range for p_2 is then

$$-\tfrac{25}{2} \leq p_2 \leq 100$$

You can demonstrate that this range is correct by showing what happens with the graphical solution. How does the objective function vary in terms of p_2?

We have described the sensitivity-analysis procedure for objective-function coefficients of variables that are basic in the optimal solution. What about nonbasic variables? For the above problem, carry out the computations assuming the objective-function coefficient of x_3 was $0 + p_3$, that is, $x_0 = 45x_1 + 80x_2 + p_3x_3$. You should be able to show that the original optimal solution will remain optimal as long as p_3 is less than or equal to the corresponding row 0 coefficient of the optimal tableau. For the Simple Furniture Company problem, we have

$$-\infty < p_3 \leq 1$$

Also,

$$-\infty < p_4 \leq 4$$

These ranges tell us that if we wanted to increase the objective-function coefficient of x_3 from its current value of 0, we could increase it to 1 before the optimal basis would change; similarly, we could increase the coefficient of x_4 by 4. Can you think of any reasons why we might want to give these slack variables a positive dollar value in the objective function? For any nonbasic variable x_j (slack variable or otherwise) the range of its p_j is bounded above by its final row 0 coefficient. This is true for any maximization linear-programming problem. What is the situation if the problem was a minimization? In either case, does the value of the objective function change when a nonbasic c_j is changed to $c_j + p_j$ for p_j within its range?

We use a similar algebraic analysis to study how changes to the right-hand side affect the original optimal basic solution. For example, what if the manufacturer is concerned with the amount of board-feet of mahogany that is available to produce the chairs and tables. There could be more or less than the assumed 400 board-feet. In either case, is the decision to make both chairs and tables the correct one? We consider the new problem:

$$\text{Maximize } x_0 = 45x_1 + 80x_2$$

subject to

$$\begin{aligned} 5x_1 + 20x_2 + x_3 \qquad &= 400 + q_1 \\ 10x_1 + 15x_2 \qquad + x_4 &= 450 \\ x_j &\geq 0 \qquad (j = 1, \ldots, 4) \end{aligned}$$

where q_1 represents a positive or negative quantity change to the original resource availability of 400 board-feet. We now carry out the simplex algorithm using q_1 as a parameter. The starting tableau is

Row	Basis	Solution	x_0	x_1	x_2	x_3	x_4	Ratio
0	x_0	0	1	−45	−80	0	0	Max
1	x_3	$400 + q_1$	0	5	20	1	0	
2	x_4	450	0	10	15	0	1	

The optimal tableau (you should fill in the missing ones) is given by

Row	Basis	Solution	x_0	x_1	x_2	x_3	x_4
0	x_0	$2200 + q_1$	1	0	0	1	4
1	x_2	$14 + \frac{2}{25}q_1$	0	0	1	$\frac{2}{25}$	$-\frac{1}{25}$
2	x_1	$24 - \frac{3}{25}q_1$	0	1	0	$-\frac{3}{25}$	$\frac{4}{25}$

The optimal solution is $x_1 = 24 - \frac{3}{25}q_1$, $x_2 = 14 + \frac{2}{25}q_1$, and $x_0 = 2200 + q_1$. If $q_1 = 0$, we have the optimal solution as before. For q_1 to take any value other than zero, and as we want the current basis to stay feasible, we must have

$$x_1 = 24 - \tfrac{3}{25}q_1 \geq 0$$

and

$$x_2 = 14 + \tfrac{2}{25}q_1 \geq 0$$

These conditions are consistent as $q_1 = 0$ satisfies them. Together these inequalities tell us, respectively, that

$$q_1 \leq 200$$

and

$$-175 \leq q_1$$

or that q_1 can range as follows:

$$-175 \leq q_1 \leq 200$$

Thus, if the amount of available board-feet varies between $400 + 200 = 600$

and $400 - 175 = 225$, the basic solution of making chairs and tables will be feasible and optimal. Of course, the number of chairs x_1 and the number of tables x_2 that would be produced varies as a function of q_1. What are the values of x_1 and x_2 if $q_1 = 100$?

You should check the graphical solution to see how the solution space is altered as q_1 takes on values within the determined range. What happens if q_1 stays inside the range? If $q_1 = 200$, the constraint set changes to make $x_1 = 0$ and $x_2 = 30$ the new optimal extreme point, with $x_0 = 2400$. The mahogany constraint has become redundant. If $q_1 > 200$, then x_1 becomes negative and we will not have a basic feasible solution. What is the optimal solution to the problem if $q_1 = 201$? Note that the mahogany constraint does not help form the convex set of solutions if $q_1 \geq 200$. Show, using the graph, what happens if $q_1 \leq -175$.

The new value of the objective function is given by $x_0 = 2200 + q_1$. Recall that the marginal value of an extra board-foot was equal to one (the dual variable for this bottleneck resource). Any change in the availability of mahogany within the range $-175 \leq q_1 \leq 200$ will change the optimal value of the objective function by q_1.

Again, we did this sensitivity analysis the hard way. Note that the coefficients of q_1 in the solution column of the optimal tableau are the corresponding column coefficients of the slack variable x_3, the variable that measures the amount of the first resource (mahogany) not used in the optimal solution. Hence, we can obtain the data for right-hand-side sensitivity analysis directly from the original final tableau. This is because the activity-analysis problem contains an explicit first basic feasible solution.

The sensitivity analysis for the second resource, labor-hours, is done by letting q_2 represent the change. The values of the basic variables are then given by

$$x_1 = 24 + \tfrac{4}{25}q_2 \geq 0$$

and

$$x_2 = 14 - \tfrac{1}{25}q_2 \geq 0$$

This yields the range

$$-150 \leq q_2 \leq 350$$

Also, $x_0 = 2200 + 4q_2$. Discuss this result graphically. Note that a degenerate basic feasible solution occurs whenever q_1 or q_2 equals a range limit.

What if a slack variable was in the optimal solution? You should be able to convince yourself that the corresponding resource could be decreased down to the value of the slack variable (it represents resource not used) and increased without limit. Here the only change to the optimal solution would be the value of the slack variable; all other variables and the value of the objective function will remain the same.

Sensitivity analysis of the technological coefficients is a bit more difficult to illustrate and accomplish. It can be systemized, but involves advanced techniques. You can certainly make a change r_{ij} to an a_{ij} and rerun the problem with the coefficient $a_{ij} + r_{ij}$. The analysis is easier if the column j corresponds to a nonbasic variable. You already know how to analyze a new product and its technological coefficients to determine if it should be considered as a candidate for the optimal solution.

17.23. Perform a sensitivity analysis on the cost coefficients and right-hand-side elements of the following problem:

$$\text{Maximize } x_0 = -x_1 + 3x_2 - 2x_3$$

subject to

$$\begin{aligned} 3x_1 - x_2 + 2x_3 + x_4 \qquad\qquad &= 7 \\ -2x_1 + 4x_2 \qquad\qquad + x_5 \qquad &= 12 \\ -4x_1 + 3x_2 + 8x_3 \qquad\qquad x_6 &= 10 \\ x_j &\geq 0 \qquad (j = 1, \ldots, 6) \end{aligned}$$

The optimal tableau can be found in two steps and is given by

Row	Basis	Solution	x_0	x_1	x_2	x_3	x_4	x_5	x_6
0	x_0	11	1	0	0	$\frac{12}{5}$	$\frac{1}{5}$	$\frac{4}{5}$	0
1	x_1	4	0	1	0	$\frac{4}{5}$	$\frac{2}{5}$	$\frac{1}{10}$	0
2	x_2	5	0	0	1	$\frac{2}{5}$	$\frac{1}{5}$	$\frac{3}{10}$	0
3	x_6	11	0	0	0	10	1	$-\frac{1}{2}$	1

Slack variable $x_6 = 11$ and thus the third equation's resource can be reduced from 10 to -1. This apparent strange happening is due to variable x_1 *producing* 16 units of this resource as $x_1 = 4$ and the corresponding technological coefficient is -4. In an activity-analysis problem, a negative coefficient means that the corresponding activity will produce a resource using other resources. The above optimal solution requires only 15 units of the third resource to make five units of x_2. The given 10 plus x_1's 16 yields 26; we have 11 units not being used.

17.24. The simplex algorithm works very well and is quite efficient in the way it searches through the set of feasible extreme points. Nobody knows why it works as well as it does. The theory upon which it is based does not indicate why it is able to limit its search to a relatively small number of solution possibilities. The simplex algorithm has been shown to be an *exponential-time algorithm*, that is, there is no reason why the time required to solve problems using the simplex algorithm does not grow exponentially with the size of the problem. Algorithms that do not have this characteristic are *polynomial-time algorithms*.

The elimination procedure for solving an $m \times m$ square set of equations is a polynomial-time algorithm in that the total number of operations (additions, multiplications, divisions) required to solve the problem is bounded above by the polynomial m^3. If a computer worked at the speed of a millionth of a second (microsecond) per operation, and $m = 50$, it would take about 0.125 of a second to solve the equations. In contrast, an algorithm that does not have a polynomial bound may have, for example, an exponential-time estimate of 3^m, where m is a measure of the problem size given in terms of the quantity of data required to encode the problem in a computer. If $m = 50$, then it could take up to 2×10^8 centuries to solve the problem on a microsecond computer! This time is a *worst-case estimate*. Fortunately, not all exponential-time algorithms require the worst case amount of time to be solved. In general, algorithms that have exponential-time bounds tend to be inefficient, while those that have polynomial-time bounds are usually efficient in terms of the total amount of time required to find a solution.

The simplex algorithm is not a polynomial-time procedure. But for most problems, the simplex algorithm requires about $2m$–$3m$ steps (iterations) to find an optimal solution (m is the number of constraints, not including the nonnegativity conditions). Each step requires about m^2 (not m^3) arithmetic operations and corresponds to the finding of an extreme point by using the elimination procedure in an efficient manner. Thus, to find $2m$ extreme points, an *empirical bound* for the number of operations for the simplex algorithm is $(2m)(m^2) = 2m^3$. If $m = 50$, then, with a microsecond computer, it would take about 0.25 seconds to solve the problem. The Russian mathematician L. G. Khachiyan has shown that the *ellipsoid algorithm* of N. Z. Shor can be used to solve linear-programming problems and it has a polynomial-time bound. This seems to be only of theoretical importance, as the applicability of the Shor–Khachiyan procedure to solve practical problems has not been demonstrated.

The worst-case situation for a linear-programming algorithm is when it would have to find all feasible extreme-point solutions. Examples have been constructed for which the simplex algorithm must do that. An upper bound on the total number of extreme points of the convex solution space to a linear-programming problem is given by the formula that expresses the total number of combinations of n things (variables) taken m at a time, that is,

$$\left(\frac{n}{m}\right) = \frac{n!}{m!\ (n - m)!}$$

For a problem with $n = 10$ variables and $m = 5$ variables this is

$$\frac{10 \cdot 9 \cdots 2 \cdot 1}{(5 \cdot 4 \cdots 2 \cdot 1)(5 \cdot 4 \cdots 2 \cdot 1)} = 252$$

For $n = 50$, $m = 20$ we get 47, 129, 212, 243, 960 possible extreme points! You might want to figure out how long the microsecond computer would take to solve that number of 20×20 square sets of equations using 20^2 as the number of operations required to find the values of the variables at one extreme point. (It is estimated that it would take 10^{11} years for our sun to run down.) We note that there are other formulas for the number of extreme points that yield smaller upper bounds than the one given above.

17.25. Our discussion of the economic interpretation of the primal–dual problems is more or less in line with the standard discussion given by economists. The equilibrium concept and interpretation of the dual variables are somewhat difficult to understand and appear to be artificial and forced. For further details you should consult the books by Baumol and Gale.

17.26. Solve the bakery problem of Section 11.11 using the simplex algorithm. The optimal solution calls for the baking of 100 batches of six English muffins and 12.5 batches of 60 cookies, for a profit of \$38.25. State the dual problem and determine the optimal dual solution from the solution to the primal. What are the units of the dual variables? Perform a complete sensitivity analysis on the primal cost coefficients and right-hand-sides. What happens if the baker decides to have a sale and sell six English muffins for \$0.29?

17.27. Solve the Simple Furniture problem of Section 16.1. The optimal solution requires the making of only chairs and tables. State and solve the dual problem, and perform a complete sensitivity analysis study on the primal.

17.28. Solve the police-scheduling problem of Section 11.14. The optimal solution requires 198 officers. Is the problem feasible if all the constraints are changed to equalities?

17.29. Find out if your school's computer system library contains the simplex algorithm. If it does, learn how to use it and solve some of the problems in this section.

Part III References

Baumol, W. J.: *Economic Theory and Operations Analysis*, 2nd Edition, Prentice-Hall, Englewood Cliffs, New Jersey, 1965.

Bland, R. G.: New Finite Pivoting Rules for the Simplex Method, *Mathematics of Operations Research*, Vol. 2, no. 2, May 1977.

Bland, R. G., D. Goldfarb, and M. J. Todd: The Ellipsoid Method: A Survey, *Operations Research*, Vol. 29, no. 6, 1981.

Dantzig, G. B.: *Linear Programming and Extensions*, Princeton Univ. Press, Princeton, N.J., 1963.

Gale, D.: *The Theory of Linear Economic Models*, McGraw-Hill, New York, 1960.

Garey, M. R., and D. S. Johnson: *Computers and Intractability*, Freeman, San Francisco, 1979.

Gass, S. I.: *An Illustrated Guide to Linear Programming*, McGraw-Hill, New York, 1970.

Gass, S. I.: Comments on the Possibility of Cycling with the Simplex Method, *Operations Research*, Vol. 27, no. 4, July–August 1979.

Gass, S. I.: *Linear Programming: Methods and Applications*, 5th Edition, McGraw-Hill, New York, 1985.

Khachiyan, L. G.: A Polynomial Algorithm in Linear Programming, *Dokl. Akad. Nauk SSSR*, pp. 1093–1096, Vol. 224 (English Translation in *Soviet Math.—Dokl* pp. 191–194, Vol. 20), 1979.

Klee, V., and G. J. Minty: How good is the Simplex Algorithm?, chapter in *Inequalities III*, O. Shisha (Ed.), pp. 159–174, Academic Press, New York, 1972.

Shor, N. Z.: Cutoff Method with Space Extension in Convex Programming Problems, *Kibernetika*, pp. 94–95, Vol. 13, (English translation in *Cybernetics*, pp. 94–96, Vol. 13), 1977.

Westlake, J. R.: *A Handbook of Numerical Matrix Inversion and Solution of Linear Equations*, Robert E. Krieger Publishing Co., Huntington, New York, 1975.

PART IV Network and Related Combinatorial Problems

Chapter 18 Network-Flow Problems 207

Chapter 19 The Traveling-Salesman Problem 218

Chapter 20 The Transportation and Assignment Algorithms 224

Chapter 21 Part IV Discussion, Extensions, and Exercises 250

In this part we extend the previous discussions and introduce you to new problem situations and procedures for solving some of them. Your approach to this material should be that of an analyst confronted with a new challenge. How would you structure the problem and find a solution? What is the model? You should give each problem serious thought before reading on. Hints, clues, and solutions are offered, but you will obtain more satisfaction and deeper insight into the process of problem formulation and solving if you use them sparingly.

When we transfer a decision problem from the real world to the pages of a text, much is lost in the translation. To be of greater value, a decision problem should be considered within the decision framework discussed in Part I. This implies a serious investigation of the problem within its decision environment, the struggle to hit upon the correct formulation, the collection of data, and many other concerns, least of which is the interaction of the analyst with the decision maker and others who will be affected by the analysis and proposed solution. Such dynamics of the decision process are difficult to capture in a text. As you work in the classroom, it would be appropriate at times to discuss a model, its algorithm, and solution in terms of the applicability of your approach in a real-world setting.

18 Network-Flow Problems

18.1 BASIC NETWORK CONCEPTS

A rather important class of linear-programming problems falls under the general heading of *network problems*. As the approach to the formulation of the related mathematical models is rather direct, we illustrate it with some small examples.

We are given a *transportation network* (pipeline system, railroad system, communication links) through which we wish to send homogeneous units (gallons of oil, cars, message units) from a particular point of the network called the *source node* to a designated destination called the *sink node*. In addition to the source and sink nodes, the network consists of a set of intermediate nodes that are connected to each other or the source and sink nodes by arcs of the network. These intermediate nodes can be interpreted as switching or *transshipment points*. We label the source node by 0 and the sink node by m and refer to the intermediate nodes by a number or a letter. A typical small-scale network, which we use in the following discussion, is

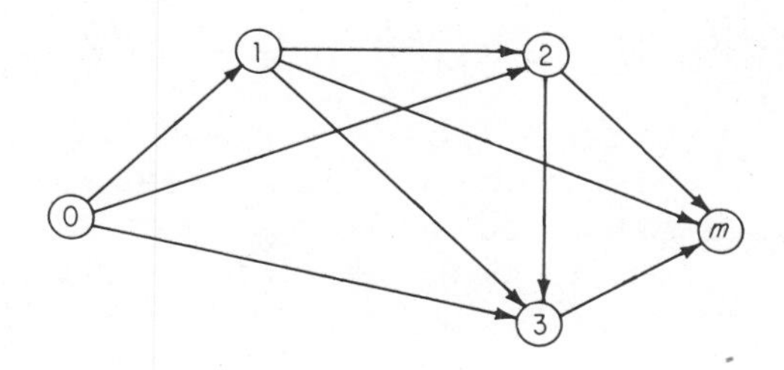

The symbol for an arc connecting nodes 2 and 3 is (2,3), or, in general, (i, j). The arcs pictured in our network are *directed arcs* in that the flow of goods along an arc is in the direction of the arrow. If goods can flow both ways, then the corresponding nodes are connected by two arcs with the flows going in the opposite directions.

Each arc of the network can accommodate only a finite amount of flow. A section of a pipeline can pass a designated number of barrels per hour; a communication link can handle so many calls per day, and so on. Thus, each arc has a specified upper bound on the flow through the arc that we designate, for example, by f_{23} for the maximum flow through arc (2,3). In the next figure we show the upper bounds on each arc, here $f_{23} = 1$. In general, f_{ij} denotes the upper-bound capacity of arc (i, j).

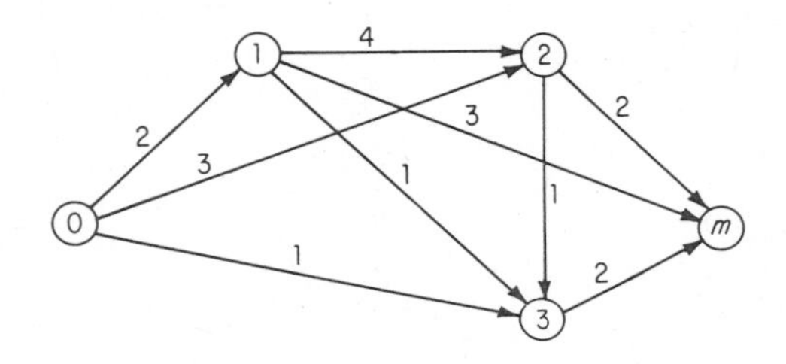

The flow of the commodity that originates at the source 0 is sent along the arcs to the intermediate nodes, then transshipped along additional arcs to other intermediate nodes, or to the sink, until all the commodity that began the trip at node 0 finally arrives at node m. That is, we impose upon the network the condition of *conservation of flow* at the nodes—what is directed into a node must be directed out. This assumption is analogous to Kirchhoff's node law in electrical networks. There are situations in which the conservation-of-flow assumption does not hold. A liquid moving through a pipeline is subject to evaporation or loss due to leaks; a gallon of gasoline that leaves one node may be only 0.9 of a gallon when it reaches the next node. If we shipped rabbits (of the opposite sex) from node i to node j, we might find that the shipment has grown. Thus, a more general assumption for networks allows for losses or gains of the commodity as it is transferred between nodes. For our problems, however, we assume conservation of flow at each node.

18.2 GETTING THE MOST: MAXIMAL FLOW

The first problem to be considered—the *maximal-flow network problem*—is concerned with sending as much of the commodity as possible through the network from node 0 to node m. We assume that there is an unlimited supply of the commodity at node 0. The only thing which restricts our sending an unlimited supply to node m is that the capacities of the arcs are limited by the given upper bounds. We wish to determine the maximum amount of flow, designated f, that can be sent from the source to the sink, along with the amount used of each arc's capacity. We wish to direct f through the network as shown in the figure below.

With the above definitions and assumptions, how can we develop a mathematical decision model? For this problem, it is easy to come by. For each arc (i, j) in the network, we define a nonnegative decision variable as

$$x_{ij} = \text{the amount of flow (goods shipped) from node } i \text{ to node } j$$

We indicate these variables, as well as the decision variable f, on the network below. Using this notation, the conservation-of-flow assumption is readily

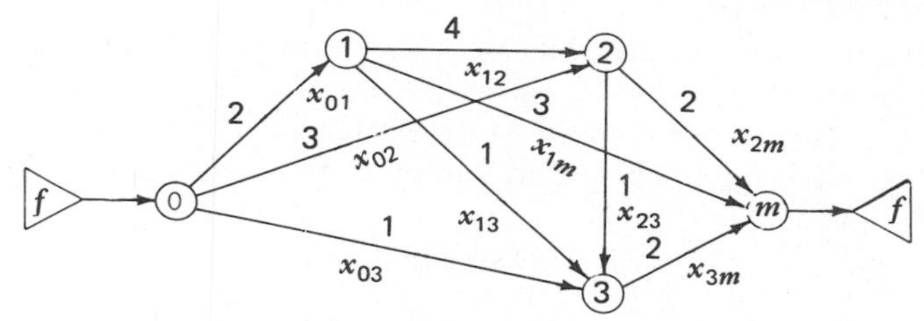

accomodated. For each node we need to express that what goes out must be equal to what comes in. For the source node 0, this is

$$x_{01} + x_{02} + x_{03} = f$$

or, since f is a variable,

$$x_{01} + x_{02} + x_{03} - f = 0$$

This equation is just a bookkeeping statement of what goes on at node 0, that is, where the commodity comes from and where it is going. Note that it is a linear expression.

The flow activity at node 1 translates into

$$x_{12} + x_{13} + x_{1m} = x_{01}$$

or

$$-x_{01} + x_{12} + x_{13} + x_{1m} = 0$$

Similarly, for nodes 2 and 3 we have, respectively,

$$-x_{02} - x_{12} + x_{23} + x_{2m} = 0$$

and

$$-x_{03} - x_{13} - x_{23} + x_{3m} = 0$$

The sink node m must receive the amount f from the shipments into m, that is,

$$f = x_{1m} + x_{2m} + x_{3m}$$

or

$$-x_{1m} - x_{2m} - x_{3m} + f = 0$$

The remaining restrictions deal with the capacities f_{ij} of each arc (i, j), that is, we must have all flows $x_{ij} \leq f_{ij}$ and, of course, $x_{ij} \geq 0$ and $f \geq 0$. The objective function is simply to maximize f. We put it all together in terms of the following linear-programming problem:

$$\text{Maximize } f$$

subject to

$$\begin{array}{rrrrrrrrrrl}
x_{01} & + x_{02} & + x_{03} & & & & & & & - f & = 0 \\
-x_{01} & & & + x_{12} & + x_{13} & + x_{1m} & & & & & = 0 \\
& - x_{02} & & - x_{12} & & & + x_{23} & + x_{2m} & & & = 0 \\
& & - x_{03} & & - x_{13} & & - x_{23} & & + x_{3m} & & = 0 \\
& & & & & - x_{1m} & & - x_{2m} & - x_{3m} & + f & = 0 \\
x_{01} & & & & & & & & & & \leq 2 \\
& x_{02} & & & & & & & & & \leq 3 \\
& & x_{03} & & & & & & & & \leq 1 \\
& & & x_{12} & & & & & & & \leq 4 \\
& & & & x_{13} & & & & & & \leq 1 \\
& & & & & x_{1m} & & & & & \leq 3 \\
& & & & & & x_{23} & & & & \leq 1 \\
& & & & & & & x_{2m} & & & \leq 2 \\
& & & & & & & & x_{3m} & & \leq 2
\end{array}$$

with $x_{ij} \geq 0$ and $f \geq 0$

This model of the maximal-flow problem is a linear-programming problem and can be solved using the simplex algorithm. The problem is certainly feasible because the constraints are satisfied if all $x_{ij} = 0$ and $f = 0$. But this is a trivial solution.

The linear structure is an outcome of the conservation-of-flow assumption and the implied physical characteristics of the network. For some applications, you might find that the amount that can flow across an arc (i, j) in a unit time period is a function of the amount already in the arc or is restricted by the nature of a node. The flow of traffic through a road system (arcs) is a function of the density of the cars in the system; the flow through a toll station (node) is limited by the number of toll booths and the traffic in the queues. More complex mathematics would be required to formulate the models for these types of relationships.

The mathematical model of the maximal-flow problem has some interesting properties. You should first note that the model, as written above, has a redundant equation. The first equation (source node) and the last equation (sink node) both define the flow variable f and, hence, one can be dropped from the model. In fact, because every variable appears twice, once with a plus sign and once with a negative sign, any one of the conservation-of-flow equations can be obtained by adding the remaining ones. For example, the node 1 constraint

$$-x_{01} + x_{12} + x_{13} + x_{1m} = 0$$

is just the negative of the sum of all the other conservation-of-flow equations; any one of the equations can be considered to be redundant and left out of the model. If you do this, then it can be shown that a basic feasible solution to the linear-programming problem will have integer values for the variables, given arc capacities f_{ij} that are also integers. This is true for all network problems having conservation of flow at the nodes. Can you figure out why this is so?

Because of their mathematical structure, most network problems can be solved by modifications of the simplex algorithm or specialized network approaches that are computationally less complex, that is, easy to implement and faster. Quite large problems, with thousands of nodes and arcs, can be handled. We describe and illustrate a special maximal-flow algorithm in Section 21.1.

How would you go about solving our small-scale maximal-flow problem? Can you obtain a bound on f by studying the flow structure and arc capacities of the network? You probably noted that no matter how much flow you tried to get through the source node 0, the capacities of the arcs out of node 0 limit the total flow to 6; while the capacities of the arcs going into the sink node m limit the flow to 7. Is there a set of arcs that limits the flow to less than 6?

For our simple network, the maximal flow $f = 6$. On the following network for this problem, we have added a number couple, for example, (0,4), to each arc (i,j), where the first number represents the flow through the arc, x_{ij}, and the second number the given upper bound for that arc, f_{ij}. Of course,

$x_{ij} \leq f_{ij}$. For the optimal solution, we then have

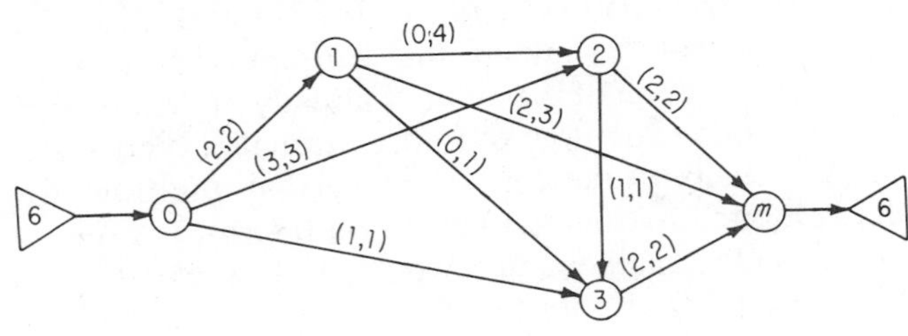

You will note that the arcs leaving node 0 are saturated—we cannot push any more through them. If we remove those arcs from the network—arcs (0,1), (0,2), and (0,3)—we end up with the source node disconnected from the sink node. Such a set of arcs is called a *cut*—a cut being a set of arcs which if removed from the network disconnects the network into two parts, with the source in one part and the sink in the other. The arcs in a cut have a certain total capacity. Here it is 6. For the cut consisting of arcs (0,3), (1,3), (1,m), (2,3), and (2,m), the total arc capacity is 8. The main theorem of networks—*the Max-flow Min-cut theorem*—states that for any network, the maximal-flow value from node 0 to node m is equal to the minimal-cut capacity of all cuts separating 0 and m. A moment's reflection will cause one to say that this theorem is intuitively obvious. An algebraic proof takes a little longer. The efficient computational schemes for the maximal-flow problem are based on the results of this theorem. These procedures also ensure that the values of the variables—the total flow and the flow through each arc—are given in terms of integers.

18.3 SPENDING THE LEAST: MINIMAL COST

Another set of significant network problems is the class of *minimal-cost network-flow problems*. As we shall see, this type of problem includes, among others, the transportation problem and the shortest-route problem.

For the minimal-cost flow problem we are given, as in the maximal-flow problem, a general network over which units of a homogeneous commodity are to be shipped from the source to the sink. Associated with each arc (i, j) is a cost c_{ij} of shipping one unit of the commodity from node i to node j. We must ship a given quantity of F units from the source to the sink so as to minimize the total cost of shipping the F units. We assume conservation of flow at the nodes and that the flow x_{ij} along any arc (i, j) is nonnegative and is bounded, that is, $0 \leq x_{ij} \leq f_{ij}$. The mathematical model for this problem is quite similar to the maximal-flow model, and we shall illustrate it for the network of the previous example. We must remember, however, that for our present problem we know what the total flow F is—in the maximum-flow problem the total flow f was to be determined. How would you modify the maximal-flow model to take care of the changes implied by the minimal-cost conditions? You need to define a new objective function. What else?

For our example let $F = 5$ (it cannot be greater than 6) and let the cost of shipping a unit of flow along an arc, the c_{ij}, and the upper bound for the arc, the f_{ij}, be shown on the network by the number couple $[f_{ij}, c_{ij}]$. For our network we have

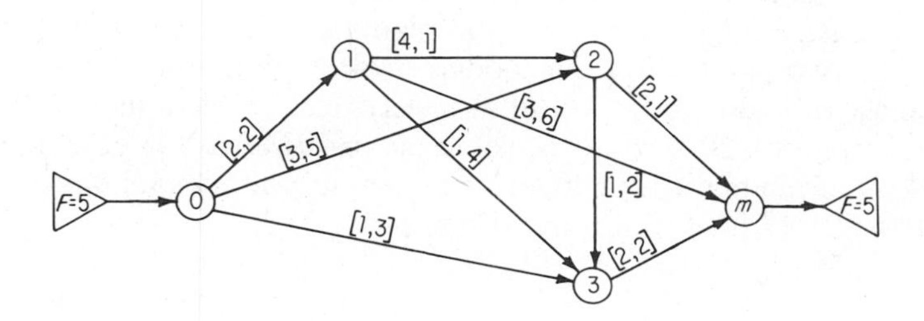

Thus, for arc (2,3), $f_{23} = 1$ and $c_{23} = 2$.

The mathematical model is to

$$\text{Minimize } 2x_{01} + 5x_{02} + 3x_{03} + x_{12} + 4x_{13} + 6x_{1m} + 2x_{23} + x_{2m} + 2x_{3m}$$

subject to

$$\begin{array}{rrrrrrrrrcr}
x_{01} & + x_{02} & + x_{03} & & & & & & & = & 5 \\
-x_{01} & & & + x_{12} & + x_{13} & + x_{1m} & & & & = & 0 \\
& - x_{02} & & - x_{12} & & & + x_{23} & + x_{2m} & & = & 0 \\
& & - x_{03} & & - x_{13} & & - x_{23} & & + x_{3m} & = & 0 \\
& & & & & - x_{1m} & & - x_{2m} & - x_{3m} & = & -5 \\
x_{01} & & & & & & & & & \leq & 2 \\
& x_{02} & & & & & & & & \leq & 3 \\
& & x_{03} & & & & & & & \leq & 1 \\
& & & x_{12} & & & & & & \leq & 4 \\
& & & & x_{13} & & & & & \leq & 1 \\
& & & & & x_{1m} & & & & \leq & 3 \\
& & & & & & x_{23} & & & \leq & 1 \\
& & & & & & & x_{2m} & & \leq & 2 \\
& & & & & & & & x_{3m} & \leq & 2
\end{array}$$

and all $x_{ij} \geq 0$.

We shall not describe the special simplex procedures for solving such problems; they have been used to solve quite large problems in a most efficient manner.

Minimal-cost network problems are also known as *transshipment problems*. The basic model given above can be generalized to allow an intermediate node to be either an additional source or sink for the goods. How would you change the model if node 1 was the source of two units and nodes 2 and 3 were sinks for one unit each? Applications of this model are found in inventory management; water-supply distribution; manpower-planning problems; financial management; routing and scheduling; transportation planning, production planning, and distribution; and in many other fields. One such application for a major U.S. automobile manufacturer determined the number of automobiles of each model to produce at its plants and how many of each model to ship from the plants to distribution centers. The network of a two-plant, three-car model, two-distribution-center problem is shown below (it is due to Glover et al.). A typical problem had 1200 nodes, 4000 arcs, and took 20 seconds to solve on an IBM 370/145 computer! Can you think of a minimal-cost network problem in your school environment? (See Section 21.17 for further discussion.)

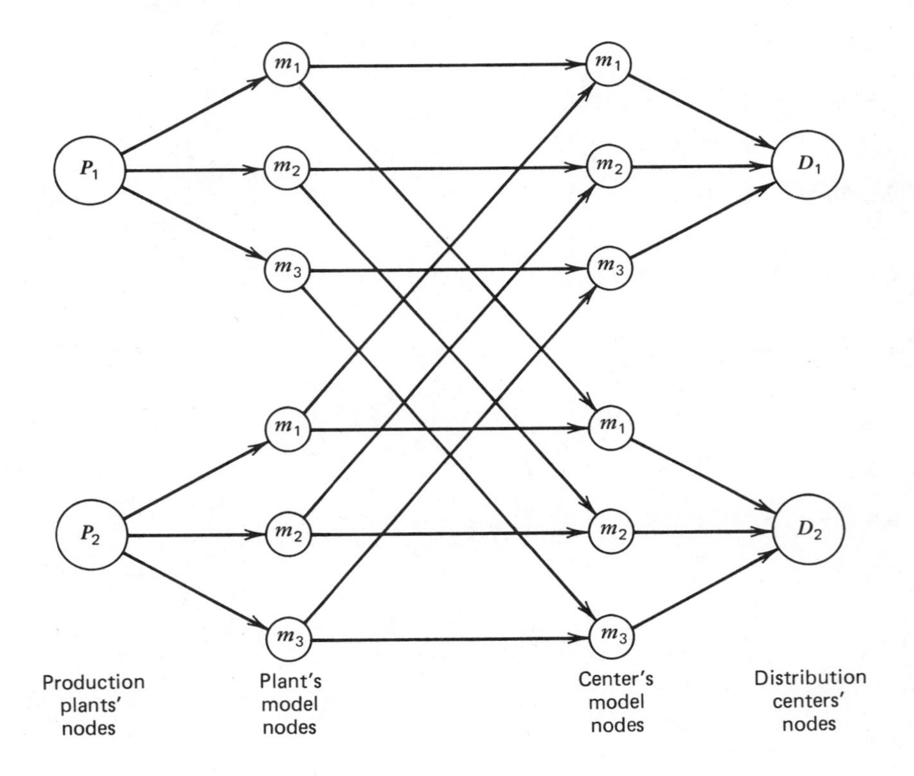

18.4 FROM HERE TO THERE: SHORTEST ROUTE

We can specialize the minimal-cost network problem to a most interesting one, one which has a rather simple solution procedure. Let the network be a road map with the source node an origin city, the sink node a destination city, and the c_{ij} the distances between cities (the intermediate nodes). We then interpret the above problem as that of finding the minimum distance between the origin and the destination by letting the amount to be shipped $F = 1$ and setting all the upper bounds $f_{ij} = 1$. We are shipping one unit from the origin to the destination so that the total distance traveled by the one unit is a minimum. This *shortest-route problem* can be solved by the regular techniques of linear programming, but more efficient algorithms exist. We shall illustrate one for the shortest-route example below.

We are given the four-city network with the distances as shown on each arc and in the distance table

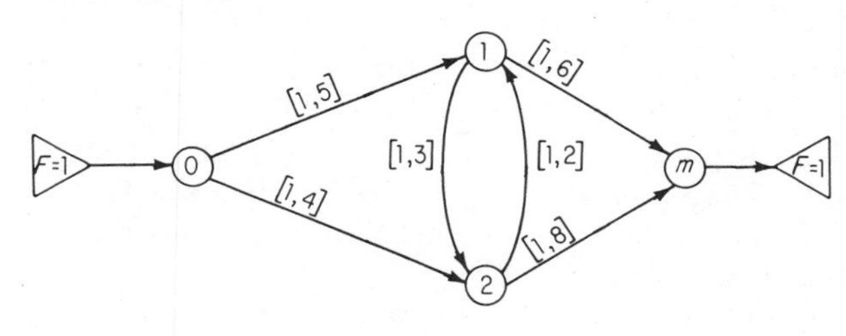

From	To		
	1	2	m
0	5	4	—
1	—	3	6
2	2	—	8

You should first develop the mathematical description of this four-city problem, that is, write the linear-programming model. For this small problem, you can determine the shortest route from node 0 to node m by trying all possible routes. Such an exhaustive approach breaks down if you have many cities connected by many arcs. Try to think of a systematic way of finding the shortest route. (*Hint:* Would it help if you knew the shortest routes of each of the intermediate cities from the origin city?) For our four-city problem, we know that the shortest distance from node 0 to any other city is the distance 4 between node 0 and node 2. What can you next say about the shortest distance of node 1 from node 0? Its shortest route will be either the direct one from node 0 or the indirect one via node 2. Try to systematize the process implied by this discussion. You will need to test it out on a more complex problem than the four-city one.

A basic shortest-route algorithm that applies the concepts hinted above is due to E. W. Dijkstra. It is known as a *node-labeling algorithm*; it finds the shortest distances and routes from the origin node to all intermediate nodes and the destination node. One form of this algorithm is given below (a more efficient version is described in Section 21.4). Other shortest-route algorithms exist for finding all shortest routes between all nodes in a network, second-best shortest routes, and shortest routes that must pass through specified cities. These algorithms require that the distances be nonnegative numbers. Of course, from a physical view of view, all distances will be positive. We could interpret the shortest-route problem as one in which we wanted to find the minimal-cost route for a truck traveling from node 0 to node m, that is, the distances represent costs. What if the trucker was paid extra for taking a particularly dangerous road or could get a free meal on that road, but in either case the value received was greater than the cost of traveling that leg of the route? For example, in the four-city problem, let the distances represent money paid by the trucker, but by traveling arc (1,2) the trucker made a profit of $3.00, that is, $c_{12} = -3$. Conceptually, what could the trucker do? Is there a minimal-cost route for this problem that gets the truck from node 0 to node m?

Shortest-Route Algorithm

STEP 1. Assign all nodes a label of the form $\{-, d_i\}$, where the first component indicates the preceding node in the shortest route and d_i indicates the shortest distance from node 0 to node i. Node 0 starts with a label $d_o = 0$, with its first component always a $-$; all other nodes start with $d_i = \infty$, any very large number, and a first component of $-$.

STEP 2. For any arc (i,j) for which $d_i + c_{ij} < d_j$, change the label of node j to $\{i, d_i + c_{ij}\}$; continue the process until no such arc can be found. In the former situation we have determined a shorter route from node 0 to node j which goes through node i; in the latter situation the process is terminated, and the node labels indicate the shortest distance from node 0 to node j.

For our problem, we have the shortest route from node 0 to node m of length 11, with the route going from node 0 through node 1 to node m. The appropriate labels are shown on the following network

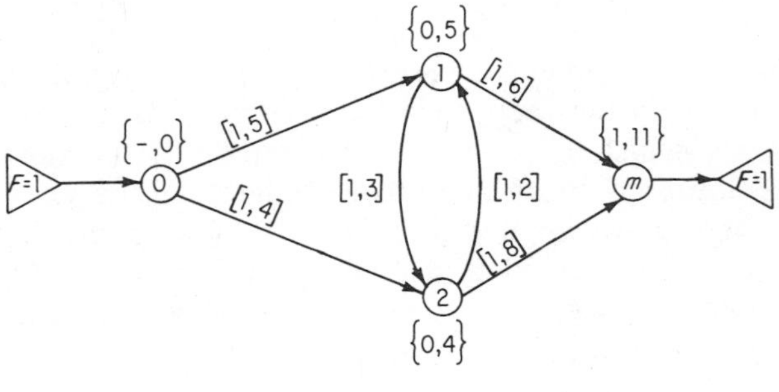

You should carry out the specific steps of the algorithm for the four-city problem. Every time you change a node number d_i, you have to check the node labels for those nodes j that have an arc (i, j). This form of the algorithm can cause the node labels to change (be reduced) many times. Thus, if the algorithm is not applied judiciously, it is rather inefficient. Can you think of a procedure that would permanently label the nodes, one at a time, with their shortest distances?

The transportation problem described in Chapter 4 can be readily interpreted as a minimal-cost network-flow problem. By referring to Fig. 4.3 of that chapter, you should be able to add source and sink nodes in a fashion which almost satisfies the requirements of the minimal-cost network-flow problem. The source node has $F = 25$ refrigerators that must be shipped—11 going to factory one and 14 to factory two. The sink node must receive 25 refrigerators—10 from store one, 8 from store two, and 7 from store three. These conditions can be made a part of the mathematical model by constraints which force shipments of 11 and 14 from the source to the respective factories, and shipments to the sink to be exactly 10, 8, and 7 from the respective stores. These types of conditions add little extra burden to the solution process and can be included for the general minimal-cost network-flow problem.

19 The Traveling-Salesman Problem[1]

There once was a farmer's daughter . . .

J. MILLER

19.1 THE EXPENSE ACCOUNT AND THE HORSE

The salesman's cry—"You got to know the territory!"—usually refers to his knowing where his prospects are and what goods and styles are preferred by his prospects and regular customers. For lo these many years, our salesman in question, Willie Simon of the Simple Furniture Company, has traveled about his wide territory, getting to know the lay of the land and building up an impressive sales record. At the same time, however, he was building up an impressive expense account. As part of the continuing management-consulting activities of Super Management Consultants, the SMC team decided to take a look at the happenings in Willie's territory. Soon Willie was getting harassed by memos and questionnaires wanting to know such things as, "What happened in Boston, Willie, on 6/30? Expense account way out of line for that date." An SMC analyst joined Willie in his travels and took copious notes and data. Willie tried to make friends with the analyst in order to influence the report. When Willie learned that his companion

[1] S. I. Gass, *An Illustrated Guide to Linear Programming*, McGraw-Hill Book Co., New York, N.Y., 1970. Reproduced with permission.

was a mathematician, he dug deep into his salesman's repertoire of stories and attempted to regale him with such tales as the following.

Horse Sense

A mathematics professor decided to take his sabbatical by resting in the country. A local farmer, learning that his new neighbor was a mathematician, approached him with a story about the farmer's horse who could add, subtract, multiply, and divide. The professor decided to humor the farmer and paid the horse a visit. The horse could do all that was claimed. He answered correctly a number of simple problems by either stomping on the stable floor or picking up a piece of chalk in his mouth and writing on a blackboard. The professor decided to work with the horse, and before long, he had the animal doing basic algebra and Euclidean geometry.

When it was time to return to the university, the professor took the horse with him and enrolled him as a special mathematics student. During the first semester, the horse managed trigonometry without any trouble. He was getting a straight A in the second semester until the lectures turned to analytic geometry. The poor horse floundered. He just couldn't make it. He was given special tutors, but nothing helped.

The mathematics department, not wanting to lose its prize student, called a special meeting of the faculty to see what could be done. They brought in a specialist in animal husbandry, psychiatrists, jockeys, but came up with no answers. They reviewed his record—excellent in arithmetic, algebra, Euclidean geometry, trig—but analytic geometry appeared to be his Waterloo.

Finally, after much contemplation, a bright assistant professor resolved the dilemma. "We should have guessed it," he said, "in trying to teach him analytic geometry, we were putting Descartes before the horse."

Willie's tales received no reaction. He fumed.

On his next visit to the home office, Willie had it out with his brother, President Simon, and threatened to quit unless his mode of operation and expense account were left alone. President Simon calmed Willie down and assured him that Super Management Consultants had given Willie a clean bill of health—except for one thing. The SMC study claimed that Willie was spending too much time and money in traveling from one city to the next. It appeared as if Willie wasn't arranging the visits to the successive cities in his territory to make the total distance traveled the shortest possible.

"All we want you to do, Willie," said President Simon, "is to visit the cities in the order recommended by the consultants. The rest of the expense account is yours."

Much relieved, Willie agreed to put the suggested routing into effect as soon as he received it—he is still waiting. It appears as if Willie has had the last (horse) laugh on the consultants. For it turned out that, although the

traveling-salesman problem can be formulated as a special integer-mathematical-programming problem, the size of Willie's territory—he covers 100 cities, towns, and hamlets—made it computationally intractable. Just what was the trouble?

19.2. CAN WE SOLVE IT?: THE PROBLEM

Let us be a little more precise and state the *traveling-salesman problem* as follows:

A salesman is required to visit each city of his territory. He leaves from a home city, visits each of the other cities exactly once, and finally returns to the home city. We are required to find an itinerary which minimizes the total distance traveled by the salesman.

There are a number of approaches to setting up this problem as an integer-mathematical-programming problem, the mathematics of which are too involved to develop here. These formulations require that the variables be restricted to integers, either 0 or 1, and a large number of equations. For Willie's problem, one formulation would require 10,000 equations.

A look at a small, five-city problem illustrates some of the difficulties. We number the cities 1, 2, 3, 4, and 5, with city 1 being the home city. We know the distances between each city, and we assume a city can be reached from any other city—that is, you can get there from here. To make the discussion more general, we also assume that the distance in going from say city 2 to city 5 is not necessarily the same distance if we reversed the trip and were going from city 5 to city 2. This nonsymmetric situation could be due to one-way bypasses, detours, and so on.

For our five cities we have the following distance table in miles:

City	1	2	3	4	5
1	—	17	10	15	17
2	18	—	6	12	20
3	12	5	—	14	19
4	12	11	15	—	7
5	16	21	18	6	—

For example, distance from city 1 to city 5, $d_{15} = 17$, while the distance from city 5 to city 1, $d_{51} = 16$.

As the salesman leaves his home city 1, he has a choice of going to any one of the remaining four cities—let us assume he goes to city 3. From city 3 he can travel to one of the remaining three—2, 4, or 5—and he selects 2. His next selection is city 5, then, of course city 4, and return to city 1.

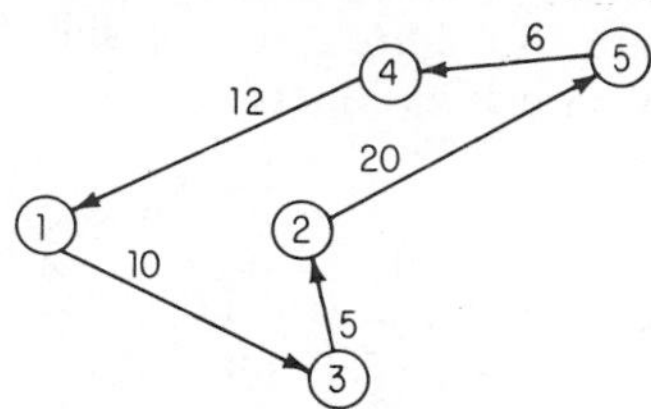

The ordering of the cities—1, 3, 2, 5, 4, 1—is called a *tour*, with a tour having a corresponding tour distance. For our tour, the total distance is

$$d_{13} + d_{32} + d_{25} + d_{54} + d_{41} = 10 + 5 + 20 + 6 + 12$$
$$= 53 \text{ miles}$$

The five-city problem has a total of $4 \times 3 \times 2 \times 1 = 24$ possible tours. For such small problems, we could enumerate all tours and their distances and select the one with the smallest number of miles. However, this approach soon becomes impractical as the number of cities slowly increases. For 10 cities there are 3,628,800 tours, while for 14 cities there are 87,178,291,200 tours.

Linear-programming approaches to this problem look quite a bit like the personnel-assignment formulation. Here we wish to assign to a tour a link connecting a city to another city, with each city being allowed the assignment of only one link emanating from it (a person can be assigned to one job, but must be assigned). The links from city one are denoted by x_{12}, x_{13}, x_{14}, x_{15}, and for a link to be assigned we have the equation

$$x_{12} + x_{13} + x_{14} + x_{15} = 1$$

A zero value of the variable means the corresponding link is not used; a value of one means that it is used. For our sample tour the assignment table is

City	1	2	3	4	5
1	0	0	1	0	0
2	0	0	0	0	1
3	0	1	0	0	0
4	1	0	0	0	0
5	0	0	0	1	0

The difficulty in treating the salesman problem like an assignment problem is that the optimal assignment, which minimizes the total distance of the assignments, might not be a tour. The following assignment table has a distance of 49 miles, less than the sample tour's distance of 53 miles, but the assignment yields two subtours, as illustrated.

City	1	2	3	4	5
1	0	0	0	1	0
2	0	0	1	0	0
3	0	1	0	0	0
4	0	0	0	0	1
5	1	0	0	0	0

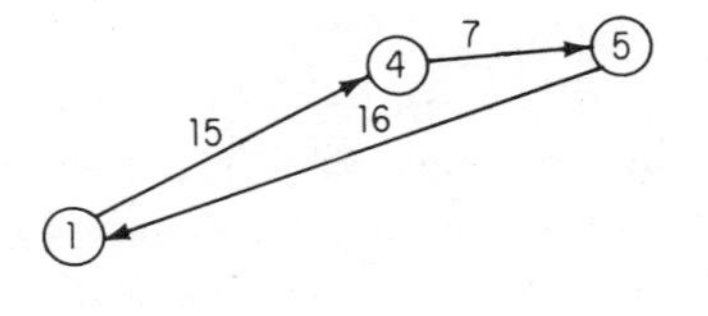

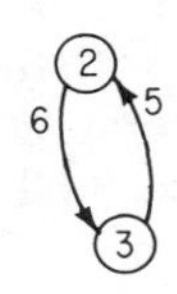

The assignment approach must be augmented by conditions which rule out the generation of assignments which are constructed of such subtours. This is the reason why the integer-programming approach requires so many equations.

Willie Simon's satisfaction was short-lived. In Willie's days (circa 1970), the largest traveling-salesman problem that had been solved was a 70-city one. Research has caught up with Willie. At this writing, the largest problem solved contains 318 cities and was solved using a special algorithm. The routine solution of large traveling-salesman problems is still an area of research. This problem is an ideal one for applying your intuition and designing a heuristic algorithm. Although we would like to find it, we are not necessarily looking for the optimal solution. We set our sights lower when we have to fall back to a nonoptimizing heuristic algorithm. A heuristic algorithm should make us feel that we tried our best to find the optimal solution and that the result is a good one.

For example, Willie's problem could be divided into smaller problems by grouping cities close together to form single-city groups. The divided parts could be linked together by short connections; then the groups redivided, and so on. Such a scheme would produce an acceptable feasible solution. You would not know if it was the optimum unless the algorithm implicitly tested and rejected all other solutions. An algorithm that does that needs to be efficient, that is, it finds a few solutions and tests them against all other solutions. The testing must demonstrate that the remaining solutions, which are not computed explicitly, would produce longer tours. An algorithm that does this, termed *branch and bound,* was first developed for the traveling-salesman problem. Variations of this algorithm have been used to solve integer-programming problems with some success. We will discuss branch and bound and a few heuristic algorithms for the traveling-salesman problem in Chapter 21. You should try to develop heuristic algorithms and use them to solve the five-city problem and the other traveling-salesman problems in Chapter 21.

20 The Transportation and Assignment Algorithms[1]

20.1 PROPERTIES OF THE TRANSPORTATION PROBLEM

We have discussed the transportation problem and its linear-programming model in Part I and in a number of exercises. In this chapter we shall develop a special adaptation of the simplex algorithm that enables us to solve transportation problems in a rather efficient manner. This procedure, the *simplex transportation algorithm,* takes advantage of the special mathematical structure of the transportation problem; it is really the standard simplex algorithm in disguise.

Up to now we have developed our algorithms in a rather informal manner. Hopefully, our discussions have given you a good understanding of model formulation and algorithms. Most of all the material covered in this book is based on mathematical theorems and proofs, for example, that a bounded linear-programming problem has an extreme-point optimal solution and the simplex algorithm will find it. Whenever you develop an algorithm that you claim finds the optimal solution, you should be able to prove that it is so.

[1] Most of the material in Sections 20.1 and 20.2 is from S. I. Gass, *Linear Programming: Methods and Applications,* 5th Edition, McGraw-Hill Book Company, New York, N.Y., 1985. Reproduced with permission.

Are you convinced that the shortest-route algorithm actually finds the shortest route? Can you prove this? To stress that what we do should be based as much as possible on mathematical foundations, we shall, in this chapter, describe the transportation problem and algorithm in a more formal manner. You should hardly notice the difference except that we shall state a few of the important theorems associated with our development.

The general statement of the transportation problem is the following: A homogeneous product is to be shipped in the amounts $a_1, a_2, \ldots, a_m$, respectively, from each of m *shipping origins* and received in amounts $b_1, b_2, \ldots, b_n$, respectively, by each of n *shipping destinations*. The cost of shipping a unit amount from the ith origin to the jth destination is c_{ij} and is known for all combinations (i, j). The problem is to determine the amounts x_{ij} to be shipped over all routes (i, j) so as to minimize the total cost of transportation.

To develop the constraints of the problem, we set up the following tableau:

		Destinations						
	(i) \ (j)	(1)	(2)	$\cdots$	(j)	$\cdots$	(n)	
Origins	(1)	x_{11}	x_{12}	$\cdots$	x_{1j}	$\cdots$	x_{1n}	a_1
	(2)	x_{21}	x_{22}	$\cdots$	x_{2j}	$\cdots$	x_{2n}	a_2
	.	.	.	$\cdots$	.	$\cdots$	.	.
	.	.	.	$\cdots$	.	$\cdots$	.	.
	(i)	x_{i1}	x_{i2}	$\cdots$	x_{ij}	$\cdots$	x_{in}	a_i
	.	.	.	$\cdots$	.	$\cdots$	.	.
	.	.	.	$\cdots$	.	$\cdots$	.	.
	(m)	x_{m1}	x_{m2}	$\cdots$	x_{mj}	$\cdots$	x_{mn}	a_m
		b_1	b_2	$\cdots$	b_j	$\cdots$	b_n	

The amount shipped from origin i to destination j is x_{ij}; the total shipped from origin i is $a_i \geq 0$; and the total received by destination j is $b_j \geq 0$. Here we temporarily impose the restriction that the total amount shipped is equal to the total amount received; that is, $\sum_i a_i = \sum_j b_j = A$. The total cost of shipping x_{ij} units is $c_{ij}x_{ij}$. Since a negative shipment has no valid interpretation for the problem as stated, we restrict each $x_{ij} \geq 0$. From the tableau we have the mathematical statement of the transportation problem: Find values for the variables x_{ij} which minimize the total cost

$$x_{00} = \sum_{i=1}^{m} \sum_{j=1}^{n} c_{ij}x_{ij} \tag{1}$$

subject to the constraints

$$\sum_{j=1}^{n} x_{ij} = a_i \qquad i = 1, 2, \ldots, m \tag{2}$$

$$\sum_{i=1}^{m} x_{ij} = b_j \qquad j = 1, 2, \ldots, n \tag{3}$$

$$x_{ij} \geq 0 \tag{4}$$

Equations (2) represent the row sums of the tableau and Eqs. (3) the column sums. In order for Eqs. (2) and (3) to be consistent, we must have the sum of Eq. (2) equal to the sum of Eq. (3); that is,

$$\sum_{i=1}^{m} \sum_{j=1}^{n} x_{ij} = \sum_{j=1}^{n} \sum_{i=1}^{m} x_{ij} = \sum_{i} a_i = \sum_{j} b_j = A$$

We note that the system of Eqs. (1)–(4) is a linear-programming problem with $m + n$ equations in mn variables.

Theorem 1. The transportation problem has a feasible solution.

Proof: Since $\sum_{i=1}^{m} a_i = \sum_{j=1}^{n} b_j = A$, we have the feasible solution $x_{ij} = a_i b_j / A$ for all (i, j). Each $x_{ij} \geq 0$, and Eq. (2) is satisfied since

$$\sum_{j=1}^{n} x_{ij} = \sum_{j=1}^{n} \frac{a_i b_j}{A} = \frac{a_i \sum_{j=1}^{n} b_j}{A} = a_i$$

and Eq. (3) is satisfied, since

$$\sum_{i=1}^{m} x_{ij} = \sum_{i=1}^{m} \frac{a_i b_j}{A} = \frac{b_j \sum_{i=1}^{m} a_i}{A} = b_j$$

For $m = 3$ and $n = 5$, let us write out the equations that correspond to (2) and (3). We then obtain the following eight (i.e., $m + n$) equations in 15 (i.e., mn) unknowns:

$$\begin{array}{lllllllllllllllll}
\text{(a)} & x_{11} + x_{12} + x_{13} + x_{14} + x_{15} & & & & & & & & & & & & & & & = a_1 \\
\text{(b)} & & x_{21} + x_{22} + x_{23} + x_{24} + x_{25} & & & & & & & & & & & & & & = a_2 \\
\text{(c)} & & & x_{31} + x_{32} + x_{33} + x_{34} + x_{35} & & & & & & & & & & & & & = a_3 \\
\text{(d)} & x_{11} & + x_{21} & + x_{31} & & & & & & & & & & & & & = b_1 \\
\text{(e)} & x_{12} & + x_{22} & + x_{32} & & & & & & & & & & & & & = b_2 \\
\text{(f)} & x_{13} & + x_{23} & + x_{33} & & & & & & & & & & & & & = b_3 \\
\text{(g)} & x_{14} & + x_{24} & + x_{34} & & & & & & & & & & & & & = b_4 \\
\text{(h)} & x_{15} & + x_{25} & + x_{35} & & & & & & & & & & & & & = b_5
\end{array}$$

If we sum Eqs. (d)–(h) and subtract Eqs. (b) and (c) from this sum, the result is Eq. (a). Hence Eq. (a) is redundant and does not need to be included in the system. Generalizing, we note that one equation from the system (2) or (3) can be eliminated, and the transportation problem reduces to $m + n - 1$ independent equations in mn variables.

Theorem 2. Construction of a basic feasible solution. A solution of at most $m + n - 1$ positive x_{ij}'s exists.

To demonstrate this theorem, we shall use the procedure for obtaining a first basic feasible solution termed the *northwest-corner rule*. We apply this scheme to the following 3×4 tableau:

x_{11}	x_{12}	x_{13}	x_{14}	a_1
x_{21}	x_{22}	x_{23}	x_{24}	a_2
x_{31}	x_{32}	x_{33}	x_{34}	a_3
b_1	b_2	b_3	b_4	

We first determine a value for the northwest-corner variable x_{11}. We let $x_{11} = \min(a_1,b_1)$; and if $a_1 \leq b_1$, then $x_{11} = a_1$, and all $x_{1j} = 0$ for $j = 2, 3, 4$. If $a_1 \geq b_1$, then $x_{11} = b_1$, and all $x_{i1} = 0$ for $i = 2, 3$. For discussion purposes, assume that the former is true; then this initial step transforms the tableau as shown in Step 1 below. Here the total left to be shipped from origin 1 has been reduced to 0, and the total left to be shipped to destination 1 is $b_1 - a_1$.

STEP 1. Assume $b_1 > a_1$.

$x_{11} = a_1$	0	0	0	0
x_{21}	x_{22}	x_{23}	x_{24}	a_2
x_{31}	x_{32}	x_{33}	x_{34}	a_3
$b_1 - a_1$	b_2	b_3	b_4	

We next determine a value for the first variable in row 2. We let $x_{21} = \min(a_2,b_1 - a_1)$. If we assume $a_2 > b_1 - a_1$, we have $x_{21} = b_1 - a_1$ and $x_{3i} = 0$. This is shown in Step 2. The total left to be shipped from origin 2 is now $a_2 - b_1 + a_1$, and the total to be shipped to destination 1 is 0.

STEP 2. Assume $a_2 > b_1 - a_1$.

$x_{11} = a_1$	0	0	0	0
$x_{21} = b_1 - a_1$	x_{22}	x_{23}	x_{24}	$a_2 - b_1 + a_1$
0	x_{32}	x_{33}	x_{34}	a_3
0	b_2	b_3	b_4	

Similarly, for the following steps, we determine a value of a variable x_{ij} and reduce to zero either the amount to be shipped from i or the amount to be shipped to j, or both.

STEP 3. Assume $a_2 - b_1 + a_1 > b_2$.

$x_{11} = a_1$	0	0	0	0
$x_{21} = b_1 - a_1$	$x_{22} = b_2$	x_{23}	x_{24}	$a_2 - b_1 + a_1 - b_2$
0	0	x_{33}	x_{34}	a_3
0	0	b_3	b_4	

STEP 4. Assume $a_2 - b_1 + a_1 - b_2 < b_3$.

$x_{11} = a_1$	0	0	0	0
$x_{21} = b_1 - a_1$	$x_{22} = b_2$	$x_{23} = a_2 - b_1 + a_1 - b_2$	0	0
0	0	x_{33}	x_{34}	a_3
0	0	$b_3 - a_2 + b_1 - a_1 + b_2$	b_4	

From Step 4, we see that $x_{33} = b_3 - a_2 + b_1 - a_1 + b_2$ and $x_{34} = b_4$. It should be noted that the values for the x_{ij} were all obtained by adding and subtracting various combinations of the a_i and b_j. Hence, if the a_i and b_j were originally nonnegative integers, then the feasible solution obtained by the above procedure would also consist of nonnegative integers. Since the northwest-corner rule of determining each x_{ij} eliminates either a row or a column from further consideration, while the last allocation eliminates both a row and a column, this feasible solution can have at most $m + n - 1$ positive x_{ij}'s. Depending on the values of the a_i and b_j and on the assumptions made in obtaining the feasible solution to the above example, we have as the six variables with possible positive values:

$$x_{11} = a_1 \qquad x_{21} = b_1 - a_1$$
$$x_{22} = b_2 \qquad x_{23} = a_2 - b_1 + a_1 - b_2$$
$$x_{33} = b_3 - a_2 + b_1 - a_1 + b_2 \qquad x_{34} = b_4$$

We apply the northwest-corner procedure to the following examples. Note that you do not need to know the costs to do this. Why could this be a drawback to the procedure? Try to develop these first basic feasible solutions yourself. You always fill in the empty cell that is the new northwest corner of the tableau.

Example 1

2	0	0	0	0	2
1	2	1	0	0	4
0	0	3	2	2	7
3	2	4	2	2	

Example 2

1	2	0	0	0	3
0	1	3	0	0	4
0	0	0	2	5	7
1	3	3	2	5	

In Example 2, only $m + n - 2 = 6$ of the x_{ij}'s are positive and we have determined a degenerate basic feasible solution. This will happen whenever the amount left to be shipped from origin i is exactly equal to the amount to be shipped to destination j, except when both $i = m$ and $j = n$. In Example 2, this is true for $x_{23} = \min(a_2 - x_{22}, b_3) = (3,3) = 3$.

As our algorithm for solving the transportation problem requires us to know exactly which $m + n - 1$ variables (positive or zero) are in a basic feasible solution, we need to identify any x_{ij}'s that are zero in a degenerate solution. This has to be done carefully—we cannot pick out any zero variables; the variables that form a basic feasible solution have to satisfy certain mathematical properties. The problem of identifying the correct zero variables occurs only when the first basic feasible solution is degenerate. When succeeding solutions are degenerate, we have a very simple rule to apply. There are two ways of selecting zero basic variables for a degenerate basic feasible solution. The first is a computational artifice; the second is based on the network structure implied by a basic feasible solution. We discuss each procedure in turn.

To avoid degenerate solutions, we have to ensure that no cell will have to ship from an origin i the same amount required by a destination j. This can be accomplished by perturbing (i.e., modifying) the values of the original a_i and b_j. The perturbation is done as follows. We set up a new problem where

$$\bar{a}_i = a_i + \epsilon \qquad i = 1, 2, \ldots, m$$

and

$$\bar{b}_j = b_j \qquad j = 1, 2, \ldots, n - 1$$

$$\bar{b}_n = b_n + m\epsilon$$

for $\epsilon > 0$.

We select an ϵ small enough that the final solution rounded to the same number of significant digits as the original a_i and b_j will yield the correct solution. This is possible since the computational procedure involves only additions and subtractions. For machine computation, it is easier always to add a 1 to each a_i and to add m to b_n in the last significant place that can be carried in the computer. (There is a mathematical proof behind this. It applies to any method for finding a first basic feasible solution that is similar to the northwest-corner technique. We allow the amount shipped from an origin i to be equal to the amount required at destination j only in the last allocation.)

Applying the ϵ procedure to Example 2, we have the solution

$$\begin{array}{ccccc|c} 1 & 2+\epsilon & 0 & 0 & 0 & 3+\epsilon \\ 0 & 1-\epsilon & 3 & 2\epsilon & 0 & 4+\epsilon \\ 0 & 0 & 0 & 2-2\epsilon & 5+3\epsilon & 7+\epsilon \\ \hline 1 & 3 & 3 & 2 & 5+3\epsilon & \end{array}$$

The basic solution now contains $x_{24} = 2\epsilon > 0$. The ϵ procedure enables the computation to proceed without any degenerate solutions. Theoretically, the selection of either x_{24} or x_{33} would have yielded a basis of $m + n - 1$ vectors. As will be seen, the computational procedure requires that we need only note which of these zero variables is selected to be in the basic solution.

The second degeneracy-avoiding method requires us to investigate the network structure implied by a basic feasible solution to the transportation problem. We illustrate it for the 3×5 problem of Example 2. The degenerate northwest-corner solution is

	i \ j	1	2	3	4	5	
		Destinations					
Origins	1	1	2	0	0	0	3
	2	0	1	3	0	0	4
	3	0	0	0	2	5	7
		1	3	3	2	5	

The positive numbers represent amounts that are shipped between the corresponding origins i and destinations j. The network for these positive shipments is represented by arcs in the following undirected *bipartite graph*:

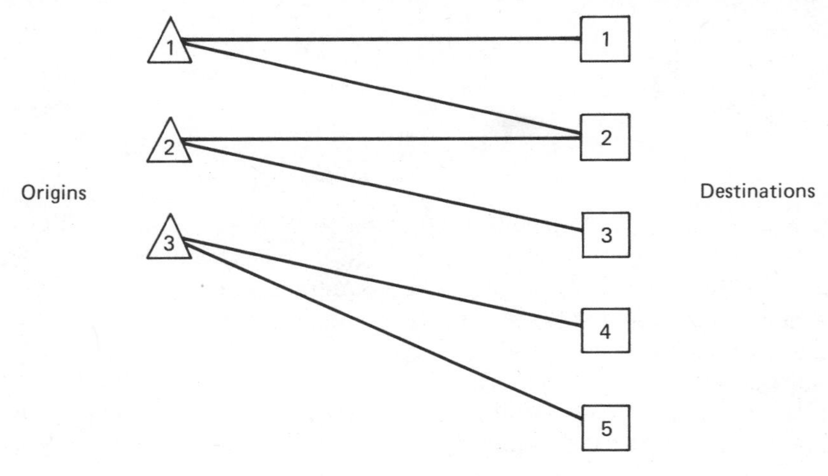

Note that the network is not *connected*, that is, you cannot reach any node from any other node. The network is split in two. A property of a basic feasible solution to a transportation problem is that its bipartite graph is connected. Any arc with zero flow that connects the graph can be used to fill out the basic feasible solution. We see that arc (2,4) or arc (3,3) will do, as well as others. A rule for choosing one is to select the arc with the lowest cost.

The connected bipartite graph for the northwest-corner solution of Example 1 is

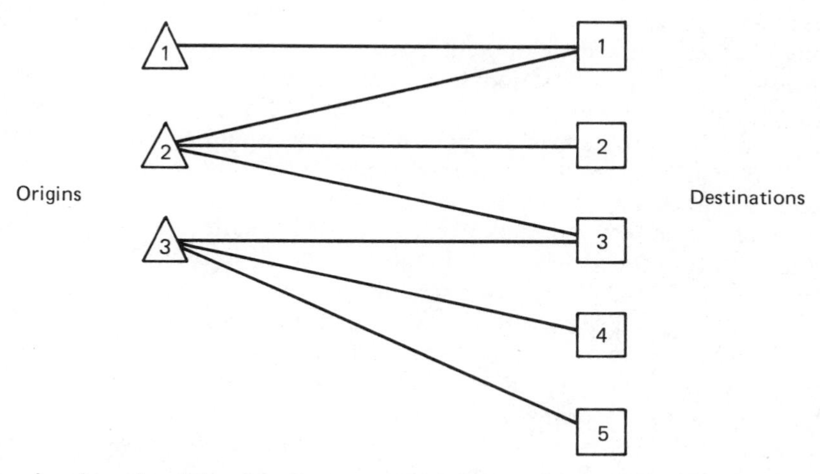

The simplex algorithm for the transportation problem changes basic feasible solutions so that the connectedness property is always maintained, even if a solution is degenerate.

Since virtually all applications of the transportation problem require the shipping of only whole units of the item being considered, it is nice to be aware of the following important property of the transportation problem:

Theorem 3. Assuming the a_i and b_j are nonnegative integers, then every basic feasible solution (i.e., extreme-point solution) has integral values.

The theorem is not too difficult to establish. It is based on the mathematical structure of the $m + n - 1$ constraints and columns that define a particular basic feasible solution. For example, you should be able to demonstrate for the northwest-corner basic feasible solution of Example 1 that the associated square set of constraints and columns can only be solved in terms of integers.

Theorem 4. A finite minimum feasible solution always exists.

The proof of this theorem is no big secret. Theorem 1 states that the problem has a feasible solution. And since each variable x_{ij} is bounded by the smaller of the corresponding a_i or b_j—you can take it from there.

20.2 THE SIMPLEX TRANSPORTATION ALGORITHM

The transportation problem can be solved as a regular linear-programming problem by direct application of the simplex algorithm. Many such large-scale problems are solved using computers and standard mathematical-programming systems. However, for hand computation, the $m + n - 1$ equations with mn variables become rather large even for reasonable values of m and n. (For example, with $m = 4$ and $n = 6$ we would have to deal with nine equations and 24 variables.) A careful observation of the simplex algorithm as it is applied to the mathematical structure of the transportation problem has led to the development of an easy-to-use special *simplex transportation algorithm*. The rationale for this algorithm is based on primal–dual problem considerations and the requirements of the simplex algorithm theory. The dual problem of the primal transportation has some nice properties that enable us to carry out the simplex algorithm without an explicit equation tableau.

The dual to the 3×5 primal problem, whose equations are given above, has the following form. For exposition purposes, it is convenient to divide the $m + n$ dual variables into two sets. We let the set of variables $u_1, \ldots, u_i, \ldots, u_m$ be associated with the corresponding m origin constraints and the set $v_1, \ldots, v_j, \ldots, v_n$ be associated with the corresponding n destination constraints. Duality theory requires that the dual be a maximization problem. Also, as the primal is in equation form, all the u_i and v_j variables are unrestricted, that is, they can take on any positive or negative values. The transportation problem's dual is then given by

$$\text{Maximize } a_1u_1 + a_2u_2 + a_3u_3 + b_1v_1 + b_2v_2 + b_3v_3 + b_4v_4 + b_5v_5$$

subject to

$$\begin{array}{lllllllll}
u_1 & & & + v_1 & & & & & \leq c_{11} \\
u_1 & & & & + v_2 & & & & \leq c_{12} \\
u_1 & & & & & + v_3 & & & \leq c_{13} \\
u_1 & & & & & & + v_4 & & \leq c_{14} \\
u_1 & & & & & & & + v_5 & \leq c_{15} \\
 & u_2 & & + v_1 & & & & & \leq c_{21} \\
 & u_2 & & & + v_2 & & & & \leq c_{22} \\
 & u_2 & & & & + v_3 & & & \leq c_{23} \\
 & u_2 & & & & & + v_4 & & \leq c_{24} \\
 & u_2 & & & & & & + v_5 & \leq c_{25} \\
 & & u_3 & + v_1 & & & & & \leq c_{31} \\
 & & u_3 & & + v_2 & & & & \leq c_{32} \\
 & & u_3 & & & + v_3 & & & \leq c_{33} \\
 & & u_3 & & & & + v_4 & & \leq c_{34} \\
 & & u_3 & & & & & + v_5 & \leq c_{35}
\end{array}$$

with all u_i and v_j unrestricted. In general, these constraints can be written as

$$u_i + v_j \leq c_{ij} \qquad \text{for all } (i, j)$$

Recall that there is a dual constraint for every variable in the primal, and a dual variable for each constraint of the primal. Since the primal transportation problem reduces to $m + n - 1$ constraints, the above dual problem for the 3×5 primal has one variable too many. In selecting the $m + n - 1$ equations for the primal, it is convenient to always leave off the first one; this implies that there is no u_1—we set $u_1 = 0$ in the dual constraints leaving us with $m + n - 1$ dual variables. For those variables x_{ij} that are in a basic feasible solution to the primal transportation problem, the primal–dual relationships require the corresponding dual constraints to be equations, that is,

$$u_i + v_j = c_{ij} \quad (\text{for } x_{ij} \text{ in the basis})$$

Thus, for the $m + n - 1$ basic variables x_{ij}, we will have a corresponding set of $m + n - 1$ dual constraints $u_i + v_j = c_{ij}$. Since $u_1 = 0$, this set of dual constraints is a square set with $m + n - 1$ variables. Not only is it square, the form of the equations is triangular. This property enables us to solve for the u_i and v_j by simple additions and subtractions. We shall illustrate this property below.

If the u_i and v_j that solve the square set also satisfy the remaining dual constraints $u_i + v_j \leq c_{ij}$ (for the x_{ij} not in the basis), then the primal–dual theory tells us that the current basic feasible solution is also optimum. If the dual constraints are not satisfied by the u_i and v_j, we have to change to a new basic feasible solution and start all over. In essence, the u_i and v_j dual variables that correspond to a primal basic feasible solution enable us to calculate the row 0 quantities of the usual simplex tableau. These quantities are just the differences $u_i + v_j - c_{ij}$. If all $u_i + v_j - c_{ij} \leq 0$, this implies optimality of the current primal solution. (The corresponding u_i and v_j will also be an optimal solution to the dual.) If some $u_i + v_j - c_{ij} > 0$, then we select a variable x_{ij} to enter the basis that corresponds to the maximum difference; this is equivalent to the usual simplex rule for selecting a variable to enter the basis in a minimization problem.

The above discussion is a brief description of the rationale and procedure of the transportation simplex algorithm. The only thing that is missing is an explanation of how do we change basic feasible solutions without the usual simplex tableau. This is described below by an example.

In the problem we have three origins, with availabilities of $a_1 = 6$, $a_2 = 8$, and $a_3 = 10$, and four destinations, with requirements of $b_1 = 4$, $b_2 = 6$, $b_3 = 8$, and $b_4 = 6$. We note that $\sum_i a_i = \sum_j b_j = 24$. The costs between each origin and destination are given in the following cost matrix:

Destinations

	1	2	3	4
Origins 1	1	2	3	4
Origins 2	4	3	2	0
Origins 3	0	2	2	1

$= (c_{ij})$

For example, the cost for shipping one unit between origin 3 and destination 2 is $c_{32} = 2$. In general, the c_{ij} can be any positive or negative numbers. Using the northwest-corner rule, we obtain the first feasible solution:

$(x_{ij}) =$

4	2			6
	4	4		8
		4	6	10
4	6	8	6	

where $x_{11} = 4$, $x_{12} = 2$, $x_{22} = 4$, $x_{23} = 4$, $x_{33} = 4$, $x_{34} = 6$, and all the other $x_{ij} = 0$. The value of the objective function is 42. Since this is a nondegenerate basic feasible solution, the first solution does not require perturbing the problem.

We are next required to determine, for those variables in the basic solution, m numbers u_i and n numbers v_j such that

$$\begin{aligned} u_1 + v_1 &= c_{11} = 1 \\ u_1 + v_2 &= c_{12} = 2 \\ u_2 + v_2 &= c_{22} = 3 \\ u_2 + v_3 &= c_{23} = 2 \\ u_3 + v_3 &= c_{33} = 2 \\ u_3 + v_4 &= c_{34} = 1 \end{aligned} \qquad (5)$$

Since we agreed to let $u_1 = 0$, we can rearrange these dual equations in the following triangular form:

$$\begin{aligned}
v_1 \qquad\qquad\qquad\qquad\qquad\qquad &= 1\\
v_2 \qquad\qquad\qquad\qquad\qquad &= 2\\
v_2 + u_2 \qquad\qquad\qquad\qquad &= 3\\
u_2 + v_3 \qquad\qquad\qquad &= 2\\
v_3 + u_3 \qquad\qquad &= 2\\
u_3 + v_4 &= 1
\end{aligned}$$

It is easy matter to solve for the u_i and v_j. We see that $v_1 = 1$ and $v_2 = 2$; with $v_2 = 2$ then $u_2 = 1$; with $u_2 = 1$ then $v_3 = 1$; with $v_3 = 1$ then $u_3 = 1$; and finally, with $u_3 = 1$ then $v_4 = 0$. This computation can be readily accomplished by setting up the following table containing the cost coefficients (in bold type) of the variables in the basic solution:

u \ v				
	1	**2**		
		3	**2**	
			2	**1**

By letting $u_1 = 0$, we can compute, as was done above, the resulting u_i and v_j and enter them in the corresponding positions as follows:

u \ v	1	2	1	0
0	**1**	**2**		
1		**3**	**2**	
1			**2**	**1**

Since equations (5) are satisfied, we have $u_i + v_j = c_{ij}$ for those x_{ij} in the basic feasible solution. We then compute the *indirect costs* $\bar{c}_{ij} = u_i + v_j$ for all combinations (i, j) and place these figures in their corresponding cells of the indirect-cost table. ($\bar{c}_{ij} = c_{ij}$ for all x_{ij} in the solution.) The indirect-cost table ($\bar{c}_{ij}$) for the first solution, along with a copy of the direct-cost table (c_{ij}), follows:[2]

$(\bar{c}_{ij}) =$

1	**2**	1	0
2	**3**	**2**	1
2	3	**2**	**1**

1	2	3	4
4	3	2	0
0	2	2	1

$= (c_{ij})$

For example, $\bar{c}_{14} = u_1 + v_4 = 0$. As shown below, the above three steps can be combined into one efficient computational step. We next compute the differences $\bar{c}_{ij} - c_{ij}$. If all $\bar{c}_{ij} - c_{ij} \leq 0$, then the solution that yielded the indirect-cost table is a minimal feasible solution. If at least one $\bar{c}_{ij} - c_{ij} > 0$, we have not found a minimum. We can, as will be described below, readily obtain a new basic feasible solution which contains a variable associated with a $\bar{c}_{ij} - c_{ij} > 0$. As in the general simplex procedure, we select for entry into the new basis the variable corresponding to $\bar{c}_{pq} - c_{pq} = \max(\bar{c}_{ij} - c_{ij} > 0)$. This scheme will yield a new basic solution whose value of the objective function will be less than the value for the preceding solution (or possibly equal to it, if the preceding solution is degenerate).

For our first solution, we find that $\bar{c}_{pq} - c_{pq} = \bar{c}_{31} - c_{31} = 2$. Hence we select $x_{pq} = x_{31}$ to be introduced into the solution. If there were ties, we would select the variable with the smaller c_{ij}. We now wish to determine a new basic feasible solution which contains x_{31}.

[2] A concise way of combining the allocation table and the indirect-cost table is to form the following tableau (shown for a 3 × 4 problem):

u_i \ v_j	v_1	v_2	v_3	v_4	
u_1	c_{11} (x_{11})	c_{12} (x_{12})	c_{13} $\bar{c}_{13} - c_{13}$	c_{14} $\bar{c}_{14} - c_{14}$	a_1
u_2	c_{21} $\bar{c}_{21} - c_{21}$	c_{22} (x_{22})	c_{23} (x_{23})	c_{24} $\bar{c}_{24} - c_{24}$	a_2
u_3	c_{31} $\bar{c}_{31} - c_{31}$	c_{32} $\bar{c}_{32} - c_{32}$	c_{33} (x_{33})	c_{34} (x_{34})	a_3
	b_1	b_2	b_3	b_4	x_{00}

Referring to the first solution matrix (x_{ij}), we introduce x_{31} into the solution at an unknown nonnegative level θ. As the row and column sums of the variables must equal the corresponding values of the a_i and b_j, we must add or subtract θ from some of the other x_{ij} in the first solution, as follows:

$4 - \theta$	$2 + \theta$			6
	$4 - \theta$	$4 + \theta$		8
θ		$4 - \theta$	6	10
4	6	8	6	

Since we put $\theta \geq 0$ in cell (3,1), we must subtract θ from x_{11}, x_{22}, and x_{33} and add θ to x_{12} and x_{23}, in order to keep the row and column sums correct. We see that the size of θ is restricted by those x_{ij} from which it is subtracted. θ cannot be larger than the smallest x_{ij} from which it is subtracted. Here θ must be less than or equal to 4 and must be greater than zero in order to preserve feasibility. Since we wish to eliminate one of the variables from the old solution and to introduce x_{31}, we let $\theta = 4$. However, since $x_{11} - \theta = x_{22} - \theta = x_{33} - \theta = 4 - \theta$, we shall eliminate three variables and obtain a degenerate solution with four positive variables. To keep a solution with exactly $m + n - 1$ nonnegative variables, we retain two of these three variables with values of zero. We select x_{11} and x_{33}, because they correspond to the smaller c_{ij}. You should construct the bipartite graph of the first basic feasible solution and convince yourself that a new connected graph results when you include in the new basic feasible solution any two of the three variables that have become zero. The new solution with $\theta = x_{31} = 4$ is

$(x_{ij}) =$

0	6			6
		8		8
4		0	6	10
4	6	8	6	

The objective function for this solution is equal to

$$42 - [\max(\bar{c}_{ij} - c_{ij} > 0)]\theta = 42 - (2)(4) = 34$$

The corresponding combined (u_i, v_j) table and $\bar{c}_{ij}$ matrix for this solution is given in the following table:

$(\bar{c}_{ij}) =$

u \ v	1	2	3	2
0	**1**	**2**	3	2
−1	0	1	**2**	1
−1	**0**	1	**2**	**1**

1	2	3	4
4	3	2	0
0	2	2	1

$= (c_{ij})$

where the bold numbers correspond to the c_{ij} for the basic variables. We see that $\max(\bar{c}_{ij} - c_{ij} > 0) = \bar{c}_{24} - c_{24} = 1$ and we next introduce $x_{24} = \theta \geq 0$ into cell (2,4).

We add and subtract $\theta = x_{24}$ from the previous x_{ij} as follows:

0	6			6
		$8 - \theta$	θ	8
4		$0 + \theta$	$6 - \theta$	10
4	6	8	6	

We have $\theta = 6$; x_{34} is eliminated; and x_{24} is introduced with a value of $\theta = 6$. The new value of the objective function is

$$34 - (\bar{c}_{24} - c_{24})\theta = 34 - (1)(6) = 28$$

The new solution is

$(x_{ij}) =$

0	6			6
		2	6	8
4		6		10
4	6	8	6	

The corresponding (u_i, v_j) table and $(\bar{c}_{ij})$ matrix is given by

$(\bar{c}_{ij}) =$

u \ v	1	2	3	1
0	**1**	**2**	3	1
−1	0	1	**2**	**0**
−1	**0**	1	**2**	0

1	2	3	4
4	3	2	0
0	2	2	1

$= (c_{ij})$

Here, all $\bar{c}_{ij} - c_{ij} \leq 0$, and this last solution ($x_{11} = 0$, $x_{12} = 6$, $x_{23} = 2$, $x_{24} = 6$, $x_{31} = 4$, $x_{33} = 6$, and all other $x_{ij} = 0$) is a degenerate minimal feasible solution. The value of the objective function is 28. We note that $\bar{c}_{13} - c_{13} = 0$ and that x_{13} is not in the solution. We can then introduce x_{13} into this last solution and obtain an alternate minimum solution. We put $\theta \geq 0$ into cell (1,3) and obtain

$0 - \theta$	6	θ		6
		2	6	8
$4 + \theta$		$6 - \theta$		10
4	6	8	6	

We then have $x_{13} = \theta = 0$ and a new degenerate basic minimum solution as follows:

$(x_{ij}) =$

	6	0		6
		2	6	8
4		6		10
4	6	8	6	

Of course, the value of the objective function is 28. For this multiple solution the value of the objective function did not change because $\bar{c}_{13} - c_{13} = 0$, while the values of the variables did not change because $\theta = 0$.

The optimality of these solutions can be seen from the following. As a constant can be added or subtracted from each element of a row or column in the cost matrix without changing the optimal allocation, the final set of u_i and v_j elements can be subtracted from the corresponding rows or columns of the original cost table. (The reason why this is so is discussed in the next section on assignment problems; try and prove it.) This results in a cost table whose elements are either zero or positive, that is,

$$c_{ij} - u_i - v_j = c_{ij} - \bar{c}_{ij} \geq 0$$

When this transformed cost table is used for our transportation problem, an optimal solution would, of course, be one that allocated shipments only to those routes with zero costs. This is just the situation encountered in the final allocation determined by the above variation of the simplex method; hence the solution is optimal.

We next summarize the simplex algorithm for solving a transportation problem.

1. Find a first feasible solution by the northwest-corner rule or an equivalent procedure, for example, row minimum (see below), and develop the solution table (x_{ij}). Let x_{00} be the current value of the objective function.

2. Develop the combined (u_i, v_j) and indirect-cost table $(\bar{c}_{ij})$; that is, solve for the u_i and v_j by requiring $u_1 = 0$ and $c_{ij} = u_i + v_j$ for x_{ij} in the basis and $\bar{c}_{ij} = u_i + v_j$ for all (i, j).

3. If all $\bar{c}_{ij} - c_{ij} \leq 0$, then the current solution is optimal. If some $\bar{c}_{ij} - c_{ij} > 0$, select the nonbasic variable x_{pq} to enter the basis corresponding to $\bar{c}_{pq} - c_{pq} = \max(\bar{c}_{ij} - c_{ij} > 0)$. (As any x_{ij} corresponding to a $\bar{c}_{ij} - c_{ij} > 0$ may be selected to enter the basis, a modified selection rule can be used.)

4. Determine the value of x_{pq} and the new basic feasible solution by entering the surrogate parameter $\theta = x_{pq}$ into position (p,q) of the current solution table (x_{ij}) and add and subtract θ from the x_{ij} to maintain row and column balances. Then for those positions in which θ is subtracted from x_{ij}, determine the minimum x_{ij}, and this value is equal to $\theta = x_{pq}$; that is, we solve the set of constraints $x_{ij} - \theta \geq 0$ and select the maximum value of θ possible. If there are ties, we obtain a degenerate solution in that more than one of the current basic variables will become zero. We drop only one of these zero variables (any one) and retain the others to form the new basic solution in order to construct the corresponding $(\bar{c}_{ij})$ table. The new value of the objective function $\bar{x}_{00}$ is given by $\bar{x}_{00} = x_{00} - (\bar{c}_{pq} - c_{pq})x_{pq}$.

5. Then apply Steps 2 to 4 to the new solution. If in the optimal solution some $\bar{c}_{ij} - c_{ij} = 0$ for a nonbasic variable, then a multiple optimal solution can be found by introducing the corresponding variable into the solution by applying Step 4 and treating the variable as x_{pq}.

In solving transportation problems by hand, you can save time and reduce errors by organizing work sheets so as to afford a ready access to the data, especially when calculating the $\bar{c}_{ij} - c_{ij}$ terms. The tableau given in the preceding footnote is a concise form for keeping the data and calculations and can be used for determining θ and the u_i's and v_j's.

Some transportation problems might have the situation where the total of availabilities, $\sum_i a_i$, is less than the total of requirements, $\sum_j b_j$. Even though we cannot satisfy all the requirements, we can still allocate the items at the origins to the destinations in a manner that minimizes the total shipping cost. For this case, we assume that we have a "fictitious" origin that has on hand a total of $\sum_j b_j - \sum_i a_i > 0$ units. The costs of shipping a unit between this fictitious or $(m + 1)$st origin and the destinations are assumed to be zero. If our problem was originally a 3×4 problem, we would set up the following 4×4 problem and solve this problem with the fictitious origin like any other transportation problem:

		Destinations 1	2	3	4	
Origins	1	c_{11}	c_{12}	c_{13}	c_{14}	a_1
	2	c_{21}	c_{22}	c_{23}	c_{24}	a_2
	3	c_{31}	c_{32}	c_{33}	c_{34}	a_3
	4	0	0	0	0	$\sum_j b_j - \sum_i a_i$
		b_1	b_2	b_3	b_4	

By letting all the $c_{m+1,j} = c_{4j} = 0$, we make the minimum value of the objective function for the fictitious problem equal to the minimum value of the original problem.

If the problem is given with $\sum_i a_i > \sum_j b_j$, we set up a similar fictitious problem. Here we assume a fictitious destination that requires $\sum_i a_i - \sum_j b_j > 0$ units. The fictitious shipping costs are again zero. For a 3×4 problem we would then solve the following 3×5 problem:

		Destinations 1	2	3	4	5	
Origins	1	c_{11}	c_{12}	c_{13}	c_{14}	0	a_1
	2	c_{21}	c_{22}	c_{23}	c_{24}	0	a_2
	3	c_{31}	c_{32}	c_{33}	c_{34}	0	a_3
		b_1	b_2	b_3	b_4	$\sum_i a_i - \sum_j b_j$	

In solving any linear-programming problem, we should, in general, expect the total number of iterations required to depend on how close the value of the objective function for the first feasible solution is to the actual minimum. Since the northwest-corner rule does not consider the size of the c_{ij}, we cannot expect the corresponding value of the objective function to be close to the minimum. A number of alternative methods have been suggested that can be adapted to machine computation. We shall illustrate three of them for the following example:

3	2	1	2	3	1
5	4	3	−1	1	5
0	2	3	4	5	7
3	3	3	2	2	

You should determine the northwest-corner solution. It yields a value of the objective function of 52.

(a) *Row Minimum.* Let the minimum element in the first row be c_{1k}. (If there is more than one minimum element, select the one with the smallest index j.) We let $x_{1k} = a_1$ if $a_1 \leq b_k$ or $x_{1k} = b_k$ if $a_1 > b_k$. In the first case, we have shipped all the a_1 units and go on to the second row after changing b_k to $b_k - a_1$. We next find the minimum element in the second row and repeat the process. In the second case, we have allocated only b_k units of the a_1; hence we change a_1 to $a_1 - b_k$ and b_k to zero and find the next smallest c_{ij} in the first row and repeat the process. Using this scheme, we have a first feasible solution as shown below:

$(x_{ij}) =$

		1			1
		1	2	2	5
3	3	1			7
3	3	3	2	2	

This is a minimal feasible solution with an objective-function value of 13.

(b) *Column Minimum.* We do a calculation similar to that using the row minimum, except that we start with the first column and proceed to the last column. Using this scheme, the first feasible solution is

$(x_{ij}) =$

		1			1
		2	2	1	5
3	3			1	7
3	3	3	2	2	

with an objective-function value of 17.

(c) *Cost Table Minimum.* Here we search the whole cost table for the smallest element and allocate accordingly. We repeat this procedure until all the units are shipped. Using this scheme on the preceding example, we obtain the minimum solution obtained by the row-minimum procedure.

Even though these schemes produce good results for the example in question, they are not foolproof. Examples have been constructed whose northwest-corner solution is much better than solutions obtained by any of the above procedures. However, experience has shown that, in general, the extra computational time required to obtain a first feasible solution by one of the above methods will be more than made up because the total iterations will be fewer. This is especially true for large problems.

20.3 THE HUNGARIAN METHOD FOR THE ASSIGNMENT PROBLEM

The mathematical model of the $m \times m$ assignment problem is to minimize

$$x_{00} = \sum_{i=1}^{m} \sum_{j=1}^{m} c_{ij} x_{ij}$$

subject to

$$\sum_{j=1}^{m} x_{ij} = 1 \qquad i = 1, \ldots, m$$

$$\sum_{i=1}^{m} x_{ij} = 1 \qquad j = 1, \ldots, m$$

$$x_{ij} = 0 \quad \text{or} \quad 1$$

Here we assume that there are m persons to be assigned to m jobs, and c_{ij} measures the cost of placing person i in job j, or the time for person i to perform job j. In our earlier discussions of the assignment problem (Chapter 9), we let c_{ij} be the value of person i in job j. We then wanted to determine the assignment that maximized total value. We address the minimization problem first as the algorithm for solving it is a bit easier to develop. It is just a question of handling minus signs, as we can convert the maximization problem to a minimization by changing the signs of the c_{ij}.

The assignment problem can be solved by the regular simplex algorithm or the simplex transportation algorithm. However, because the extreme-point solutions tend to be highly degenerate (many basic variables are zero), these algorithms take many iterations to converge to an optimum. There are other ways of solving this problem. One that has proved very effective and is not troubled by degeneracy was developed by the mathematician Harold Kuhn. He named it the *Hungarian method* as its theoretical basis rests on a theorem first proved by the Hungarian mathematician E. Egérváry. (The theorem was originally stated by another Hungarian mathematician D. König.)

Let us consider the following minimal-cost assignment problem and associated discussion due to a third Hungarian mathematician, B. Krekó. Assume that a company has five garages in each of which is stationed a bus. The busses, which are of the same type, have to go to five destinations—one bus to each destination. We need to find an assignment of the busses to destinations that yield the minimal total cost of the assignment. If we assume that the cost of a truck is proportional to the distance traveled, we can deal with the following distance table:

Garage \ Destination	1	2	3	4	5
1	5	3	4	2	6
2	8	3	5	5	4
3	2	5	3	6	8
4	4	2	8	3	6
5	3	6	9	5	3

Before going on with this problem, let us develop a few ideas and concepts that will lead you to an algorithm for solving assignment problems. Consider the following cost table for a 3 × 3 minimization problem:

		Jobs		
		1	2	3
	1	5	0	2
Persons	2	0	1	4
	3	2	3	0

It is clear that if we assign 1 to 2, 2 to 1, and 3 to 3, that we have a mininal-cost assignment with an objective-function value of zero. Wouldn't it be nice if all assignment cost tables had this feature? Can we make it so for any cost table? How do you think you can go about doing it and why can you get away with it? It can be done!

There are two parts to this process. One is based on a simple property and mathematical structure of the assignment problem, and the other is based on the Hungarian theorem. Let us consider another cost table that does not have an obvious solution:

	1	2	3
1	5	4	7
2	6	7	3
3	8	11	2

The objective function of this problem is to minimize the sum:

$$\begin{aligned} x_{00} = \; & 5x_{11} + 4x_{12} + 7x_{13} \\ & + 6x_{21} + 3x_{22} + 3x_{23} \\ & + 8x_{31} + 11x_{32} + 2x_{33} \end{aligned}$$

What if we subtract four units from all the costs in the first row? What does this do to the objective function? Doing this explicitly we have

$$\begin{aligned}(5 - 4)x_{11} &+ (4 - 4)x_{12} + (7 - 4)x_{13}\\ + 6x_{21} &+ \quad 3x_{22} + \quad 3x_{23}\\ + 8x_{31} &+ \quad 11x_{32} + \quad 2x_{33}\end{aligned}$$

or, after combining terms,

$$x_{00} - 4(x_{11} + x_{12} + x_{13})$$

Since $x_{11} + x_{12} + x_{13} = 1$, the transformed objective function is just $x_{00} - 4$. Our problem is exactly the same except that the value of x_{00} will be decreased by four units. Thus, the solution to the original problem, that is, the optimal assignments, will be the same as the optimal assignments of the problem whose costs have been reduced by 4. As long as we subtract the same amount from all costs in a row, or all costs in a column, the same analysis holds. Hopefully, the problem with reduced costs will be easier to solve; maybe even by inspection. Look what happens if we subtract the minimal row elements from the corresponding rows for the above 3×3 table:

	1	2	3	
1	1	0	3	4
2	3	4	0	3
3	6	9	0	2
				Min

We introduce additional zeros by subtracting the minimal element from each column:

	1	2	3	
1	0	0	3	4
2	2	4	0	3
3	5	9	0	2
	1	0	0	Min

The problem with this reduced cost table is the same as the original, except that the objective function will be 10 units less. But this table does not have an assignment that uses just the zero cells. What do we do?

First, we need some new terminology. Let the cost elements in the table be called *points*, and the rows and columns be termed *lines*. A set of points is said to be *independent* if none of the lines contain more than one point of the set. The set of zero points in the following table is independent:

	1	2	3
1	5	0	2
2	0	1	4
3	2	3	0

The set of zero points in the next table is not independent:

	1	2	3
1	0	0	3
2	2	4	0
3	5	9	0

A single point is independent. If we want to add another point to the set, then it must not be on the line (row or column) of the first point. Following this prescription, an independent set containing m points can be found. You can always pick out m independent points. The trick is to pick out the set of m that has minimum cost. To proceed, we need to combine the above reduction process with the result of the following theorem.

König-Egérváry (Hungarian) Theorem.

The maximum number of independent zero points in a square table is equal to the minimum number of lines required to cover all zeros in the table.

If we use dashed lines to indicate that a row or column is a covered line, we have the following situations for our two 3 × 3 examples:

```
—5—0—2—
—0—1—4—
—2—3—0—

       |
—0—0—3—
       |
 2  4  0
       |
 5  9  0
       |
```

In the first problem it takes three lines to cover all the zeros; in the second it takes two. The theorem then tells us that the three zeros in the first problem represent an independent set of points and an assignment; the four zeros in the second example do not. Let us proceed with the second example.

We want to introduce at least one additional zero into the tableau by subtracting the correct amounts from the rows and columns while keeping all the zeros we have intact. Think of how this can be done; you want all the reduced costs to stay positive or zero. The second problem's reduced costs are

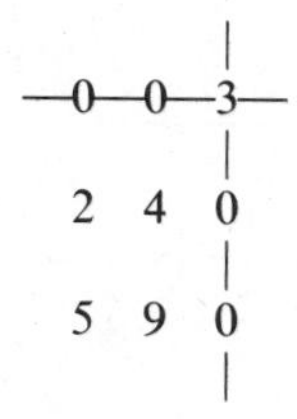

We select the minimum cost to be reduced to zero that is not covered by a line. There is, of course, no advantage in choosing an element already covered as we want to add to the independent set of zeros. We denote this minimum reduced cost by $\bar{c}_{ij}$. We subtract this cost, here $\bar{c}_{ij} = 2$, from all uncovered costs. But to maintain the balance implied by the assignment equations, we add this cost to any element that is covered by two lines. Conceptually, what this step really does is to subtract the minimum element from *all* costs. Then to bring back all the covered zeros, we readd $\bar{c}_{ij}$ to the elements of all rows and columns that were covered. This causes a cost that was covered twice to have $\bar{c}_{ij}$ added to it, and any element covered once is not changed. The new reduced table is

—0—0—5—
—0—2—0—
—3—7—0—

We now need three lines to cover all the zeros and thus there are three independent zeros and a corresponding optimal assignment. Can you pick them out? How do you figure out the value of the assignment?

The subtracting of $\bar{c}_{ij}$ is summarized as follows:

1. The costs that are not covered are reduced by $\bar{c}_{ij}$.
2. The costs that are covered once do not change.
3. The costs that are covered twice are increased by $\bar{c}_{ij}$.

Linear-programming theory shows that you can perform this reduction a finite number of times and eventually you will be able to find an independent set of m zeros and an optimal assignment. If the table ever gets reduced to all zeros, you have an obvious optimal solution. Note that each reduction reduces the value of the objective function by a positive amount. The value of the objective function for the optimal assignment needs to be calculated directly. It is equal to the sum of the first row and column reductions, plus a function of the $\bar{c}_{ij}$ reductions. That function is a little complicated, but you should be able to figure it out. (*Hint:* It depends on how many independent zeros need to be found.)

Let us return to the 5 × 5 garage-truck example:

Destinations

		1	2	3	4	5
Garages	1	5	3	4	2	6
	2	8	3	5	5	4
	3	2	5	3	6	8
	4	4	2	8	3	6
	5	3	6	9	5	3

We first subtract out the row minimums:

	1	2	3	4	5	
1	3	1	2	0	4	2
2	5	0	2	2	1	3
3	0	3	1	4	6	2
4	2	0	6	1	4	2
5	0	3	6	2	0	3
						Min

Then we subtract the column minimums to obtain the first reduced table:

	1	2	3	4	5	
1	3	1	1	0	4	2
2	5	0	1	2	1	3
3	0	3	0	4	6	2
4	2	0	5	1	4	2
5	0	3	5	2	0	3
	0	0	1	0	0	Min

The total cost reduction is the sum of the minimum elements and equals 13.

The resultant zeros can be covered by only four lines as shown:

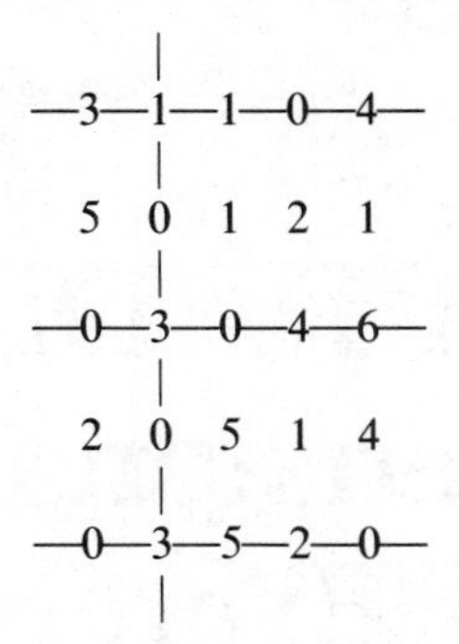

We do not have an independent set of five zeros. The minimum uncovered element is equal to 1. We subtract 1 from all uncovered elements and add it to those elements that are covered twice to obtain the second reduced cost table:

3	2	1	0	4
4	0	0	1	0
0	4	0	4	6
1	0	4	0	3
0	4	5	2	0

If you check this table, you will see that you will require five lines to cover all the zeros. Thus, we can select a set of five independent zeros from the set of 10 zeros that corresponds to an optimal assignment. You should be able to "eyeball" the table to pick out the correct zeros. The minimal cost is 14.

21 Part IV Discussion, Extensions, and Exercises

21.1. We first present a labeling method for solving the maximal-flow problem, and then illustrate its use by solving the problem of Section 18.2. It will appear to be complicated at first reading, so you should not be concerned if it has to be reread. After you understand the process, you should try to rewrite the algorithm in a simpler fashion.[1]

The algorithm converges in a finite number of steps if the arc capacities f_{ij} are all rational numbers. Its validity is based on the Max-flow Min-cut theorem. The steps of the algorithm are designed to find a path over which a positive flow can be sent from source to sink. The steps are repeated until no such path can be found.

Maximal-Flow Algorithm

STEP 1. Find an initial feasible solution for the network. We may, for example, begin with all $x_{ij} = 0$.

STEP 2. Start with the source node 0 and give it a label $[-,\infty]$. The general label for any nodes i and j is indicated by $[i\pm,v_j]$, with v_j being a positive number representing a change in flow between i and j, $i+$ representing an increase of flow by the amount v_j from i to j, and $i-$ representing a decrease

[1] The material in this section is from S. I. Gass, *Linear Programming: Methods and Applications,* 5th Edition, McGraw-Hill Book Co., New York, N.Y., 1985. Reproduced with permission.

of flow by the amount v_j from j to i. The source label $[-,\infty]$ indicates that an unlimited amount of commodity is available at the source for shipment to the sink.

STEP 3. Select any unlabeled node i. Initially only node 0 is labeled.

a. For any unlabeled node j for which $x_{ij} < f_{ij}$, assign the label $[i+,v_j]$ to node j, where $v_j = \min(v_i, f_{ij} - x_{ij})$. This limits the amount sent from i to j to either the amount already sent to i or the remaining capacity of the arc connecting i to j, whichever is smaller.

b. To all nodes j that are unlabeled and such that $x_{ji} > 0$, assign the label $[i-,v_j]$, where $v_j = \min(v_i, x_{ji})$. This enables us to reroute a flow going into i away from i. [For example, in Fig. 21.1 if we started with the initial solution $x_{01} = 2$, $x_{12} = 1$, $x_{20} = 1$, $x_{1m} = 1$, $x_{2m} = 0$, we have, starting with the origin label $[-,\infty]$, that the label for node 2 should be $[0-,1]$ since $v_2 = \min(\infty, x_{20} = 1) = 1$.] As node j receives a label, it is processed through Step 3 until all labeled nodes have been looked at with respect to the unlabeled nodes connecting them. Either this process will bring us along a path to the sink node which then gets a label (Step 4), or we cannot find such a path which leads to the sink (Step 5).

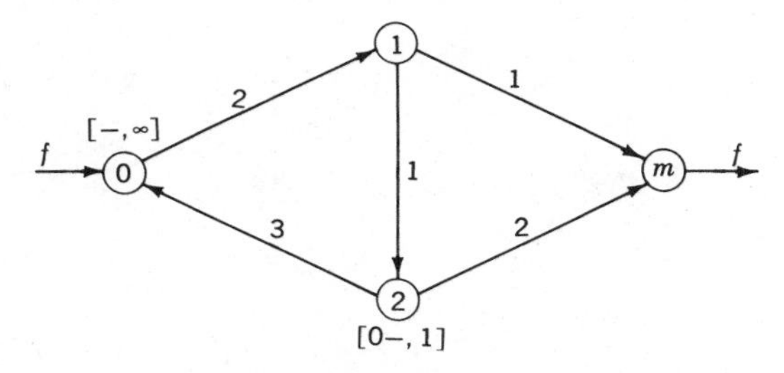

Figure 21.1

STEP 4. The sink node has received a label $[i+,v_m]$ with $i \neq m$ and $v_m > 0$ by the selection rule for v_m. Hence we have found what is termed a flow-augmenting path and can add or subtract v_m to the arcs of the path leading from 0 to m. The new feasible solution is

$$x'_{ij} = x_{ij} + v_m \quad \text{for } (i, j) \text{ in path and } j \text{ has label } [i+,v_j]$$

$$x'_{ji} = x_{ji} - v_m \quad \text{for } (j,i) \text{ in path and } j \text{ has label } [i-,v_j]$$

$$x'_{ij} = x_{ij} \quad \text{for } (i, j) \text{ not in path}$$

The labels are all erased, and we repeat the process starting with Step 2 and the new feasible flow x'_{ij}.

STEP 5. The process terminates in that we cannot find a path from source to sink. The maximal flow f is equal to the sum of all the v_m generated in the applications of Step 3, assuming initial solutions of all $x_{ij} = 0$.

We illustrate the algorithm by finding the value of the maximal flow for the network of Fig. 21.2. We let the initial feasible solution be $x_{ij} = 0$ for all i, j. We indicate the current feasible solution and the arc capacities by the ordered pair of numbers (x_{ij}, f_{ij}) attached to each arc of the network as shown in Fig. 21.2. The node labels are indicated by $[i\pm, v_j]$. Along the flow-augmenting path $(0,1,2,m)$ of Fig. 21.3 we can send an additional two units of flow ($v_m = 2$) which yields a new feasible solution as shown in Fig. 21.4. We again find an unsaturated path as shown by the labels of Figure 21.4. Here $v_m = 1$. The new feasible solution which sends a total of three units from source to sink is shown in Fig. 21.5, along with a new path over which an additional unit can be sent. The resulting feasible solution is shown in Fig. 21.6, along with a new unsaturated path for which $v_m = 2$. Note that the finding of the labeled path of Fig. 21.6 required an application of the rerouting Step 3*b*. The corresponding feasible solution is shown in Fig. 21.7. This solution has a flow of six units and is optimum as we cannot label any node except the origin. The corresponding cut is given by arcs (0,1), (0,2), (0,3) and has a cut capacity of 6.

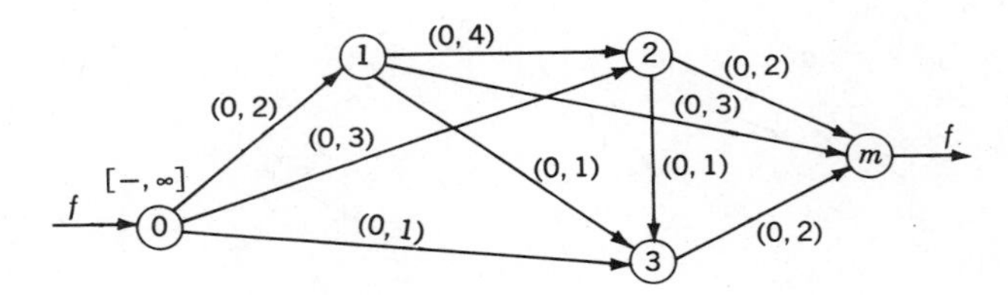

Figure 21.2

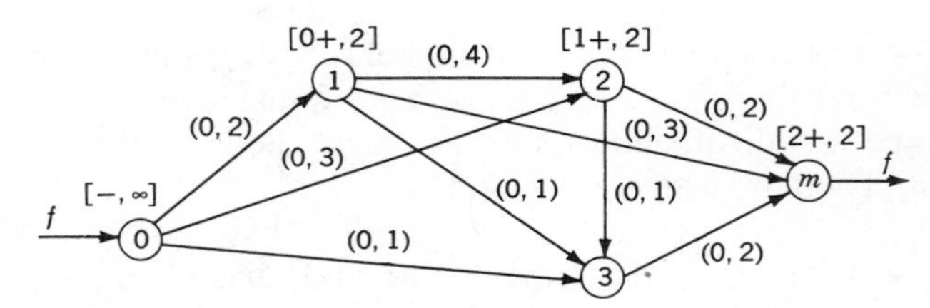

Figure 21.3

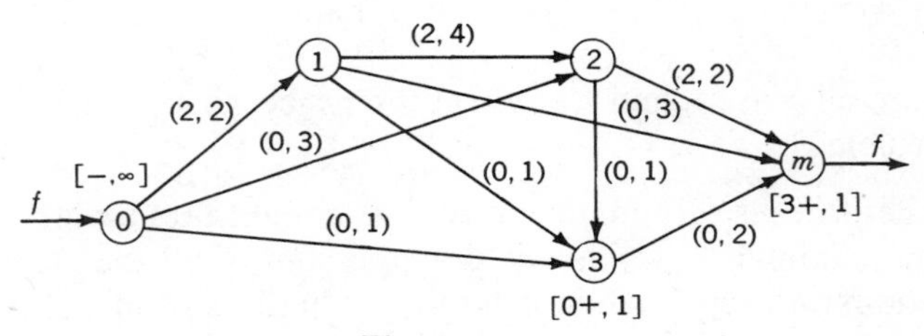

Figure 21.4

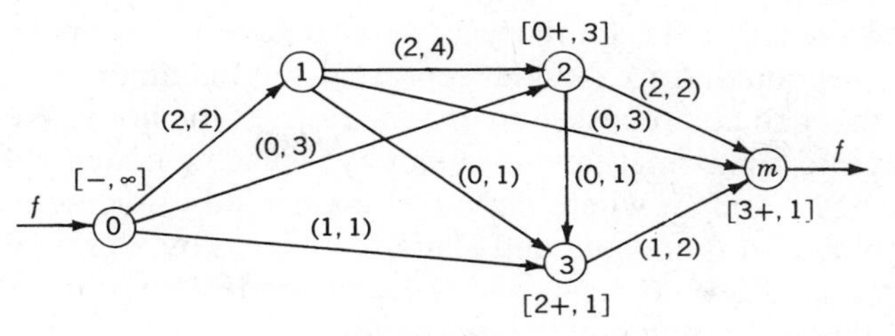

Figure 21.5

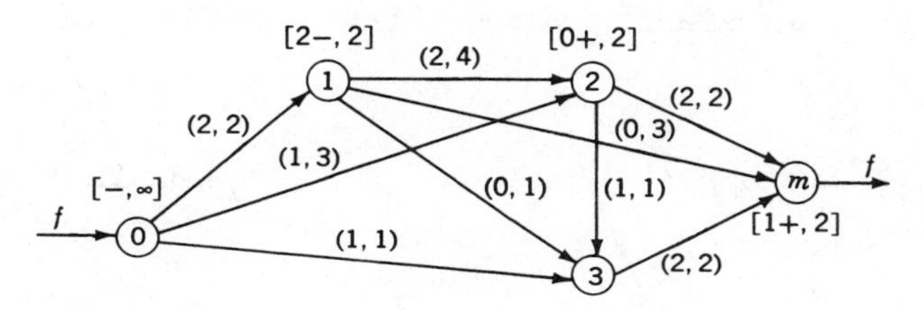

Figure 21.6

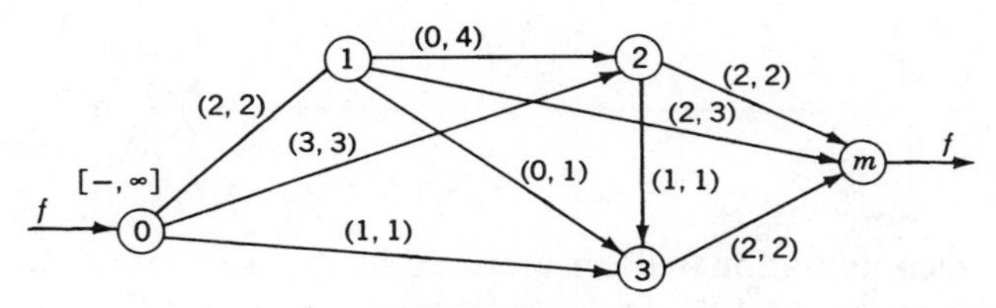

Figure 21.7

21.2. (1) Write out the explicit equations of the maximal-flow network shown below and (2) solve the problem using the maximal-flow algorithm.

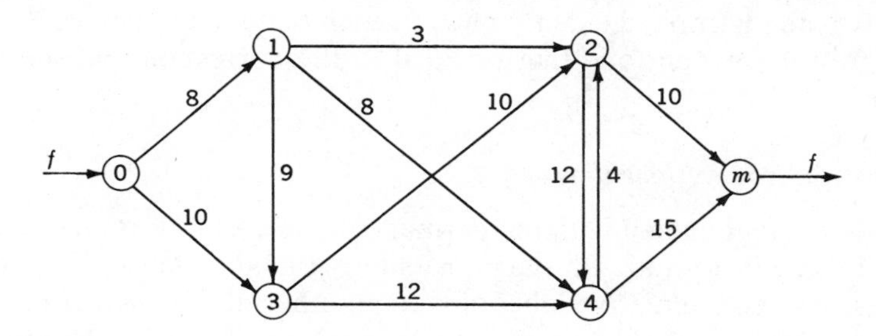

The maximal flow f is 18.

21.3. A number of modified maximal-flow network problems can be formulated in terms of the basic network model. For example, if a network has two (or more) sources, it can be recast into a single-source network by joining the given sources to a third source by artificial arcs which have infinite capacity, and similarly if we have multiple sinks. Another variation is when in addition to the capacities on the arcs we also have node capacities. These latter capacities could, for example, represent the maximum amount of the commodity that can be processed by the facilities at the node. We transform this problem into a maximal-flow problem by replacing node i with capacity k_i by two nodes i' and i'', where node i' is connected to the network by the same arcs going into i, nodes i' and i'' are connected by a directed arc from i' to i'' with capacity k_i (i.e., $f_{i'i''} = k_i$), and node i'' is connected to the network by the arcs going out of i.

Redraw the following maximal-flow network so that it has a single source and single node; also assume that node 3 has a capacity of k_3.

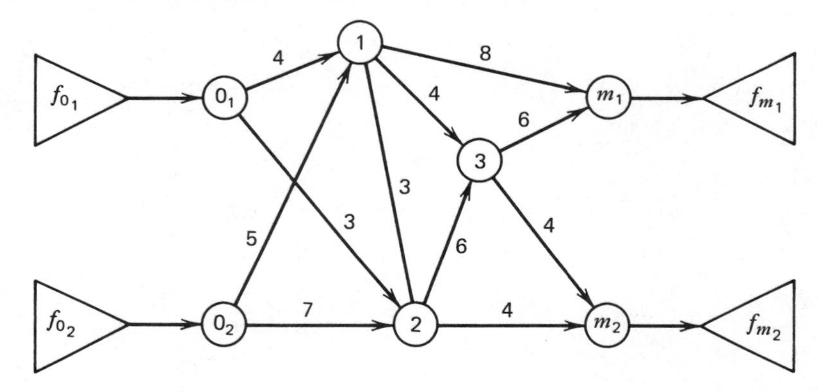

21.4. The computational effort required for the shortest-route algorithm described in Section 18.4 depends on how you organize the computations. The following statement of Dijkstra's algorithm is more efficient for finding the shortest routes from all nodes to the origin node.

The algorithm first assigns tentative labels to all nodes (these are upper bounds on the shortest distances from node 0 to all nodes). Then, a simple iterative step is applied exactly once to each node that changes the tentative label to a permanent one that is equal to the corresponding shortest-route distance.

Shortest-Route Algorithm

STEP 1. Label node 0 with the permanent value of 0. For all nodes that are linked directly to node 0, assign a tentative label equal to their direct distances. Assign a tentative label of ∞ to all other nodes. Select the minimum of the tentative labels and make it permanent. This label represents the shortest distance from node 0 to the new permanently labeled node. We now have two nodes with permanent labels—node 0 and a node connected directly to node 0.

STEP 2. For each node with a tentative label that is linked directly to any of the permanently labeled nodes, compare its tentative label with the sum of the permanent label plus the direct distance to the corresponding node. If the sum is less, make this value the tentative label. Do this for all tentatively labeled nodes that are linked to the permanently labeled nodes. Select the minimum tentative label and make it permanent. We have now added an additional node to the permanently labeled set. If this set contains the sink node, the process stops and we have found the shortest distance from node 0 to node m. If not, repeat Step 2. If we want the shortest routes from all nodes to the source node, continue the process until all nodes have permanent labels.

The arcs in the shortest-route can be determined if you keep track of the node from which each permanently labeled node was labeled. For a node j, you can add to your label a node indicator i that tells you that to get to j along the shortest route you must come from i.

We illustrate the above algorithm by solving the following shortest-route problem for an undirected network (from Phillips et al.):

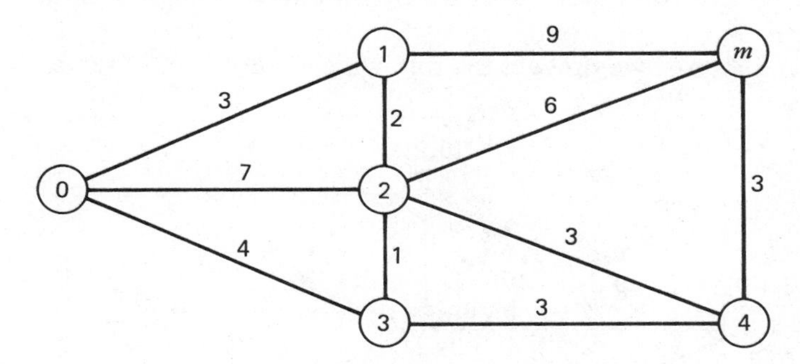

We shall use a two-part label $[i,d]$. The first part of the label will indicate the preceding node in the shortest path for a permanently labeled node, and the second number is either the tentative or permanent distance label. We initiate Step 1 and determine the following labeled network (an asterisk indicates a permanent label):

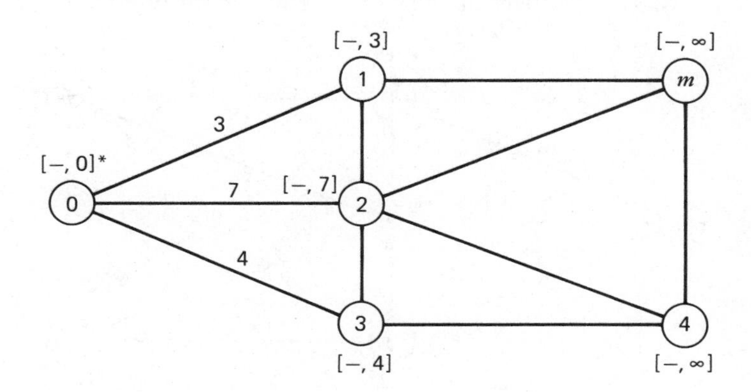

Selecting the minimum of the tentative labels to be made permanent, Step 1 concludes with

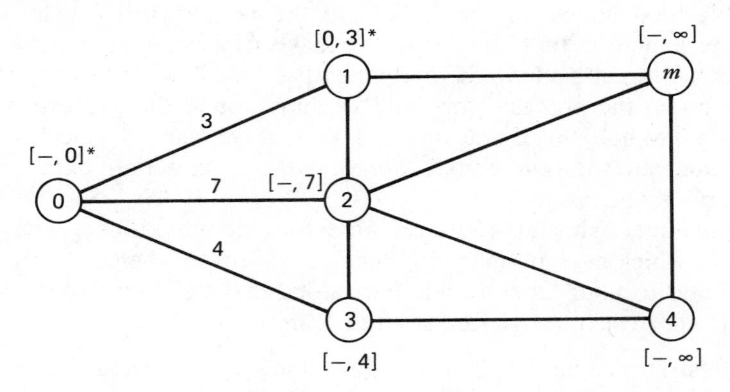

It should be clear that there is no other route from node 0 to node 1 that is less than 3.

Applying Step 2, we first get the following changed set of tentative labels (do you see why?):

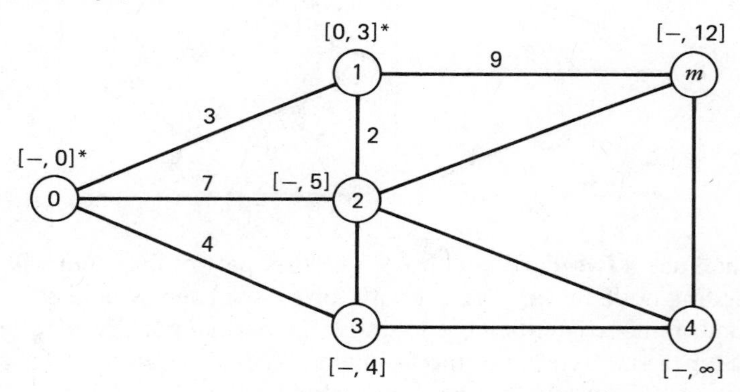

Completing Step 2, we permanently label node 3 with [0,4]:

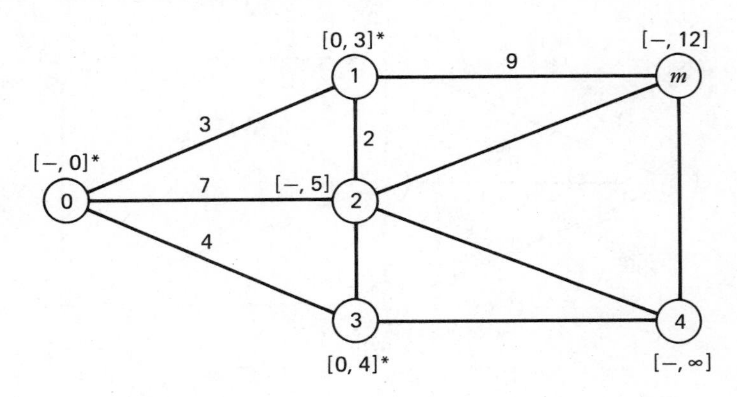

Repeating Step 2 yields the labeled network

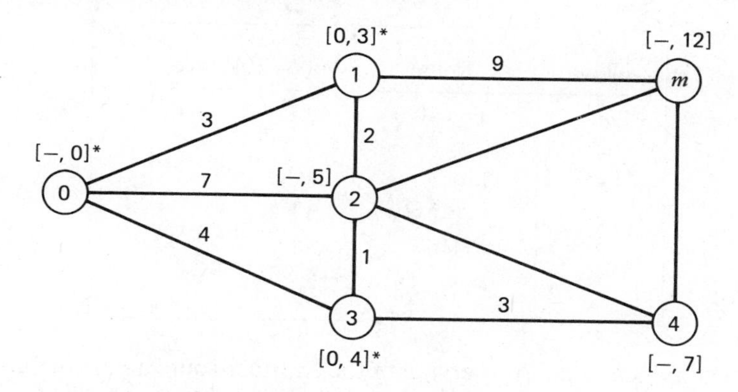

We now permanently label node 2 and note that there are two shortest routes from 0 to 2—(0–1–2) and (0–3–2), both with a distance of 5.

The new network with the permanent and tentative labels is now

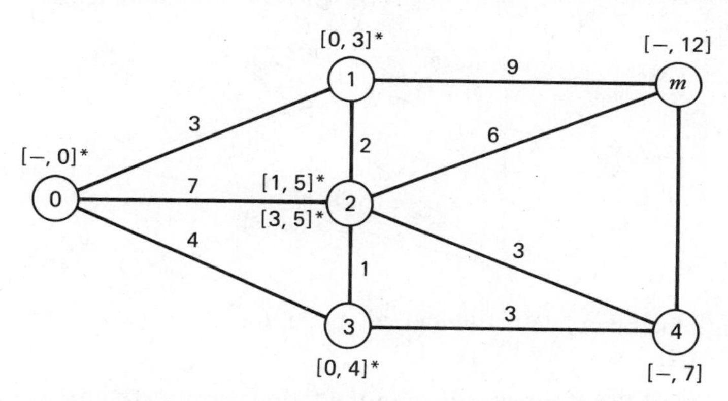

Step 2 yields only a change in the label of node m making it $[-,11]$. Node 4 is permanently labeled as its distance is 7. The final network with all permanent labels is

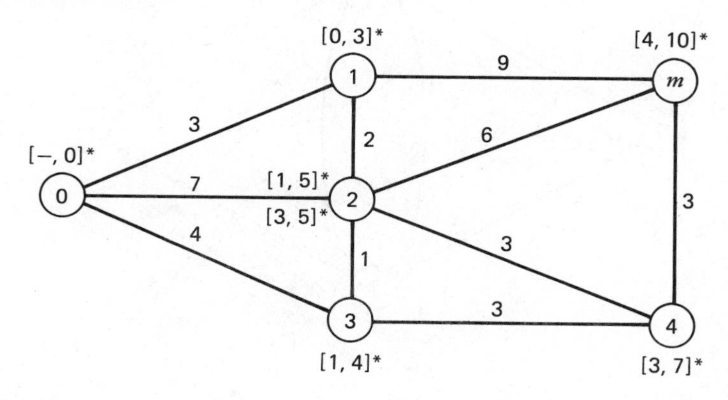

All the shortest routes are given below.

Node	Route	Distance
m	m–4–3–0	10
4	4–3–0	7
3	3–0	4
2	2–1–0	5 } multiple routes
2	2–3–0	5 } multiple routes
1	1–0	3

Are you convinced that this version of the shortest-route algorithm does find the shortest route?

21.5. Find the shortest-distance route from the origin to the destination in the following network:

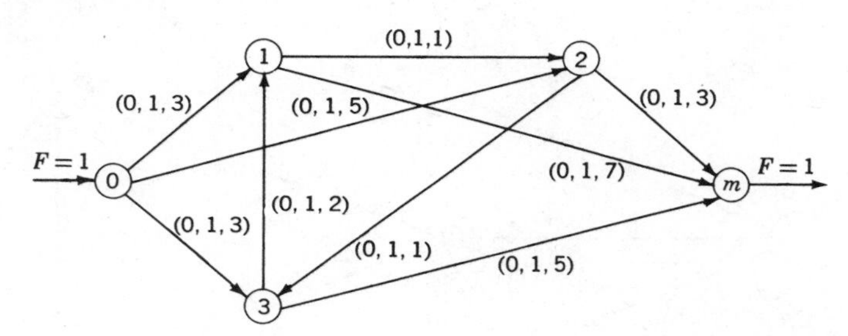

Formulate the complete mathematical model for this network. The shortest distance is 7.

21.6. Find the shortest routes in the following directed networks:

(a)

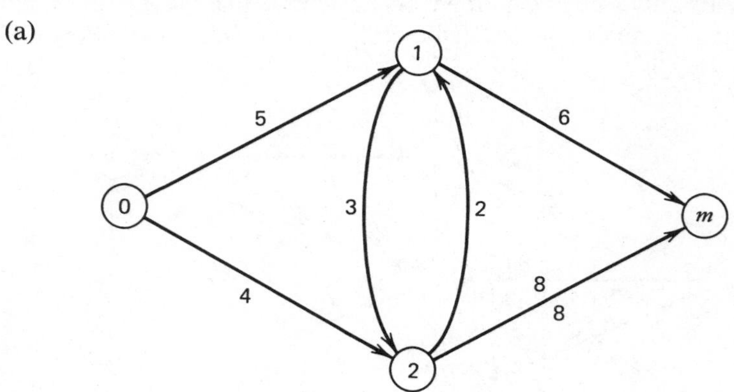

Shortest distance is 11.

(b)

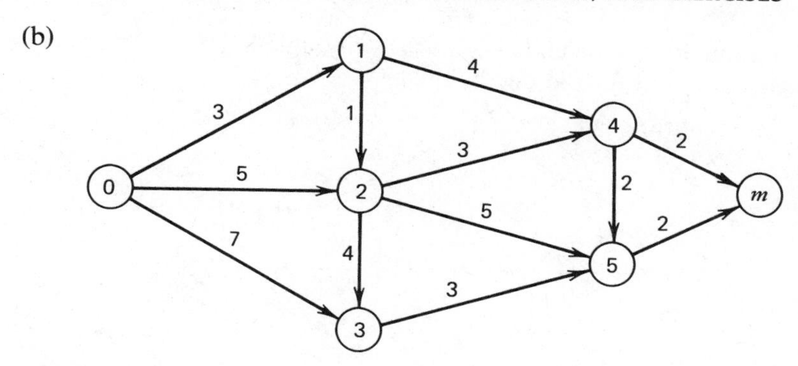

Shortest distance is 9.

21.7. Many applications can be analyzed by a special network termed an *acyclic network*. An acyclic network is a directed network in which there are no cycles, that is, once you leave a node you cannot return to it. The network

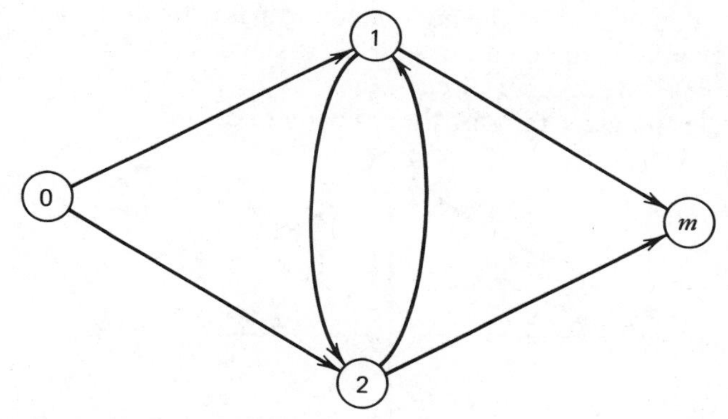

is not acyclic as it has the cycle 1–2–1. A key property of an acyclic network is that if the source node is numbered 0, you can find a sequential numbering scheme for all the other nodes such that every arc leads from a lower numbered node to a higher numbered node. Here we do not distinguish the sink node with the letter m; the sink node would be numbered with the highest number in the sequence. This property is illustrated in the following figure:

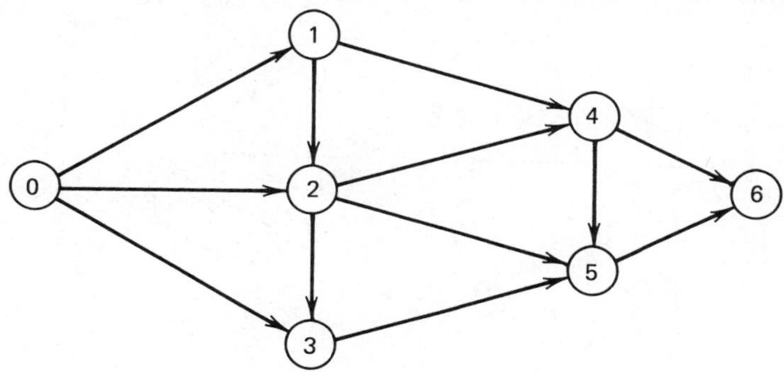

Such a numbering scheme is not necessarily unique. It is unique for the above figure, but not for the one below:

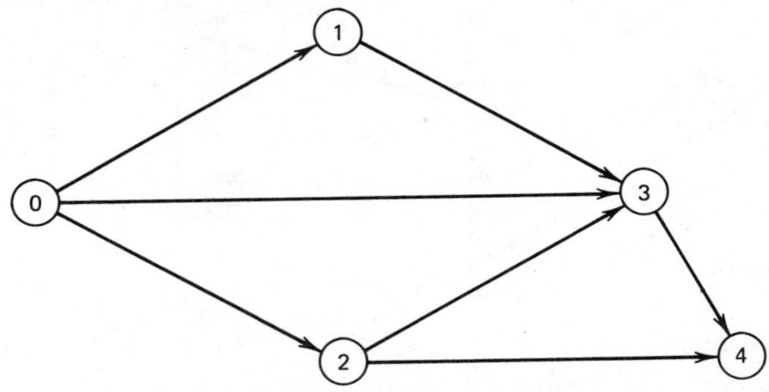

Node numbers 1 and 2 can be interchanged.

If a shortest-route problem is defined by an acyclic network, then the algorithm for finding all shortest routes from the source node 0 is greatly simplified. You can adapt the node-labeling algorithm to an acyclic network by noting that you calculate the shortest distances from node 0 in sequence by node number 1, 2, 3, . . . , each time giving the node a permanent label. We illustrate the algorithm for the following network:

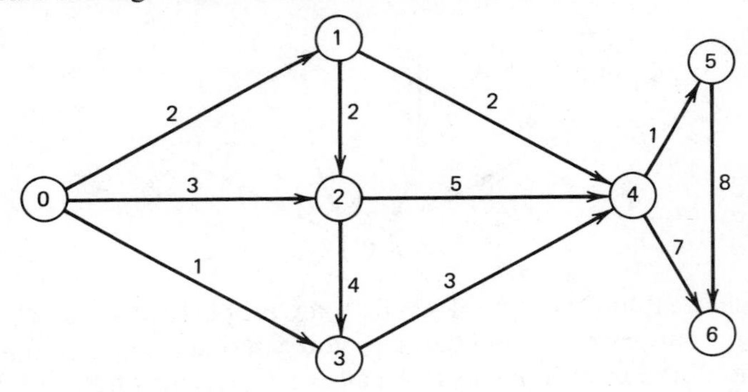

The permanent node labels are

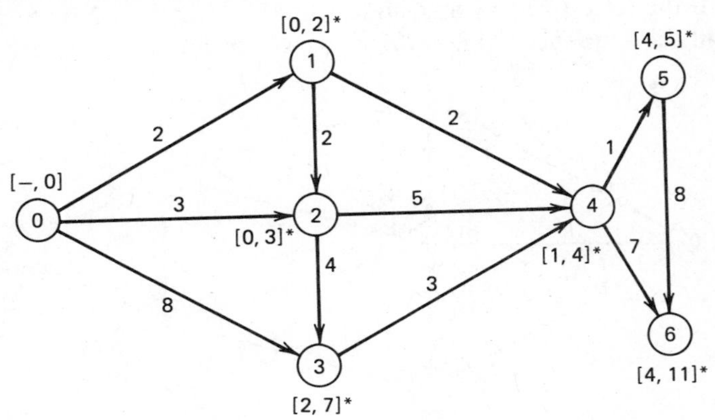

Note that the label for node 1 is immediate; the label for node 2 is derived by comparing the direct-route distance from node 0 to node 2 (value of 3) to the distance along the indirect route from node 1 (value of 2 + 2 = 4), and similarly, for node 3. The label at node 4 is obtained by comparing the three distances for the 1–4 route, 2–4 route, and 3–4 route, that is, we compare (2 + 2), (3 + 5), and (7 + 3), with the minimum being 4 along the route from node 1. We continue in this manner until we reach the last node 6. The shortest distances of each node from the source node is given by the second component of each label; the first component indicates the preceding node in the shortest route. The answers are summarized in the following table:

Node	Route	Distance
6	6–4–1–0	11
5	5–4–1–0	5
4	4–1–0	4
3	3–2–0	7
2	2–0	3
1	1–0	2

Write the algorithmic steps for finding all shortest routes in an acyclic network problem.

21.8. Number the nodes of the following network to show that it is acyclic. Apply your acyclic shortest-route algorithm to find the shortest routes from all nodes to the source node. The route from the source node to the sink node has a total distance of 9. (The problem is from Wagner.)

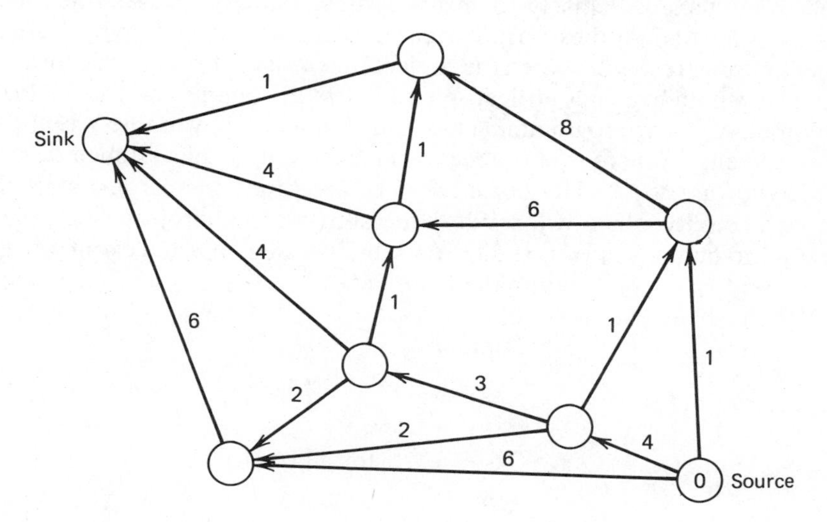

21.9. There are important applications that require the finding of the longest route in an acyclic network. You should be able to convert the acyclic network shortest-route algorithm so it finds the longest routes from all the nodes to the origin node. Describe the steps of the modified algorithm and apply it to the network of Section 21.11.

21.10. A particular application which considers the longest route is that of the management techniques termed PERT (program evaluation and review technique) or CPM (critical-path method), which are used to analyze the scheduling of construction, manufacturing, and other complex projects. The building of a house is a good example. A project is divided into well-defined, independent activities (put up walls, lay shingles) and important events or milestones (painting finished, roof completed). Each activity takes a certain length of time, and the activities are ordered in that the walls must be put up before they are painted. If we picture the activities leading from one event to another and let activities be denoted by directed arcs and events by nodes, then, due to the ordering requirements of the activities, the total project can be viewed as an acyclic network. The following figure is a PERT network, with the distances interpreted as the time to complete an activity, node 0 the start of the project, node m the completion of the project, and the intermediate nodes specific events.

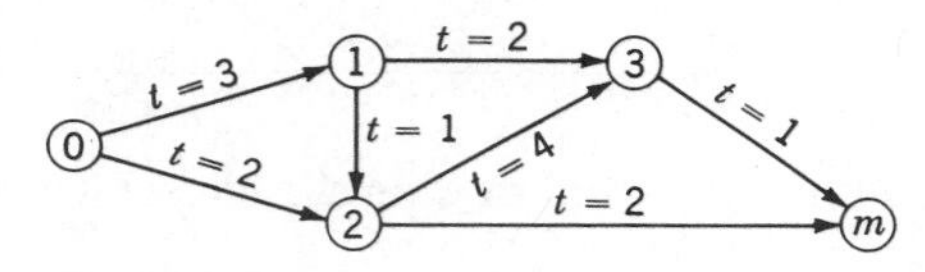

The minimum time required to complete the entire project is the largest total time for any path from node 0 to node m. The corresponding path is termed the *critical path* and is the longest route (time) from node 0 to node m. The activities on the critical path represent the set of activities which need to be managed closely to ensure against slippage in the project-completion time. Also, if these critical activities can be speeded up by a greater expenditure of resources, then the project time will be decreased accordingly until a new path becomes critical. We can view the timing aspects of a PERT network as a linear-programming model as follows. Let variable x_i represent the occurrence time of event (node) i and let $t_{ij} \geq 0$ be the assumed duration of activity (arc) (i, j). The objective is to determine the earliest start time for each activity which will result in the shortest total project time; that is, we want to minimize the time difference between event 0 and event m which is given by $x_m - x_0$. The problem is then:

$$\text{Minimize} \quad x_m - x_0$$

subject to

$$x_j - x_i \geq t_{ij} \qquad \text{for all } (i, j)$$

with the $x_j \geq 0$. As the start time of node 0 is arbitrary, it can be set equal to zero. The dual of this problem is the finding of the longest path in the network from node 0 to node m, which corresponds to the critical path. By the duality theorem the value of the longest path will be equal to the minimum value of x_m. We illustrate the above formulation for the PERT network figure, where the t on each arc represents the activity time:

$$\text{Minimize } x_m$$

subject to

$$\begin{aligned} x_1 \quad\quad\; &\geq 3 \\ x_2 \quad\quad\; &\geq 2 \\ x_2 - x_1 &\geq 1 \\ x_3 - x_1 &\geq 2 \\ x_3 - x_2 &\geq 4 \\ x_m - x_2 &\geq 2 \\ x_m - x_3 &\geq 1 \\ x_j &\geq 0 \end{aligned}$$

The optimal answer is $x_1 = 3$, $x_2 = 4$, $x_3 = 8$, $x_m = 9$, with the critical path of (0,1), (1,2), (2,3), (3,m). State and solve the corresponding dual problem.

21.11. Assume the following acyclic network describes a building project. Number the nodes and find the critical (longest) path.

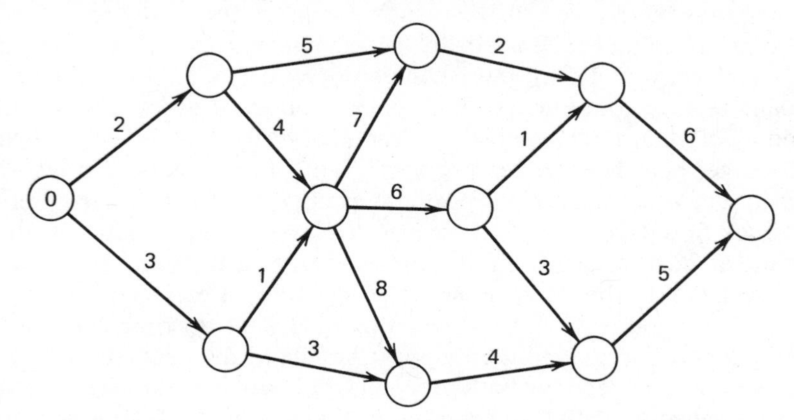

21.12. Describe a project of your choice in terms of activities and events, and construct the corresponding PERT network. You might consider all the steps that a team of analysts employ to solve a decision problem, or the steps required to produce a play put on by the drama department.

21.13. The operations officer at an air base wants to find a flight path across which an attack bomber can be sent to the target. Each leg of the path is exposed to mobile enemy surface-to-air missiles. Using daily intelligence information, the operations officer can estimate, on an increasing scale of 1 to 10, the level of danger of each possible leg. The officer has organized the legs and exposure levels into the following acyclic network. Your problem is to number the nodes correctly and find the minimum exposure path. The bomber is restricted to certain turns and directions, and thus, not all legs between the nodes are possible. How would you get the bomber back to the base?

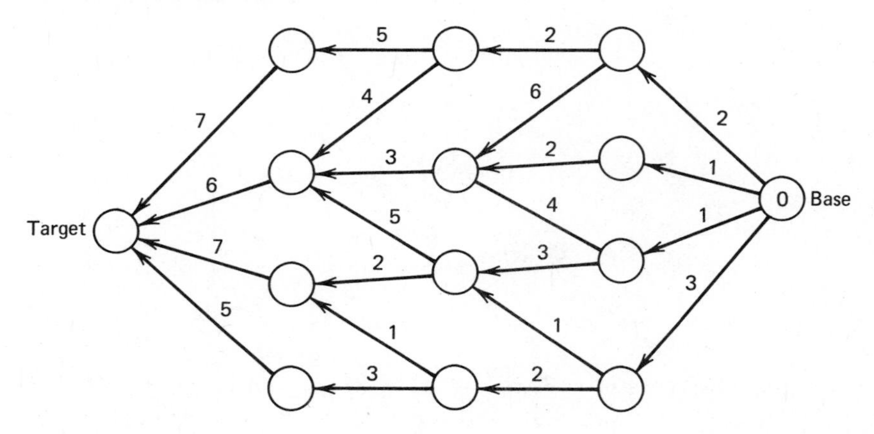

21.14. Formulate the following problem as a shortest-route acyclic-network problem. A trucking company wishes to determine a leasing plan so it will always have a certain size truck available during the next 5 years. The company can meet its needs by leasing a new truck at the beginning of year 1 and keeping it until the beginning of year $t \leq 5$. If $t < 5$, then the company replaces the truck at the beginning of year t and keeps it until the beginning of year $t' \leq 5$, and so on. The cost c_{ij} for $1 \leq i < j \leq 5$ includes the rental fee plus the expected operating maintenance costs of a truck leased at the start of year i and replaced at the start of year j. The aim is to determine the "truck flow" from year 1 to year 5 so as to minimize total cost, that is, the minimal-cost leasing plan. The nodes represent the years replacements could occur. You should have an arc going from an early-year node to all the later-year nodes. Could you use this model to determine a 5-year replacement plan for an auto that you just bought at the beginning of year 1? What data do you need for both problems and how accurate are such data? Outline a sensitivity study for the truck-leasing model. How does the data change if the company plans to buy a truck instead of leasing and there is a trade-in value that depends on when a truck is replaced? (The replacement problem is described in Wagner, and Phillips et al.)

21.15. Construct an acyclic network that takes you from your home to school. Collect two sets of data: (1) the arc distances in actual miles; and

(2) average driving or walking or biking time across each of the arcs. Solve both problems.

Many highway-traffic studies are concerned with the movement of individuals from node positions outside of the central business district (CBD) (source nodes) to nodes within the CBD (sink nodes). Describe how you can decompose your town into a number of sink and source nodes with interconnecting directed arcs. You need to assume everyone is trying to find the shortest-time routes. How could you use such a network to study a new highway link? Traffic assignment problems deal with two broad extremal principles that were first stated by the engineer J. G. Wardrop. They are:

1. Flow distributes to produce an equilibrium condition in which no user can reduce his or her transport cost by a unilateral change of route.
2. Flow through a transportation network distributes to minimize the sum total of network costs.

The first principle deals with traffic patterns that are user optimized while the second is for system-optimized patterns. These principles lead to the study of two network-flow problems: the finding of the cheapest routes on a network (user-oriented shortest-route problem) and the minimization of total network cost (systems-oriented minimal-cost network problem). To study these problems in correct detail for a highway, bus, or subway system, we would have to determine the number of people (users) that must flow from each source node to each sink node, worry about the flow rates changing as volume increases, determine the effects of capacities of the arcs (traffic jams), and so on. Then, you might have to solve such problems for different hours of the day. One relationship between travel volume V and travel time T is given by Oppenheim:

$$T_k = T_0 \left(1 + 0.15 \frac{V_k}{C}\right)^4$$

or

$$T_k = 2T_0 \frac{V_k}{C}$$

where T_0 is the minimum travel time corresponding to no traffic, V_k the current volume at time k, and C the maximum capacity of the arc. The second formula is an approximation of the first obtained by dropping low-order terms. Check it out. How would you go about determining the accuracy of this formula? How would you use it to solve a two-time-period minimal-cost flow problem? Traffic analysts use the term origin and destination ($O - D$) for source and sink.

21.16. The following is a traffic network through which 9 units of flow must go from the origin node 0 to the destination node m.

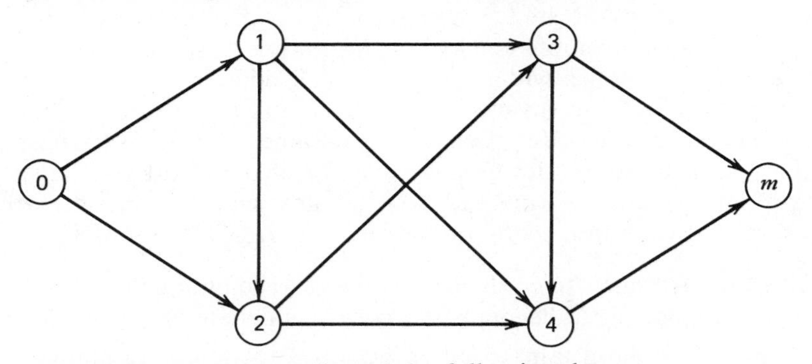

The traffic engineers have collected the following data:

Arc	Arc capacity	Arc cost/unit flow	Optimal solutions	
01	6	1	6	6
02	4	5	3	3
12	4	1	0	0
13	6	3	6	6
14	1	7	0	0
23	3	1	3	3
24	4	6	0	0
34	6	3	6	5
3m	4	6	3	4
4m	6	3	6	5

Write out the system-wide, minimal-cost network-flow model and check to see if the two optimal solutions listed satisfy your flow and capacity constraints. Calculate the total system cost. (This problem is from Potts and Oliver. Using your formulation, solve the problem if you have a computer-based simplex algorithm.)

21.17. For the automobile manufacturer problem of Section 18.3, we are given the following data:

Plant	Plant production/cost			Total
	Model 1	Model 2	Model 3	
P_1	300/$4000	200/$5000	100/$6000	600
P_2	400/$4025	150/$4800	150/$5900	700

Distribution center	Demand			Total
	Model 1	Model 2	Model 3	
D_1	400	175	130	705
D_2	300	180	115	595

	Transportation costs (same for all models)	
Plant	D_1	D_2
P_1	\$200	\$225
P_2	\$180	\$240

Write out the network-flow model, assuming that there are no lower and upper bounds on arc capacities except those implied by the data. (Solve the problem if you have a computer-based simplex algorithm.) Modify your formulation to determine how much of each model should be produced at each plant. Also, how could you determine where you should open or close a plant or distribution center, or increase or decrease production capacity?

21.18. The following is one way of writing a mathematical model of the traveling-salesman problem. Define a tour as a sequence of cities starting at city 1 that goes through all other cities exactly once and returns to city 1 at the end. Let

$x_{ij} = 1$ if a tour requires the salesman to travel from city i directly to city j

$x_{ij} = 0$ otherwise

The salesman is to enter each city exactly once:

$$\sum_{i=1}^{n} x_{ij} = 1 \qquad j = 1, \ldots, n$$

The salesman is to leave each city exactly once:

$$\sum_{j=1}^{n} x_{ij} = 1 \qquad i = 1, \ldots, n$$

Also, the $x_{ij} = 1$ must form a tour. We want to minimize

$$x_0 = \sum_{i=1}^{n} \sum_{j=1}^{n} c_{ij} x_{ij}$$

This formulation assumes that the salesman can reach any city from any other city. The distance table can be asymmetric, that is, we can have some $c_{ij} \neq c_{ji}$ for $i \neq j$. There are other formulations that define explicitly the condition that the $x_{ij} = 1$ must form a tour. If the tour requirement is left out, the model is an assignment problem.

If the distance table is asymmetrical, then there are $(n - 1)!$ different tours (all starting at city 1); if the table is symmetrical, then the total number of different tours is reduced to $(n - 1)!/2$. Many traveling-salesman-type problems occur in production settings, for example, the transporting of goods between factories, warehouses, depots, and customers; the movement of materials; and the routing of watchmen. Other application areas include the routing of delivery vehicles, messenger delivery and pickup, special bus routes, and so on. Can you think of a traveling-salesman problem that you need to solve? How about shopping tours?

21.19. In telling the story about the mathematical horse, Willie Simon might have been aware of the arithmetical feats of "Clever Hans" and his friends, as reported in "Significa," from *Parade* Magazine, June 1982, by Irving Wallace, David Wallechinsky, and Amy Wallace.

THE AMAZING CALCULATING HORSES

How long would it take you to subtract $\sqrt{12{,}769}$ from $\sqrt{15{,}876}$—without a calculator? Muhamed, an Arabian stallion, could tap out the correct answer (13) with his hooves before any of his human observers could solve the problem.

In 1895, Wilhelm von Osten of Elberfeld, Germany, started studying the mental powers of animals. His first student, a bear, was a dull pupil. But in 1900, he purchased a Russian stallion that soon earned the nickname "Clever Hans." Pawing with his hooves, Hans could add, subtract, divide, do fractions, and even spell and form sentences using a special tapping code. The horse's feats caused enormous controversy. A committee of experts found nothing to criticize.

Oskar Pfungst, member of a second investigating committee, felt differently. He published a book denouncing von Osten's claim that animals could be educated and argued that Hans was merely responding to subtle movements by his owner.

Von Osten died in 1909, willing Hans to Karl Krall, a manufacturer. Krall added two Arabian stallions, Muhamed and Zarif, and a Shetland pony named Hänschen and brought their abilities to new levels. A fifth horse, Berto, was blind but could perform like the others—thus dashing Pfungst's theory of visual cues. Despite all the neigh-saying, no one could prove that the five horses were a hoax.

21.20. Construct a 10-city random distance table and corresponding map by the following procedure. On a piece of graph paper, draw the usual horizontal and vertical axes with the origin (0,0) as far into the lower left-hand corner as possible. Using a $\frac{3}{4}$-inch per-mile scale, mark off 0–10 miles on each axis. Write the numbers 1–10 on ten separate, similar pieces of paper and put them into an envelope. Draw a number from the envelope 20 times, recording the number and replacing it after each draw. Use the numbers of the first two draws as coordinates of city 1, the next two for city 2, and so on. This process enables you to select 10 points at random to be used as cities for a traveling-salesman problem. Randomly drawn problems are used to test out algorithms.

Plot the points and construct the distance table. You will need to measure 45 distances, as your table will be symmetric. Use this problem to test out any algorithm you develop. You should exchange random-city problems with other students to obtain more information on how good your algorithm is.

21.21. Even though we may not have an exact algorithm for determining an optimal solution, there are a number of advantages gained by structuring the decision problem as a mathematical model. The explicit form of the set-covering museum-problem model highlights the possible feasible solutions and enables us to choose guard positions that can be combined to cover all the walls. For the traveling-salesman problem, the mathematical model does not help much except to indicate its close relationship to the assignment problem. As we try to develop heuristic algorithms for the traveling-salesman problem, we automatically try to satisfy the assignment conditions.

The traveling-salesman problem, the shortest-route problem, and many other decision problems are solved on a regular basis by people unschooled in optimization techniques. Certainly, the letter carrier, newspaperperson, and others who encounter such problems apply heuristic procedures based on their experiences; good decisions are derived and applied. They probably do not even think about the situation as a decision problem. When decision problems are not too complex or do not involve many alternatives, we find that experienced persons are able to manage the problem without resorting to optimizing algorithmic procedures. The human mind is a great problem solver. In many instances, when algorithms are applied to ongoing situations to find an optimal solution, the heuristic solution being used turns out to be a very good one. But when problems need to be solved repeatedly, and costs and profits are of concern, routinized algorithms that guarantee the optimum—or at least a very good solution—can be of great value to the managers of the system.

How have you done with your traveling-salesman heuristic algorithms? You should discuss them in class and demonstrate their efficacy by solving your random-city problem. The most simple of heuristic algorithms for the traveling-salesman problem is the following, termed the *nearest-neighbor algorithm*; it is really a greedy-type algorithm. It systematically selects nodes one at a time until a tour is formed. It can be applied to any distance table. Did you think of it?

Nearest-Neighbor Algorithm

STEP 1. Select any node. This is the first node of the set that will define the tour.

STEP 2. Find the node closest to the last node added to the set. Add this node in order to the set.

STEP 3. Repeat Step 2 until all nodes are included in the set; then join the first and last nodes to form a tour.

Try this procedure out on the five-city problem and your random 10-city problem. This algorithm depends quite a bit on the starting node. Can you think of a decision rule to select the starting node? The total number of computational steps is of the order of n^2. A variation of the nearest-neighbor algorithm is to replace Step 2 by a procedure that selects the best node to be inserted into the subtour formed by the nodes already selected. We leave it to you to think about this idea and to construct an insertion algorithm and test it out. Write the steps of your insertion algorithm.

Convex-Hull Algorithm

Take your random 10-city map and connect the outside city points, that is, form a boundary so that all points are either on the boundary or interior to the polygon formed by the boundary (see figure below). The boundary is called the *convex hull* of the points. The convex hull is an initial subtour. There is a theorem that states that the order in which the points on the boundary are connected is the same as in the optimal traveling-salesman tour. This theorem holds for a symmetric distance table and when the distances are measured in the usual Euclidean-plane sense. Given the convex hull, your problem is to design an algorithm that joins the interior points to the subtour formed by the boundary, keeping the order of the boundary points. For example, city 2 or 3 would not come between cities 5 and 1. You would insert city 4 between cities 5 and 1 if $c_{14} + c_{45} - c_{15}$ is less than the corresponding distance changes if we insert 4 between 2 and 5, or 2 and

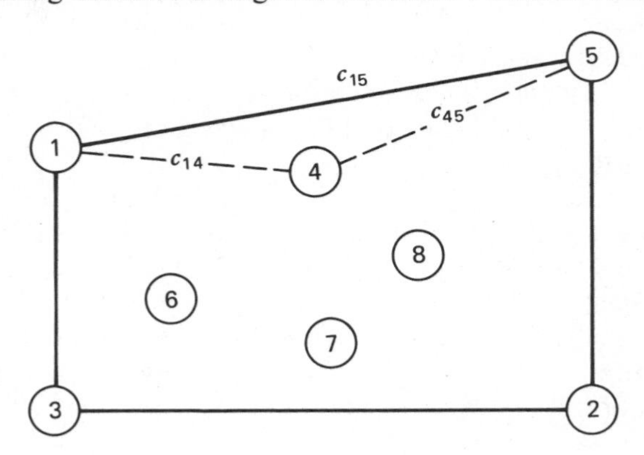

3, or 1 and 3. Write the steps for your convex-hull algorithm and apply it to your 10-city random problem.

Most traveling-salesman problems assume a symmetric distance table, that the cities (points) lie on a plane, and that the triangle inequality holds, that is, for cities i and j, the direct-route distance c_{ij} is less than an indirect route passing through any other city k. For the following figure this means that $c_{ij} \leq c_{ki} + c_{jk}$.

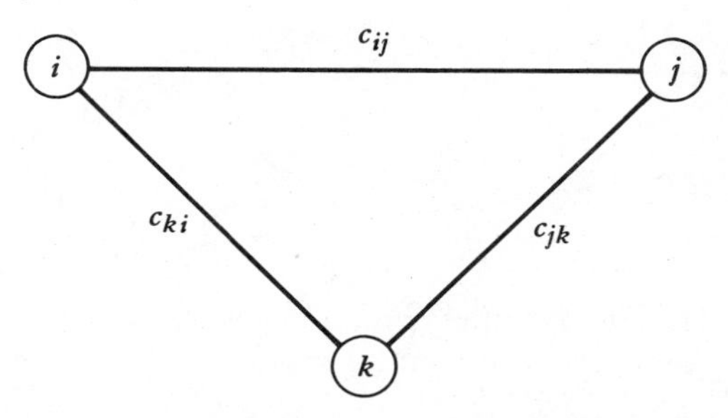

A traveling-salesman problem with Euclidean distances has a symmetric distance table and the distances satisfy the triangle inequality. The optimum tour for such a problem does not intersect itself, that is, you cannot have a tour like

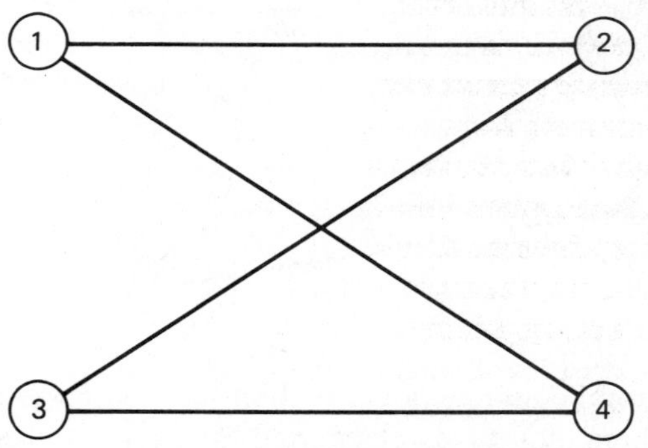

Can you prove why? You also should be able to prove why the points on the convex hull have to appear in the same order in the optimal tour. The proofs rely on the triangle-inequality assumption. If the "distances" are travel times between the cities, we could have an asymmetric table. (If the distance matrix is symmetric and satisfies the triangle inequality, the greedy solution will be less than twice the distance of the optimal tour, see Tucker.)

Tour-Improvement Algorithm

This algorithm for symmetric distance problems starts with any tour (e.g., a random solution or one found by the nearest-neighbor algorithm). It then tries to exchange arcs that connect nodes in the tour with other arcs not in the tour whose distance sum is less than the arcs being removed from the tour. A tour is always maintained. For example, if we start with the five-city tour (1–2–3–4–5–1), we might start by removing any two arcs that do

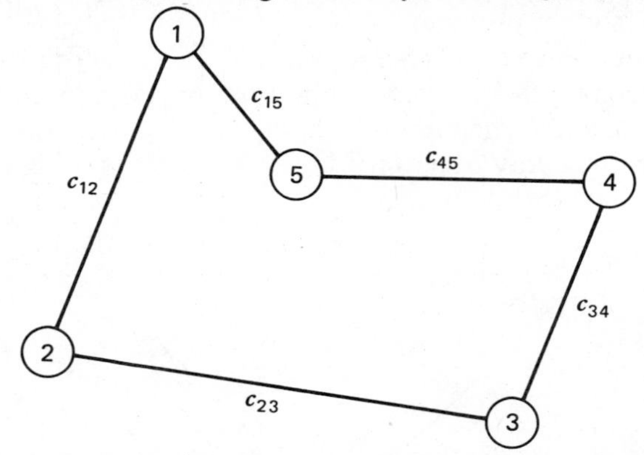

not meet at a node, for example, arc (1,5) and arc (3,4), and reconnect the tour by inserting appropriate arcs (1,4) and (3,5) as shown.

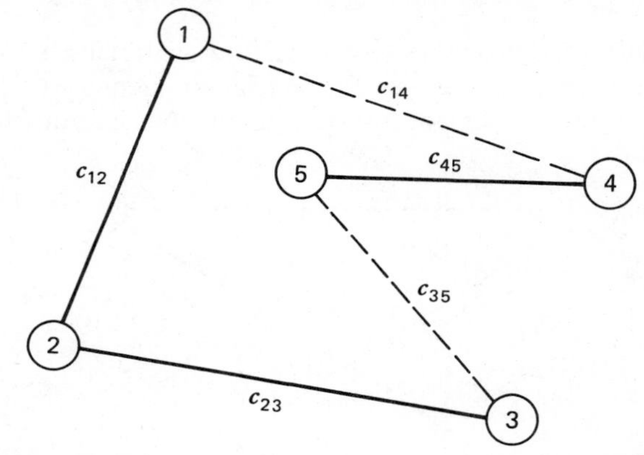

If the distance savings $s = c_{15} + c_{34} - c_{14} - c_{35} > 0$, we have found a new tour whose total distance is s less than the original tour's distance. An exchange of the arcs then takes place, and the process is continued until all two-arc exchanges have been evaluated. Write up the steps of a two-arc exchange algorithm and apply it to your random 10-city problem.

21.22. Use one of your algorithms to find a traveling-salesman's optimal route for the following symmetric four-city problem. The salesman lives in city 1.

City	1	2	3	4
1	—	120	140	180
2	—	—	70	100
3	—	—	—	110

You can show that the optimal solution is (1,2,4,3,1) with a distance of 470 by enumerating all possible routes. How many are there? (This problem is from Ore.)

21.23. Solve the five-city traveling-salesman problem defined by the following distance table:

City	1	2	3	4	5
1	—	11	6	6	4
2	7	—	10	6	12
3	11	8	—	11	13
4	4	10	6	—	8
5	4	9	8	4	—

The optimal tour is (1,5,4,3,2,1) with a distance of 29.

21.24. Consider the following 10-city traveling-salesman problem defined by the following nonsymmetric distance table:

City	1	2	3	4	5	6	7	8	9	10
1	—	24	18	22	31	19	33	25	30	26
2	15	—	19	27	26	32	25	31	28	18
3	22	23	—	23	16	29	27	18	16	27
4	24	31	18	—	19	13	28	9	19	27
5	23	18	34	20	—	31	24	15	25	8
6	24	12	17	15	10	—	11	16	21	31
7	28	15	27	35	19	18	—	21	21	19
8	13	24	18	13	13	22	25	—	29	24
9	17	21	18	24	27	24	34	31	—	18
10	18	19	29	16	23	17	18	31	23	—

$= (c_{ij})$

If you tried to solve this by treating it as an assignment problem, the optimal solution would consist of three subtours: (1,6,7,2), (3,9), and (4,8,5,10) with a total distance of 140. The subtours are connected as follows:

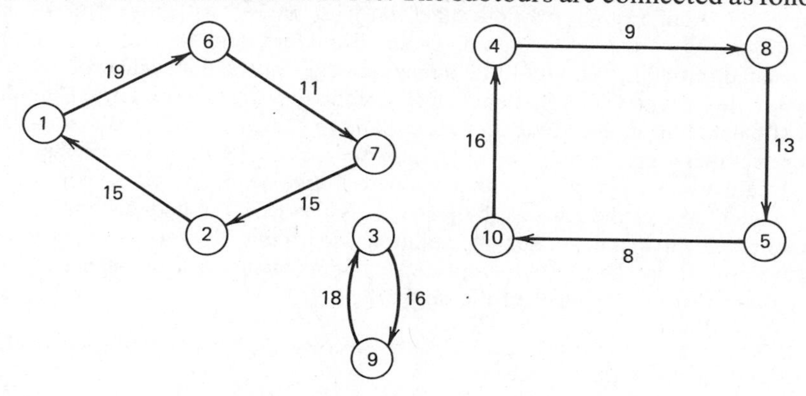

Using these three subtours, develop a procedure for linking them together in the best manner. Systemize the decisions that have to be made. The optimal tour is (1,3,9,4,8,5,10,6,7,2,1) with a total distance of 146. (This problem is from Lawler and Wood.) Try your algorithm using the two subtours (1,4,5) and (2,3) of the five-city problem:

City	1	2	3	4	5
1	—	17	10	15	17
2	18	—	6	12	20
3	12	5	—	14	19
4	12	11	15	—	7
5	16	21	18	6	—

21.25. A traveling-salesman's tour that passes through each city once is also called a *Hamiltonian circuit,* named after the mathematician Sir William Hamilton. He first studied such circuits by determining one that passes through the 20 vertices of a dodecahedron, with the only allowable routes being along the edges of the figure. In his book *Graphs and Their Uses*, the mathematician Oystein Ore relates the following story about Hamilton's interest in the problem:[1]

> In the year 1859 the famous Irish mathematician Sir William Rowan Hamilton put on the market a peculiar puzzle. Its main part was a regular dodecahedron made of wood. This is one of the so-called regular Platonic bodies, a polyhedron having regular pentagons for its 12 faces, with three edges of these pentagons meeting at each of the 20 corners.

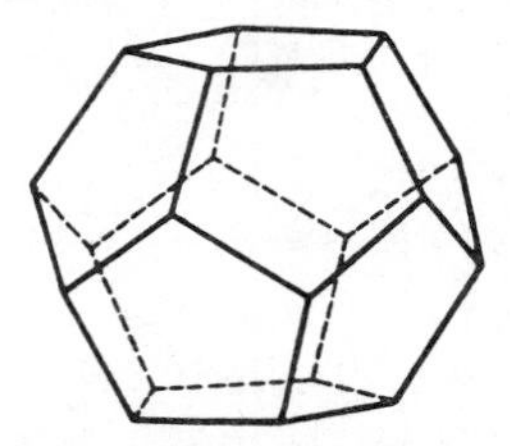

> Each corner of Hamilton's dodecahedron was marked with the name of an important city: Brussels, Canton, Delhi, Frankfurt, and so on. The puzzle consisted in finding a travel route along the edges of the dodecahedron which passed through each city just once; a few of the first cities to be visited should be stipulated in advance to render the task more challenging. To make it easier to remember which passages had already been completed, each corner of the dodecahedron was provided with a nail with a large head, so that a string could be wound around the nails as the journey progressed. The dodecahedron was cumbersome to maneuver so that Hamilton also produced a version of his game in which the polyhedron was replaced by a planar graph isomorphic to the graph formed by the edges of the dodecahedron.

[1] O. Ore, *Graphs and Their Uses,* Random House, New York, N.Y., 1963. Copyright Yale University. Reproduced with permission.

There is no indication that the Traveller's Dodecahedron had any great public success, but mathematicians have preserved a permanent memento of the puzzle: A Hamilton line in a graph is a circuit that passes through each of the vertices exactly once. It does not, in general, cover all the edges; in fact, it covers only two edges at each vertex.''

What are the differences between Hamilton's problem and the usual traveling-salesman problem? Note that Hamilton's puzzle is not a decision problem. Construct the two- and three-dimensional versions of Hamilton's puzzle.

21.26. The following rules have been suggested for trying to find a Hamiltonian circuit (Tucker):

1. If a node i has two arcs, both arcs incident at i must be part of any Hamiltonian circuit.
2. No subcircuit, that is, a circuit not containing all vertices, can be formed when constructing a Hamiltonian circuit.
3. Once the Hamiltonian circuit has passed through a node i, all other (noncircuit) arcs incident at i can be deleted since they cannot be used later in the circuit.

Use these rules to show that the following networks have no Hamiltonian circuits (from Tucker):

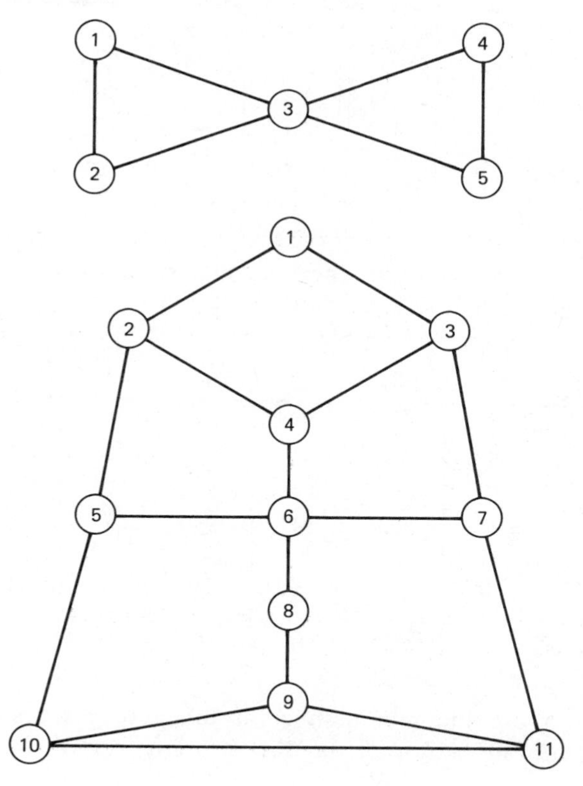

21.27. There is a simple way to find a lower bound for the value of the minimal-distance tour of a traveling-salesman problem. (We already know how to obtain one lower bound by solving the assignment problem associated with the traveling-salesman problem.) We illustrate the approach using the following distance table whose optimal-tour length is 20 (the problem is from Garfinkel and Nemhauser):

D_1 =

City	1	2	3	4	5	6
1	—	7	3	12	5	8
2	4	—	2	10	9	3
3	6	7	—	11	1	7
4	7	3	1	—	8	8
5	2	10	2	7	—	3
6	4	11	7	6	3	—

First, subtract the smallest number in a row from all other numbers in that row; do it for all rows to obtain the transformed table

D_2 =

City	1	2	3	4	5	6	Min
1	—	4	0	9	2	5	3
2	2	—	0	8	7	1	2
3	5	6	—	10	0	6	1
4	6	2	0	—	7	7	1
5	0	8	0	5	—	1	2
6	1	8	4	3	0	—	3

If possible, do the same for the columns. Here we can only do it for columns 2, 4, and 6 to obtain the table

D_3 =

City	1	2	3	4	5	6	Min
1	—	2	0	6	2	4	3
2	2	—	0	5	7	0	2
3	5	4	—	7	0	5	1
4	6	0	0	—	7	6	1
5	0	6	0	2	—	0	2
6	1	6	4	0	0	—	3
Min	0	2	0	3	0	1	$\Sigma = 18$

The sum of these minimum numbers is 18 and represents a lower bound for the optimal tour. Any tour must have a length of at least 18. (Why?)

Solving the traveling-salesman problem using the distance table D_3 will be equivalent to solving the original problem using table D_1, except the length of the D_3 tour will be 18 units less. Explain why this reduction process works and why the solutions to D_1 and D_3 are equivalent. (*Hint:* The length of any Hamiltonian circuit of D_3 must be greater than or equal to zero. Each node will have to be left for some other node and each node will have to be reached from some other node.) Using D_3, you should be able to find a tour that has length 2 and thus corresponds to an optimal tour for D_1 with length 20. There are two such optimal tours.

21.28. Here is a very easy traveling-salesman problem, with asymmetric distances (from Vajda). The tour starts and returns at city 1.

City	1	2	3	4
1	—	1	4	5
2	2	—	1	2
3	2	4	—	3
4	5	2	6	—

To find a lower bound for all possible tours, we subtract the smallest nonzero entry from each row first, then from each column to obtain the reduced distance table; the lower bound is 7.

City	1	2	3	4
1	—	0	3	3
2	2	—	0	0
3	0	2	—	0
4	3	0	4	—

You should now try to construct a tour from city 1 along arcs that have a zero reduced distance. For example, if we start with arc (1,2), we know we then have to go from 2 to 3 or 4, return to 1, and that arc (2,1) will not be used, that is, we make sure we do not allow subtours to be formed. We can then eliminate row 1 and column 2 from the reduced table; this gives us the smaller reduced distance table

City	1	3	4
2	—	0	0
3	0	—	0
4	3	4	—

Note that if we did not go from 1 to 2, it would have increased the lower bound by 3 as we would have to go to 3 or 4. We can now subtract 3 from the last row and obtain the following reduced table. The lower bound is increased by 3 to 10.

City	1	3	4
2	—	0	0
3	0	—	0
4	0	1	—

We can now complete the tour with zero reduced distance arcs, thus not adding any more to the lower bound. The tour proceeds from 2 to 3, 3 to 4, and 4 to 1, with a total distance of 10. If we did not go from 2 to 3, the tour would be increased by one unit as we would have had to use arc (4,3). Can you systemize the above approach into an algorithm that will guarantee your finding the minimum tour? This is an example of the *branch and bound algorithm* in which you eliminate possible tours by knowing that they have larger lower bounds than the tour(s) that are left. The branch and bound procedure forms a network, treelike figure in which the nodes indicate the value of the current lower bound and the arc being included in the tour. The tree branches out from the nodes to indicate possible tours that need to be investigated. The above example has the following tree:

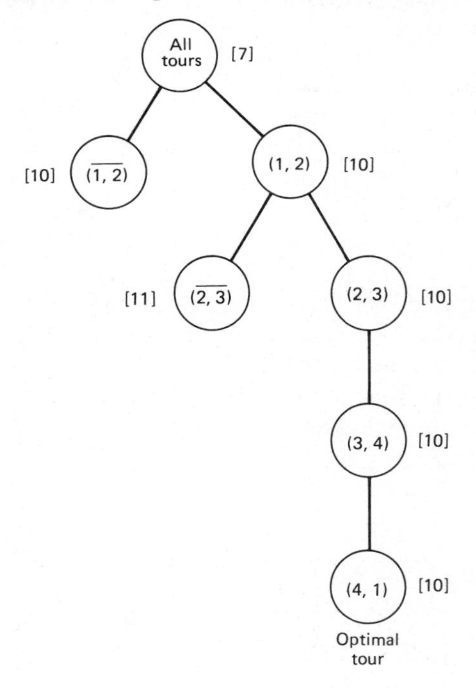

The current lower bounds are in square brackets. The notation $(\overline{i, j})$ means that any tours coming from that branch would *not* include arc (i, j). Thus, node $(\overline{2,3})$ represents all tours that include (1,2) but not (2,3). These tours have a lower bound of 11. The final node (4,1) represents all the arcs contained in the nodes leading to it. To complete this analysis, you have to investigate the node $(\overline{1,2})$. It has a lower bound of 10 that is equal to the optimal tour. There is another optimal tour that has a value of 10. Can you find it?

In picking out nodes (arcs to include in a tour), there is a good rule that enables you to set the lower bounds more efficiently. We usually have a choice in selecting an arc that corresponds to a zero distance in the reduced matrix. Consider an arc (i, j) that has a cost $c_{ij} = 0$ in the reduced matrix, and let us look at the tours that do not include (i, j). Since city i must connect with some city, these tours must incur at least the cost of the smallest element in row i, excluding c_{ij}. Also, since city j must connect to some city, these tours must incur at least the cost of the smallest element in column j, excluding c_{ij}. We add these two numbers and denote this sum by $\bar{c}_{ij}$. We choose as the next arc (i, j) to be in the tour the one that has $c_{ij} = 0$ and the largest $\bar{c}_{ij}$. We know that the lower bound for those tours that do not include arc (i, j) will be increased by $\bar{c}_{ij}$. We determine a new lower bound for those tours that do include arc (i, j) by crossing out row i and column j in the last reduced distance table; the corresponding change in the lower bound is found by subtracting the minimum distances from the rows and columns as before. The process is repeated until the distance table is reduced to a 2×2 table that yields the final arcs in the tour. See if you can work the process out for the above four-city problem.

The original branch and bound example of Little et al. has the distance table of

City	1	2	3	4	5	6
1	—	27	43	16	30	26
2	7	—	16	1	30	25
3	20	13	—	35	5	0
4	21	16	25	—	18	18
5	12	46	27	48	—	5
6	23	5	5	9	5	—

and the following branch and bound tree. Can you figure out how it was done? Write the steps of your branch and bound algorithm for solving the traveling-salesman problem. Note that here we are dealing with asymmetric distances. Also, we do not allow branches that would cause us to form a subtour (cycle) out of those cities that have already been allowed into the partial tour.

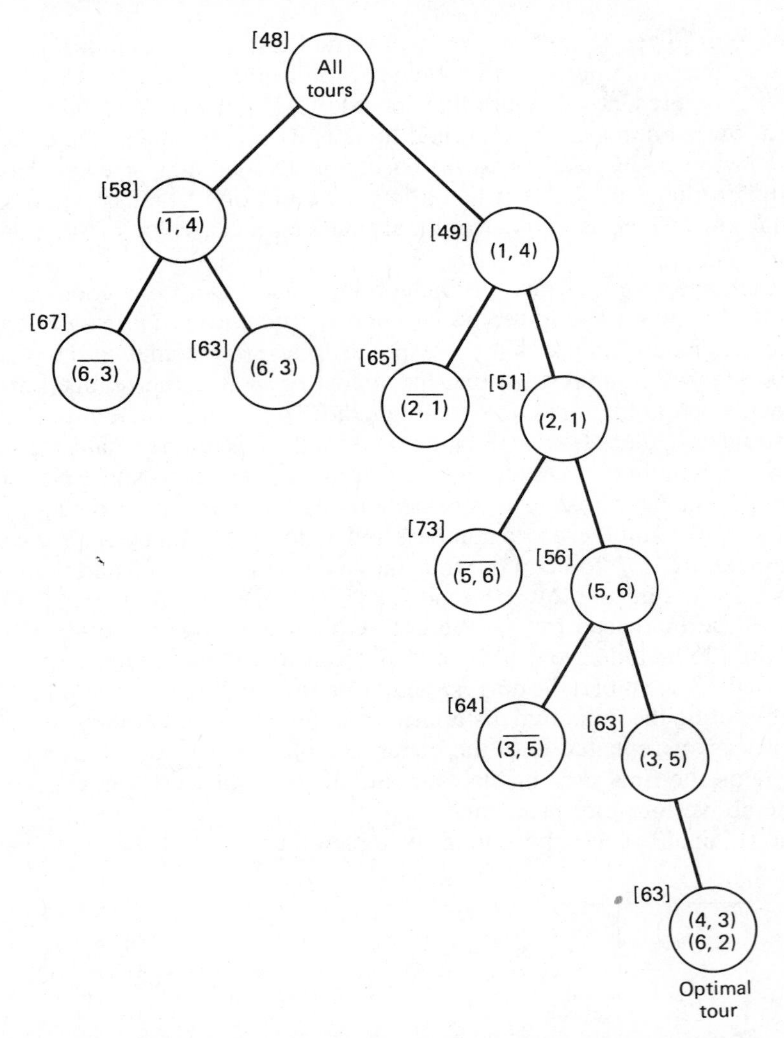

21.29. For the following distance table, find the shortest traveling-salesman tour using the branch and bound method and a heuristic algorithm. Does the city you start with make any difference? The solution is 3–1–5–4–2–3, with a distance of 53. (The problem is from Vajda.)

City	1	2	3	4	5
1	—	16	25	27	30
2	20	—	1	25	7
3	7	16	—	5	8
4	23	9	19	—	6
5	30	18	7	6	—

21.30 If the (6×6) city problem of Section 21.28 is first solved as an assignment problem, you obtain the subtours of (1,4,2,1) and (3,5,6,3) with an objective function value of 54. How would you use this information to initiate a branch and bound procedure? Use your subtour-combining algorithm to join these subtours into a Hamiltonian tour. Is it optimal?

21.31. Solve the following traveling-salesman problem that starts out in city 1. First, construct the distance table; note that it is not complete, that is, not all arcs are allowed. The Hamiltonian-circuit solution is (1–2–5–4–6–3–1), with a minimum cost of 12. (The problem is from Minieka.)

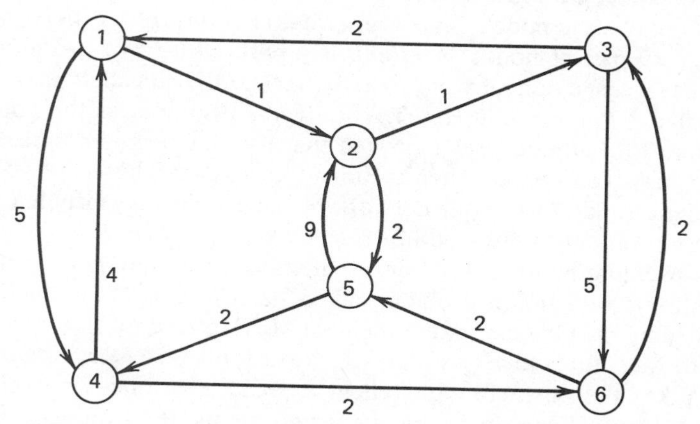

21.32. How would you schedule a cargo airplane that had to deliver goods to Cleveland, Cincinnati, St. Louis, Atlanta, and Washington, D.C., and return to its starting point, New York City. Construct the required symmetric distance table by consulting a suitable atlas. A nonsymmetric "cost" table for this problem could consist of flight times, assuming that these times vary by direction. Organize a flight-time table for the above cities.

21.33. There are a number of problems similar to the traveling-salesman problem whose study form part of the mathematical field of *graph theory*. The earliest is the Königsberg Bridge Problem, another type of routing puzzle that was solved by the mathematician Leonhard Euler. The old city of Königsberg is located on the banks and on two islands of the river Pregel. The various parts of the city were connected by seven bridges:

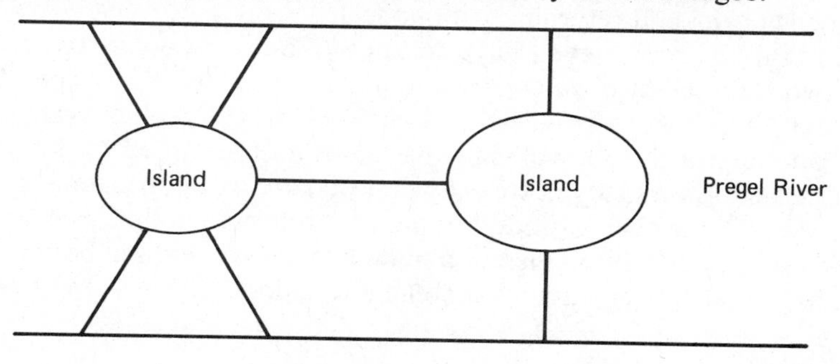

The problem is to determine if it is possible to start a walk from home and return after having crossed each bridge just once. Euler showed that it could not be done without retracing your steps. Can you see why? Draw a schematic network that represents the Königsberg problem. (*Hint*: Let the banks and islands be nodes and the bridges arcs.) Unlike the traveling-salesman problem (which always has a solution, although sometimes difficult to find), the city–bridge network does not let you go directly from any point (node) to any other point.

To investigate such problems, we need to define the following graph (network) concepts. A *path* is a sequence of distinct edges (arcs) (i, j), (j, k) . . . (q,r), (r,s) connecting nodes i and s. A graph is *connected* if there is a path between every pair of nodes. A *circuit* is a path with $i = s$. A circuit that traverses every edge of a network exactly once and visits each node at least once is called a *Euler circuit* (*tour*). The *degree* of a node is the number of arcs incident to the node. Euler proved that if a network is connected and each node has even degree, then a Euler tour exists. These are necessary and sufficient conditions. Your network of the Königsberg problem should show that all the nodes have odd degree.

Euler's problem is not a decision problem, but it is intimately related to an interesting optimization problem called the *Chinese Postman Problem* (first studied by the Chinese mathematician Mei-Ko Kwan). In this problem, we want to find, for a given (street) network, the shortest or minimal-cost tour that takes the postman along each street at least once and returns to the starting point. There are many similar problems: the routing of garbage collection, street cleaning, and snow-removal crews; the design of patrol beats for police-patrol cars; tourists walking through a city; road-repair crews; and so on. In their paper, Edmonds and Johnson develop algorithms for solving the following variations of the postman problem: (1) the network consists entirely of undirected arcs (all streets are two-way); (2) the network consists entirely of directed arcs (all streets are one-way); and (3) the arcs are mixed (both one- and two-way) but the degree at each node is even. For the undirected network, if there is a Euler tour, then it solves the postman problem; the postman does not have to repeat an arc—the optimal-tour length is just the sum of all the individual arc distances, each added only once. Can you develop an algorithm that finds a Euler tour for a network all of whose nodes have even degree? Try it out on the following street network, starting and returning with node 1.

Note that we do not even have to know the distances, as the nodes of the network are all of even degree and a Euler tour solves it. The solution is the tour (1–2–3–6–9–8–7–4–2–5–6–8–5–4–1). (The problem is from Minieka.) One approach to developing an algorithm (described by Minieka) is to start at node 1 and move across any unused arc to another node until you are forced to come back to node 1. You must return to node 1 as every node has even degree and every visit to a node leaves an even number of unused arcs at that node. Every time you come into a node, there is an unused arc

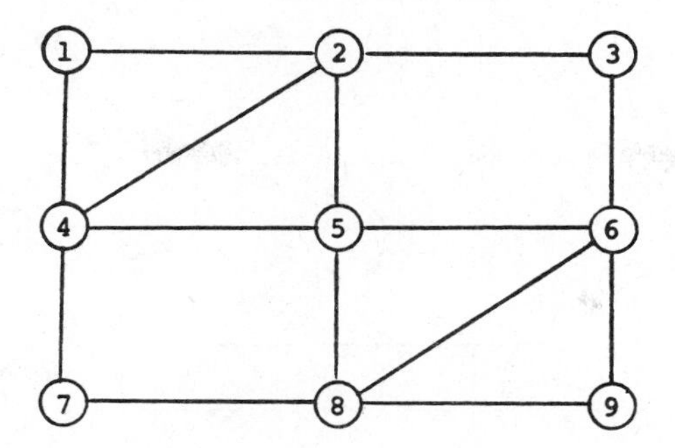

for leaving. The traversed arcs form a subtour, T_1. If T_1 contains all arcs, you have a Euler tour. Otherwise, generate another subtour T_2 of unused arcs starting with any unused arc. In a similar manner, continue to develop subtours $T_3, T_4, \ldots$, until all arcs have been used. Next, you need to splice all the subtours together. You should be able to figure out how to do that. There could be different Euler tours for the same network.

If the undirected network has odd-degree nodes, then the algorithm is more complex. Try to develop an algorithm and try it out on the following problem:

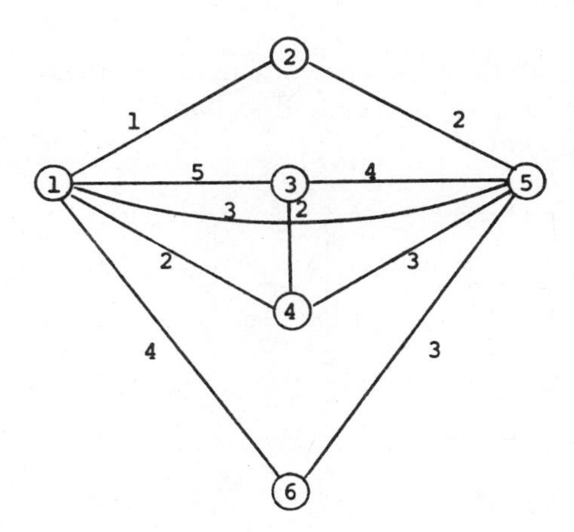

The optimal tour is (1–2–5–6–1–3–5–4–3–4–1–5–1) with a distance of 34. It is five greater than the sum of all the arcs as you have to traverse arcs (1,5) and (3,4) twice. (The problem is from Minieka.)

21.34. Select a many-block and -street area in your town and develop a routing for garbage collection or snow-removal crews. Also, locate on your map the locations of 15–20 mailboxes and figure out how a truck should be routed to collect the mail.

21.35. Find the northwest-corner solution to the following transportation problems:

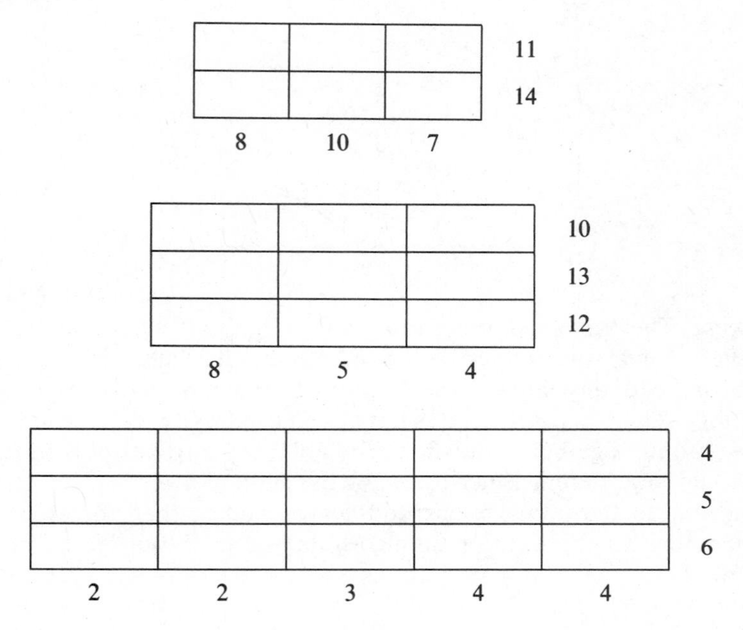

21.36. Apply both degeneracy-avoiding methods in applying the northwest-corner procedure for finding a first basic solution for the following transportation problems:

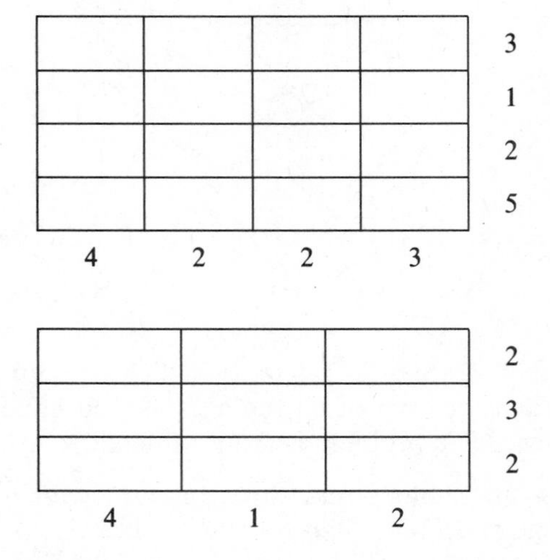

21.37. Solve the following transportation problems using the transportation simplex algorithm and any procedure for finding a first basic feasible solution:

(a)

6	8	10	11
5	9	7	14
10	8	7	

This is the refrigerator problem discussed in Chapter 4. The solution is $x_{11} = 10$, $x_{12} = 1$, $x_{13} = 0$, $x_{21} = 0$, $x_{22} = 7$, and $x_{23} = 7$; $x_{00} = 170$.

(b)

1	2	4	5
3	2	1	10
8	5	2	

Minimum: $x_{00} = 26$.

(c)

4	3	2	10
1	5	0	13
3	8	6	12
8	5	4	

(d)

3	6	3	1	1	4
2	4	3	2	7	5
1	1	2	1	2	6
2	2	3	4	4	

The solution is $x_{15} = 4$, $x_{21} = 1$, $x_{24} = 4$, $x_{31} = 1$, $x_{32} = 2$, and $x_{33} = 3$; $x_{00} = 23$.

(e) For problem (d), find and compare the northwest-corner row minimum, column minimum, and cost-table minimum solutions.

21.38. For the transportation problem example solved in Chapter 20, subtract the final u_i and v_j from the cost elements in the corresponding rows (i) and columns (j) and show that the optimal solution has positive x_{ij}'s only in cells that are zero in the transformed cost table.

21.39. You can perform sensitivity analysis on the c_{ij} and the a_i and b_j data of the transportation problem, but the process is a bit complicated. However, you should be able to figure out the following.

For a variable x_{ij} *not* in the final optimal solution, we want to see how much its cost c_{ij} can be changed before the current optimal solution is no longer optimal. Let p_{ij} be the change that could occur to the cost, that is, the cost of x_{ij} is considered to be $c_{ij} + p_{ij}$. In the original problem, $p_{ij} = 0$. The only condition that needs to be satisfied is that for the *nonbasic* x_{ij} we have

$$u_i + v_j - (c_{ij} + p_{ij}) \leq 0$$

Letting $\bar{c}_{ij} = u_i + v_j$, we see that any value of p_{ij} must satisfy

$$\bar{c}_{ij} - c_{ij} - p_{ij} \leq 0$$

or

$$\bar{c}_{ij} - c_{ij} \leq p_{ij} < +\infty$$

Thus, you can let p_{ij} vary in that range and the current solution will remain optimal. If $p_{ij} < \bar{c}_{ij} - c_{ij}$, this would mean that you can improve (decrease) the value of the objective function by making the now nonbasic x_{ij} a basic variable. Find the ranges for the p_{ij} for the nonbasic variables in the example discussed in the Chapter 20.

For a variable in the optimal basic feasible solution, it is a bit more difficult to determine what happens if its cost is changed to $c_{ij} + p_{ij}$. What you need to do is to substitute $c_{ij} + p_{ij}$ for c_{ij} in the final indirect-cost table and trace the role of p_{ij} in satisfying the dual relations $u_i + v_j = c_{ij}$ for all x_{ij} in the basis. Give it a try for the example in Chapter 20.

21.40. Solve the following transportation problem:

2	3	11	7	6
1	0	6	1	1
5	8	15	9	10
7	5	3	2	

The optimal solution is $x_{12} = 5$, $x_{13} = 1$, $x_{23} = 1$, $x_{31} = 7$, $x_{33} = 1$, and $x_{34} = 2$; $x_{00} = 100$.

	5	1		6
		1		1
7		1	2	10
7	5	3	2	

Perform a sensitivity analysis on the costs for the nonbasic and basic variables. Partial answer: $-1 \leq p_{11} < +\infty$, $-2 \leq p_{14} < +\infty$, $-\infty < p_{12} \leq 1$, $-1 \leq p_{13} \leq 1$. (The problem is from Wagner.)

21.41. The transportation problem can be viewed as a minimum-cost network-flow model whose basic structure is a *bipartite network,* that is, a network in which the nodes can be divided into two subsets such that the arcs of the network join the nodes of one subset to the other. The network for the 2×3 transportation problem follows.

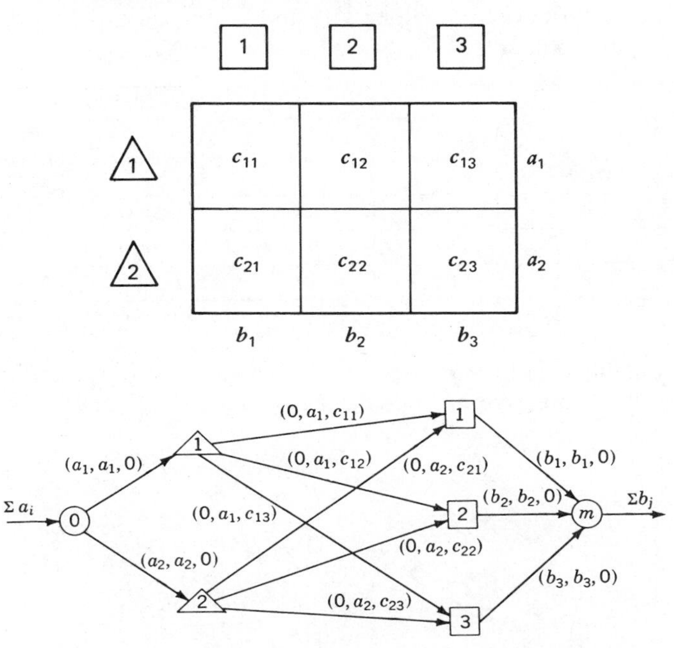

The first number of each arc is the lower bound of the flow across the arc; the second is the upper bound; and the third is the cost per unit flow. Write the constraints of the network problem and compare them to the usual transportation constraints.

21.42. Solve the following transportation problem using the transportation simplex algorithm.

The problem calls for shipping a supply of an item from a group of Air Force depots to a group of receiving stations. Each depot has a limited amount of the item. There are many routings which will supply each station with exactly the required amount of the item. The problem is to determine the routing which not only fulfills the requirements, but also minimizes some measure of the cost. For example, the objective might be to minimize one of the following: the total dollar cost, the total number of miles of the shipping schedule, or the total time the items are in transit.

Assume that Lockbourne Air Force Base (AFB) at Columbus, Ohio, has been testing a large item of equipment, weighing a ton, for the B-52. It is now desired that this equipment be tried at other bases. Five of each are required by March AFB at Riverside, California; Davis-Monthan AFB at

Tucson, Arizona; and McConnell AFB at Wichita, Kansas. Pinecastle AFB at Orlando, Florida, and MacDill AFB at Tampa, Florida, each need three. To supply these 21 items, Lockbourne AFB at Columbus, Ohio, can ship 8; Oklahoma City Depot has 8; and Warner-Robins AFB at Macon, Georgia, has 5. The items of equipment are to be airlifted to their destinations. All the foregoing information, together with approximate air distances, are given in the following table:

City	Items available	Required items: MacDill 3	March 5	Davis-Monthan 5	McConnell 5	Pinecastle 3
Oklahoma City	8	938	1030	824	136	995
Macon	5	346	1818	1416	806	296
Columbus	8	905	1795	1590	716	854

Distance in miles

Here the objective is to *minimize the total ton-miles*. The optimal solution in terms of this objective has been computed and is given below.

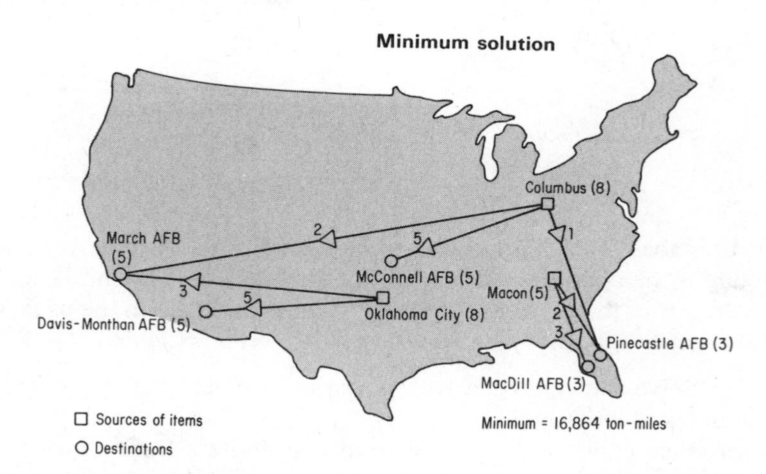

21.43. Solve the following minimum assignment problem (from Kwak):

	1	2	3	4
1	30	25	26	28
2	26	32	24	20
3	20	22	18	27
4	23	20	21	19

There are three optimal solutions with a total cost of 86.

21.44. Discuss the formulation of the following problem as an assignment problem. Five graduate assistants are available to assist in five different undergraduate courses, one for each course. On a scale of 0 to 100, each assistant rates him- or herself as to their value in each of the classes, with a 100 being the highest value. Do you see any problems with this procedure? How could an assistant attempt to be assigned to an easy workload course? What other procedure could you use to develop the value table? Remember, most graduate assistants get A's in all their courses or they do not remain assistants very long.

21.45. Show that a maximization assignment problem can be solved by the minimizing algorithm discussed in Chapter 20 if you transform the c_{ij} to $\bar{c}_{ij} = r - c_{ij}$, where r = maximum c_{ij}. Solve the Able, Baker, and Charlie personnel-assignment problem (Chapter 9) using this transformation.

21.46. The costs c_{ij} of an assignment problem need not be positive. Apply the minimizing algorithm to solve this problem:

	1	2	3	4	5
1	2	3	5	1	4
2	−1	1	3	6	2
3	−2	4	3	5	0
4	1	3	4	1	4
5	7	1	2	1	2

The optimal assignment is $x_{12} = 1$, $x_{21} = 1$, $x_{35} = 1$, $x_{44} = 1$, and $x_{55} = 1$; $x_{00} = 5$. (The problem is from Bazaraa and Jarvis.)

21.47. Solve the following maximizing assignment problem:

	1	2	3	4
1	8	7	9	9
2	5	2	7	8
3	6	1	4	9
4	2	3	2	6

The optimal assignment is $x_{11} = 1$, $x_{23} = 1$, $x_{34} = 1$, and $x_{42} = 1$; $x_{00} = 27$. (The problem is from Kuhn.)

21.48. The Hungarian algorithm can also be used to solve the transportation problem. Any ideas on how to go about doing it?

21.49. Solve the following minimum assignment problem (from Balinski and Gomory):

	1	2	3	4
1	2	3	1	1
2	5	8	3	2
3	4	9	5	1
4	8	7	8	4

The minimum value is 13.

21.50. It turns out that many linear-programming problems can be viewed as minimal-cost network problems, shortest-route problems, or transportation problems even though their problem statements give no hint of a network structure. The caterer problem is a case in point. In what follows, we reintroduce the caterer problem using notation that enables us to write a more concise model of the problem. Then, using the tea-party example, we show how this problem can be interpreted as a transportation problem and a minimal-cost network problem.[1]

A caterer knows that, in connection with the meals to be served during the next n days, $r_t \geq 0$ fresh napkins will be needed on the tth day, with $t = 1, 2, \ldots, n$. Laundering normally takes p days; that is, a soiled napkin sent for laundering immediately after use on the tth day is returned in time to be used again on the $(t + p)$th day. However, the laundry also has a higher-cost service which returns the napkins in $q < p$ days (p and q integers). Having no usable napkins on hand or in the laundry, the caterer will meet early needs by purchasing napkins at a cents each. Laundering expenses are b and c cents per napkin for the normal slow and high-cost (fast) services, respectively. We assume $a > b$ and $a > c > b$. How does the caterer arrange matters to meet the needs and minimize the outlays for the n days?

Define

$0 \leq x_t =$ the number of new napkins bought on day t
$0 \leq y_t =$ the number of soiled napkins sent to the slow laundry on day t
$0 \leq z_t =$ the number of soiled napkins sent to the fast laundry on day t
$0 \leq s_t =$ the number of soiled napkins not sent to any laundry on day t

We assume all purchases, uses, and shipments happen simultaneously at time t, $t = 1, 2, \ldots, n$, and all the variables are integer-valued. Also, as we want to minimize the total cost of maintaining the proper inventory of clean napkins, it does not pay to buy more napkins than needed for the current day, or to send out a napkin to be laundered unless it will be used at a future time. For discussion purposes we shall develop the equations for

[1] From S. I. Gass, *Linear Programming: Methods and Applications,* 5th Edition, McGraw-Hill Book Co., New York, N.Y., 1985. Reproduced with permission.

the set of tea-party requirements $r_t = (5,6,7,8,7,9,10)$, with $a = 25$ cents, $b = 10$ cents, and $c = 15$ cents. We next develop the 14 equations, two for each day, corresponding to the daily requirements for clean napkins and disposition of dirty napkins:

Day	Clean napkins	Dirty napkins
1	$x_1 = 5$	$z_1 + y_1 + s_1 = 5$
2	$x_2 = 6$	$z_2 + y_2 + s_2 = 6 + s_1$
3	$x_3 + z_1 = 7$	$z_3 + y_3 + s_3 = 7 + s_2$
4	$x_4 + z_2 + y_1 = 8$	$z_4 + y_4 + s_4 = 8 + s_3$
5	$x_5 + z_3 + y_2 = 7$	$z_5 + s_5 = 7 + s_4$
6	$x_6 + z_4 + y_3 = 9$	$s_6 = 9 + s_5$
7	$x_7 + z_5 + y_4 = 10$	$s_7 = 10 + s_6$

For example, for day 1 we must purchase five new napkins and, once used, they are distributed to the fast or slow laundries, or left in a dirty inventory pile; for day 4, we must have eight clean napkins which can come from either new purchases, the shipment sent to the fast laundry on day 2, or the shipment sent to the slow laundry on day 1, with the resultant eight dirty napkins and the previous day's dirty pile available for laundering or storage; finally, we see that dirty napkins are inventoried at the end of the time horizon in that the laundries would not be able to return them in time to be used.

The objective function is to minimize the total cost function of

$$25 \sum_t x_t + 10 \sum_t y_t + 15 \sum_t z_t.$$

The optimal solution has a total cost of \$8.80 with the nonzero variables being $x_1 = 5$, $x_2 = 6$, $x_3 = 7$, $x_4 = 3$, $z_3 = 1$, $z_4 = 3$, $z_5 = 5$, $y_1 = 5$, $y_2 = 6$, $y_3 = 6$, $y_4 = 5$, $s_5 = 2$, $s_6 = 9$, and $s_7 = 10$.

The general statement of the caterer problem is given by

$$\text{Minimize} \quad a \sum_t x_t + b \sum_t y_t + c \sum_t z_t$$

subject to

$$y_t + z_t + s_t - s_{t-1} = r_t \qquad t = 1, 2, \ldots, n$$

$$x_t + y_{t-p} + z_{t-q} = r_t \qquad t = 1, 2, \ldots, n$$

with $x_t \geq 0$, $y_t \geq 0$, $z_t \geq 0$, and $s_t \geq 0$; $y_{t-p} = 0$ if $t - p \leq 0$, $z_{t-q} = 0$ if $t - q \leq 0$, $y_t = 0$ if $t + p > n$, and $z_t = 0$ if $t + q > n$ (all variables integer-valued).

To interpret the caterer problem as a transportation problem, we define x_{ij} to be the number of dirty napkins sent to a laundry (slow or fast) on day i to be returned as clean napkins for use on day j. Here we are allowed to launder napkins in advance and do not store dirty napkins unless they will not be laundered at all. Let x_{i0} be the number of dirty napkins retired from future service on day i. Thus, each day i represents an origin of dirty napkins with availability of r_i to be shipped (laundered or retired), and we must have

$$\sum_{j=1}^{n} x_{ij} + x_{i0} = r_i \qquad i = 1, 2, \ldots, n$$

Similarly, each day j represents a destination for clean napkins, and r_j napkins must be received from previous shipments or from new purchases. Let x_{0j} be the number of new napkins purchased on day j. We then have

$$\sum_{i=1}^{n} x_{ij} + x_{0j} = r_j \qquad j = 1, 2, \ldots, n$$

As the process starts out with no napkins on hand and ends with all napkins retired, we have

$$\sum_{j=1}^{n} x_{0j} = \sum_{i=1}^{n} x_{i0}$$

The above equations represent a transportation problem with the origins being each day's dirty napkins and the supply of new napkins, and the destinations being each day's requirement of new napkins and the pile of retired dirty napkins. Of course, many of the x_{ij} are zero, as they represent impossible shipments and can be considered to have a very high cost, for example, a shipment x_{31} of dirty napkins on day 3 which arrives on day 1. The transportation tableau has $n + 1$ rows and $n + 1$ columns. To form the necessary balanced structure between the origin availabilities of napkins and the destination requirements, we let

$$\sum_{j=1}^{n} x_{0j} + x_{00} = \sum_{i=1}^{n} x_{i0} + x_{00} = \sum_{i=1}^{n} r_i$$

where x_{00} represents new napkins which would have had to be purchased if there were no laundries. The above equations, for the previous numerical problem, are represented in the following tableau. The optimal solution is also shown.

Origins \ Destinations	Day 1	Day 2	Day 3	Day 4	Day 5	Day 6	Day 7	Dirty napkins retired	r_t
Day 1			[15] 0	[10] 5	[10] 0	[10] 0	[10] 0	[0] 0	5
Day 2				[15] 0	[10] 6	[10] 0	[10] 0	[0] 0	6
Day 3					[15] 1	[10] 6	[10] 0	[0] 0	7
Day 4						[15] 3	[10] 5	[0] 0	8
Day 5							[15] 5	[0] 2	7
Day 6								[0] 9	9
Day 7								[0] 10	10
Napkins purchased	[25] 5	[25] 6	[25] 7	[25] 3	[25] 0	[25] 0	[25] 0	[0] 31	52 ($\sum r_i$)
r_t	5	6	7	8	7	9	10	52 ($\sum r_i$)	

(Rows for Days 6–7 are marked q; columns for Days 1–2 are marked q.)

To transform the caterer problem to a minimal-cost network-flow problem we assume a source node of new napkins and a sink node for retired dirty napkins, with intermediate nodes representing the days in which clean napkins must be on hand and dirty napkins must be redistributed to the laundries and dirty-napkin storage. The network for the above seven-day example is shown below. We have labeled each arc with the corresponding variable or daily-napkin requirement. You will note how the network assumption of conservation of flow through a node forces the caterer-problem equations to be satisfied. Also, to obtain the proper balance of flow at the sink node, we need to assume a total of $F = \sum_i r_i$ napkins at the source node and that any of the F not purchased are sent to the sink node. We leave as an exercise the development of each arc's lower and upper bounds and associated cost, and the development of the conservation-of-flow equations. Note that both the transportation model and the network model guarantee that the optimal solution will be in integers. Which of the appropriate algorithms would you use to solve a caterer problem?

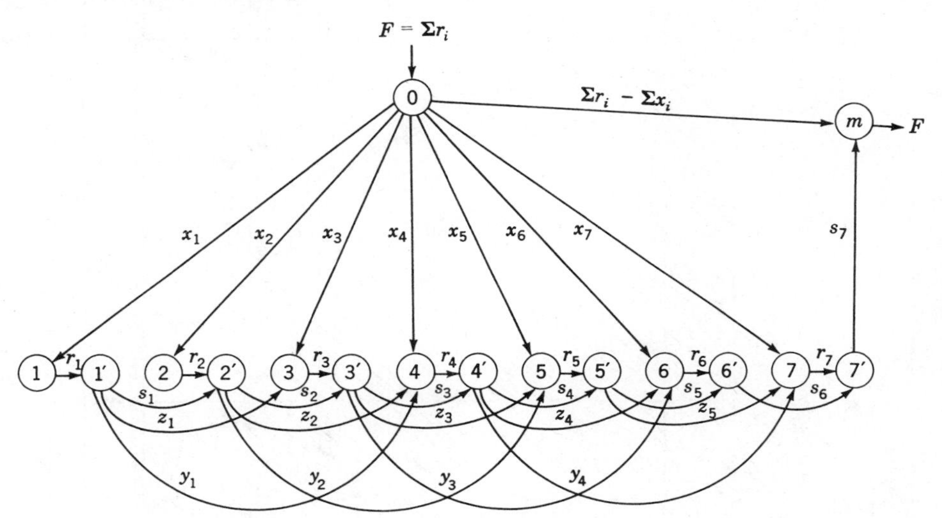

21.51. Whenever we are faced with an optimization problem, our natural inclination is to make sure we select a feasible solution which is also an optimal solution.[1] We want to do our best no matter what—or do we? As the computational procedures of linear programming search for the optimum by first finding a solution, then a better one, then an even better one, and so on, we might question the need to actually wring out of our computational washing machine the last penny of savings. The labor and computational costs to do so might cost more than the savings afforded by improving upon the solution at hand. Do we really want the minimal-cost diet or just a good diet which satisfies the daily requirements? Can we live with Sid Simon's solutions to his activity-analysis problems, or do we really want to make the last dollar?

For some problems a good solution would suffice, but, in general, if the economics of the computation are favorable, the optimal solution should be found and used. There are some problems, however, where the use of an

[1] The material in this section is from S. I. Gass, *An Illustrated Guide to Linear Programming*, McGraw-Hill Book Co., New York, N.Y., 1970. Reproduced with permission.

optimal solution is a must. In particular, with the governmental contract-awards problem the need for determining the lowest-cost solution is dictated by the U.S. Congress.

The Armed Forces Procurement Act requires the Armed Forces to purchase supplies on a competitive basis. Contracts must be advertised, sealed bids submitted, and the awards made so as to minimize the total cost to the government. Once the contracts have been awarded, the procurement office must be ready to demonstrate—especially to the satisfaction of the losing bidders and their Congressional representatives—that the total cost was the least possible.

The award of the contracts can be a rather complicated decision-making problem as constraints imposed by the bidders can be quite complex. As all the bids are made public after an award is made, the contract evaluators have to be very astute in juggling and analyzing the data. With the advent of linear programming however, this heavy burden—and some ulcers—have been removed. How does linear programming fit in? Let's take a closer look at the problem.

For a given procurement, the manufacturers submit bids on which they state:

- □ The price per unit of article or articles.
- □ The maximum or minimum quantity of each item that can be produced at the stated price.
- □ Any other conditions they wish to impose.

The bids reflect the manufacturers' desire for profit, their guesses about each other's bids, and their limitations.

In evaluating the bids, the procurement office must add shipping and other related costs to each bidder's quoted prices. Similarly, any savings that could be effected by agreeing to certain conditions are subtracted; for example, a discount may be allowed for payments made within a certain time. The basic contract-awards problem can be formulated as a transportation problem, but certain conditions imposed by the bidders require special handling by the problem formulator. For example, a bidder might require the manufacture of at least 500 units and no more than 1000 units, or have different prices for each lot of 1000 units to reflect quantity discounts. The following example is an actual contract-awards problem as encountered by an army purchasing agency. We shall only present the facts of the case and leave it to you to develop the equations and approach to solving it.

The quartermaster depots listed below require the stated amount of packages of copy machine paper. The amounts required are separated into domestic and export categories as the special packaging for export dictates a higher cost per unit, Table 1.

Four manufacturers submit bids. Bidder 1 requires an award of at least 11,500 packages and allows for a maximum award of 33,145 packages for domestic and a maximum of 3510 for export. Bidder 2 has no minimum or

TABLE 1. Requirements

	Requirements	
Depots	Domestic	Export
Columbus	10,395	
Richmond	12,420	
San Antonio	10,395	
Schenectady	9,720	39,555
Utah	3,240	
Sharpe	5,535	
Auburn	3,645	
Atlanta	10,330	3,510
Totals	65,680	43,065
Grand total	108,745	

maximum conditions. Bidder 3 offers a maximum of 60,000 packages, while bidder 4 bids only for the domestic contract and has no other conditions. This information, along with the cost per package for each bidder–depot combination, is summarized in Table 2. The given prices include packaging and shipping costs; for example, it costs $0.6868 to ship one package from bidder 2 to the Columbus depot.

TABLE 2. Costs and Amounts Bid

		Maximum Number Offered				
		Bidder 1	Bidder 1			
		Domestic	Export	Bidder 2	Bidder 3	Bidder 4
City	Number Required	33,145	3,510	108,745	60,000	65,680
Columbus	10,395	$0.7289	—	$0.6868	$0.6574	$0.6832
Richmond	12,420	0.7398	—	0.7058	0.6489	0.6724
San Antonio	10,395	0.7229	—	0.7204	0.6904	0.7227
Schenectady domestic	9.720	0.7406	—	0.7075	0.6318	0.6627
Schenectady export	39,555	—	0.7749	0.7319	0.6452	—
Utah	3,240	0.7247	—	0.7358	0.6944	0.7306
Sharpe	5,535	0.7276	—	0.7389	0.6973	0.7339
Auburn	3,645	0.7297	—	0.7389	0.6973	0.7339
Atlanta domestic	10,330	0.7325	—	0.7049	0.6646	0.6917
Atlanta export	3,510	—	0.7663	0.7291	0.6816	—

The solution which satisfies the conditions imposed by the bidders and minimizes the total cost to the purchasing agency is given in Table 3. This award results in a total cost to the government of $72,953.1935. The linear-programming procedure guarantees that no better result is possible. How

TABLE 3. Contract Award Minimum Solution

City	Bidder 1	Bidder 2	Bidder 3	Bidder 4
Columbus	—	—	—	$10,395
Richmond	—	—	—	12,420
San Antonio	—	$10,395	—	—
Schenectady domestic	—	—	$ 4,515	5,205
Schenectady export	—	—	39,555	—
Utah	—	—	3,240	—
Sharpe	—	—	5,535	—
Auburn	—	—	3,645	—
Atlanta domestic	—	—	—	10,330
Atlanta export	—	—	3,510	—

would you demonstrate to a losing bidder that the linear-programming solution was correct?

Historical Note

The relationship between the contract-awards problem and the transportation problem was first discovered and exploited by mathematicians of the U.S. National Bureau of Standards (NBS), working with personnel from the Philadelphia quartermaster depot. Prior to the advent of linear programming, each problem was solved by submitting the bids to a series of evaluations conducted by different analysts. When no change could be found, the successful bidders would be announced, and everyone would hope for the best. When the first operational tests were conducted using the NBS computer, the SEAC,[1] the quartermaster group continued to solve the problems with their analysts in order to build up the necessary confidence in the linear-programming approach. The computer solutions were always better than, or at least as good as, the quartermaster solutions. In fact, this test again demonstrated that experienced personnel can do a good job in finding the optimal solution, if the data and constraints are not too involved. However, as many such problems have to be solved on a daily, routine basis, the computerized linear-programming approach has other economic advantages, along with its ability to determine the low-cost solution.

21.52. We have shown how the caterer problem can be formulated as a transportation problem. A number of other production–inventory problems can also be treated as transportation problems. Here are two of them. You should draw the bipartite networks before you try to write any constraints.

(a) A manufacturer of an appliance has just completed a new design and wants to determine the monthly production schedule for the first quarter of the year. The marketing department expects sales demand to be 80 units in January, 95 units in February, and 125 units in March. The production department has a fixed production schedule of 110 units a month. The production cost for an item is $25 in January and increases $3.00 a unit in each succeeding month. Any unit not assigned to meet demand during the month it is manufactured is put into inventory starting with the month after it is made. Inventory costs are $1.00 a month per unit. Any production not used to meet the first-quarter's demand ends up in inventory to be used for the next quarter. Formulate this as a transportation problem. Can you solve this problem by inspection? Try the column-minimum, first basic feasible solution.

(b) Consider the problem in (a) above, but the marketing department has come in with an optimistic demand forecast of 150 units in January, 160 units in February, and 200 units in March. The only way demand can be met is to allow overtime production. This will add $5.00 per unit to the cost;

[1] The SEAC operated from June 20, 1950 to April 23, 1964.

65 units can be produced each month on overtime. Formulate this as a transportation problem. Can you solve this problem by inspection?

(c) Formulate both problems (a) and (b) as regular production–inventory type linear-programming problems.

21.53. An important aspect of solving any decision problem is getting the answer accepted and putting it to work, that is, the selling and implementation of the analysis. Can you think of ways to present the results of a transportation problem so they are understood readily by management? What do you think of the map illustrating the routes for the Air Force problem (Section 21.42)? You could contrast the minimum solution with a similar map that shows the solution before the analysis was made (like the one below).

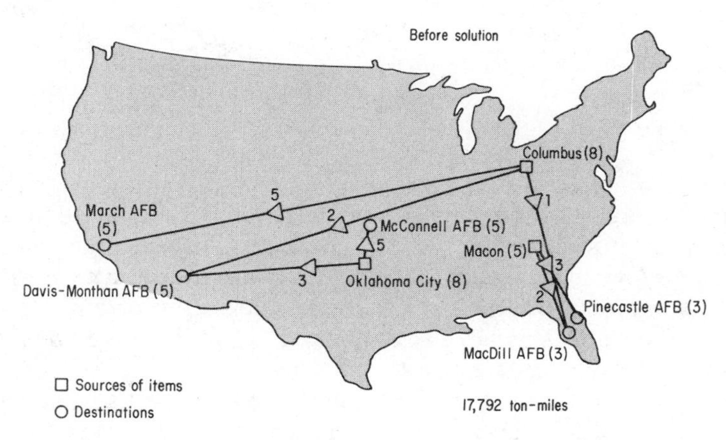

21.54. A scientist is usually not content with just solving a particular problem. The first thing he or she tries to do is to vary the assumptions of the problem and generalize and extend the solution procedure so that it solves the new problem. This tendency to look for new mountains to climb is very strong in persons trained in the mathematical sciences. Sometimes after they have finished, you can hardly recognize that the impetus for the analysis was just a simple practical problem. But much valuable theory originates from this type of investigation. You should develop this inquiring attitude and not be complacent with knowing you have solved a specific problem—and just leaving it at that. Many of the decision problems described so far have intriguing and practical variations. Let us look at some.

Up to now we have considered the traveling-salesman problem to be just that—only one salesman is traveling. What if there is more than one salesman?

The problem can be stated as follows: There are m salesmen at the origin (home) city and their are $n - 1$ (customers') cities to be visited, with each city to be visited exactly once by one of the salesmen. We need to determine

m different tours (that start with the origin city) so that the total distance traveled by all the salesmen is a minimum. For a $n = 10$ city problem and $m = 3$ salesmen we could have the following situation (with city 1 being the home city):

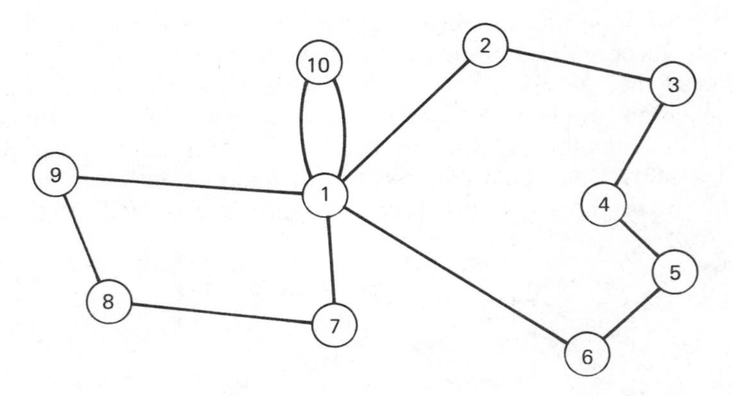

This type of problem is encountered by a crew of bank messengers who must pick up deposits at branch banks and deliver them to the central office. You can think of a number of other applications, for example, Brink's armored trucks, mailbox pickups, and so on. It would be nice if we could convert the m-salesmen problem into a one-salesman problem and thus be able to use the algorithms already developed. Well, it can be done! Give it some thought before reading ahead.

The approach is to reproduce the home city $m - 1$ times so that it appears exactly once for each of the m salesmen; renumber these origins $0_1, 0_2, \ldots, 0_m$. For each i $(1 \leq i \leq m)$, 0_i is connected to the $n - 1$ cities in the same manner as origin city 1, that is, with the same arcs and distances. Each 0_i is connected, conceptually, to the other 0_j by an arc of infinite distance. If we solve this $m + n - 1$ city problem as a single-salesman problem, using 0_1 as the origin city, we would never use an arc that connected an 0_i to an 0_j. For the 10-city problem, we might end up with the sequence $(0_1,2,3,4,0_2,5,6,7,8,0_3,9,10,0_1)$. As all the 0_i's are really the same, the three tours are (1,2,3,4,1;1,5,6,7,8,1;1,9,10,1). We just collapse all the 0_i's into the original origin city 1.

Let the original n-city distance table be

	1	2	$\cdots$	n
1	—	c_{12}	$\cdots$	c_{1n}
2	c_{21}	—	$\cdots$	c_{2n}
$\vdots$	$\vdots$		$\vdots$	$\vdots$
n	c_{n1}	c_{n2}	$\cdots$	—

Then the expanded $(m + n - 1)$-city distance table is

	O_1	O_2	$\cdots$	O_m	2	$\cdots$	n	
O_1	—	∞	$\cdots$	∞	c_{12}	$\cdots$	c_{1n}	Row one repeated m times
O_2	∞	—	$\cdots$	∞	c_{12}	$\cdots$	c_{1n}	
$\vdots$	$\vdots$		$\vdots$	$\vdots$	$\vdots$			
O_m	∞	∞	$\cdots$	—	c_{12}	$\cdots$	c_{1n}	
2	c_{21}	c_{21}	$\cdots$	c_{21}	—	$\cdots$	c_{2n}	
$\vdots$	$\vdots$	$\vdots$		$\vdots$	$\vdots$		$\vdots$	
n	c_{n1}	c_{n1}	$\cdots$	c_{n1}	c_{n2}	$\cdots$	—	

Column one repeated m times

The following example (from Larson and Odoni), illustrates how to set up the expanded distance table for a two-salesmen problem:

Original Distance Table (Miles)

	1	2	3	4	5
1	—	28	57	20	45
2	28	—	47	46	73
3	57	47	—	76	85
4	20	46	76	—	40
5	45	73	85	40	—

Expanded Distance Table

	O_1	O_2	2	3	4	5
O_1	—	∞	28	57	20	45
O_2	∞	—	28	57	20	45
2	28	28	—	47	46	73
3	57	57	47	—	76	85
4	20	20	46	76	—	40
5	45	45	73	85	40	—

The optimal two-tour solution is (1,2,3,1;1,4,5,1) and has a total distance of 237. Can you prove that this solution is optimal? Use this table to set up a three-salesmen problem and solve it.

The solution to the above two-salesman problem yielded tours each of which have only two cities to visit, that is, each salesman has the same workload. (However, the first travels 132 miles, while the second travels 105 miles.) There is nothing in our algorithm that could prevent a solution in which one salesman visits only one city and the other the rest. This could happen as we are only interested in minimizing the combined total distance traveled by all salesmen. Is that the correct measure of effectiveness? How would you approach the situation in which you wanted to minimize the total distance, but have each salesman visit the same number of cities, or as close to the same number as possible? We investigate aspects of this problem next.

21.55. We began our discussion of the traveling-salesman problem with Willie Simon and his run in with the management analysts; it is fitting that we end our discussion with Willie's latest problem. He has been promoted to sales manager of the Simple Furniture Company's textile distributing division. The division's main activity is the buying of upholstery material from different mills and selling it to large and small upholstery and decorator shops throughout the country. The latest Fall line of goods has just been announced and Willie's staff has decided on the materials it wants to market and has prepared sample books with swatches of each item. It is traditional for these books to be distributed by having a salesman drop them off at each shop. A customer gets at least one book, but the actual number is based on a customer's volume of purchases—the bigger the customer the more books received. The actual numbers have been determined by Willie and he has told his staff how many sample books to make up. Each of Willie's salesmen works out of the home city and drives the same type of station wagon. Past experience has shown that the wagon can carry up to 100 books and that a salesman has difficulty managing the distribution of more than 100 books. Willie's problem is to figure out a set of traveling-salesman tours (routes) such that the total number of books required by the customers on each tour is less than or equal to 100, all customers are visited and receive the correct number of books, and the total sum of the distances traveled on all tours is minimized. Based on his experiences with Super Management Consultants, Willie feels as if he can solve this problem. Here is how he went about it.

First, a few assumptions: (1) a customer is visited by and receives books from only one salesman; (2) no customer requires more than 100 books; and (3) Willie has ordered enough books to satisfy the requirements of each customer. Willie has discovered that, to date, there is no optimization algorithm available to solve reasonable-size problems of this type. Exact solutions have been found for problems involving up to 30 customers. Willie's and our interest here is in developing heuristic algorithms that are efficient and that can be used to solve problems with many customers.

For discussion purposes, we can assume that Willie has n customers (cities) indexed by $i = 1, 2, \ldots, n$ and the ith customer has a demand for books denoted by d_i, with the total demand $d = \sum_i d_i$. The distances c_{ij}, between all combinations of customers i and j are known and given in a distance table. We denote the home city by the index $i = 0$; the distances c_{0i} for $i = 1, \ldots, n$ are also known. Finally, we let k be the capacity of the station wagons, that is, the number of sample books each salesman can deliver is the same and equal to $k = 100$.

We first ignore the optimization aspect of the problem and see how we can obtain feasible solutions. Willie would like the number of tours to be as small as possible as a tour corresponds to sending out a salesman in a station wagon—a very expensive activity. (Note that if $k \geq d$, then the problem reduces to the single-traveling-salesman problem in that one salesman can handle the total demand.) The worst situation is to send one salesman to service exactly one customer. This would require n salesmen with each tour to city i having a distance of $(c_{0i} + c_{i0})$ for $i = 1, \ldots, n$. For a symmetric distance table $(c_{ij} = c_{ji})$, the total distance for the n tours is given by $2 \sum_{i=1}^{n} c_{0i}$; for an asymmetric distance table the total distance is $\sum_{i=1}^{n} (c_{0i} + c_{i0})$. For $n = 5$, we would have the following situation on five tours:

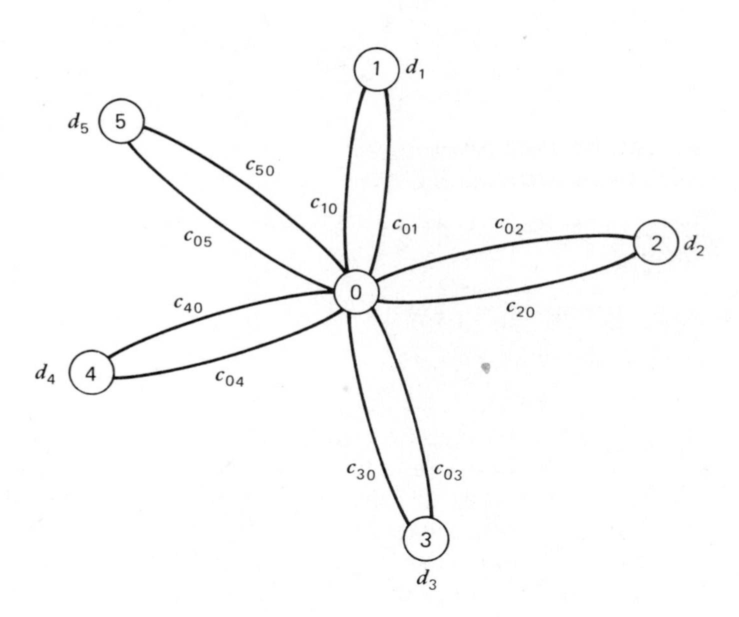

Depending on the values of the d_i, this could be the optimal solution. Can you think of a set of d_i that would force Willie to use this solution? What if each $d_i = 51$, so that a salesman could not transport enough books to service more than one customer? The d_i play a rather important part in the algorithm;

we simply cannot form a tour without ensuring that the resulting sum of the d_i's is less than or equal to k.

It seems reasonable to try to see if these one-city tours can be combined in ways that reduce the total distance traveled and still yield tours that satisfy the capacity limit of k per salesman (tour). For example, in the five-city problem we could try to link city 1 with city 2 as follows:

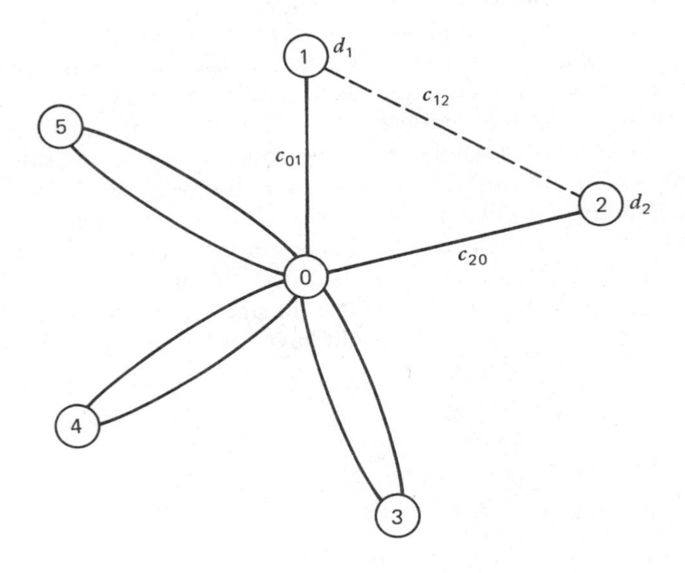

The change (savings) in the distance is given by

$$s_{12} = (c_{01} + c_{10} + c_{02} + c_{20}) - (c_{01} + c_{12} + c_{20})$$
$$= (c_{10} - c_{12} + c_{02})$$

If $s_{12} > 0$, *and* $d_1 + d_2 \leq k$, then we form the new tour (0,1,2,0). The larger a savings s_{ij}, the better off we are as far as reducing the total distance. This type of exchange is the basis of a heuristic algorithm due to Clarke and Wright. You start with the feasible set of n tours and calculate the savings $s_{ij} = c_{i0} - c_{ij} + c_{0j}$ for every pair of (i,j) customers. Next, we rank the s_{ij} with the largest at the top. Starting with the largest s_{ij}, we form a tour that includes cities i and j if the total demand for this new tour is less than or equal to k. This exchange can take place as follows: (1) if both cities i and j were not already in the same tour, a tour is formed that now includes both i and j; or (2) if either i or j is already in a tour, then this tour is changed to include arc (i,j) and both cities i and j. We have to watch out for one thing, however. For example, what if we have formed the following tours:

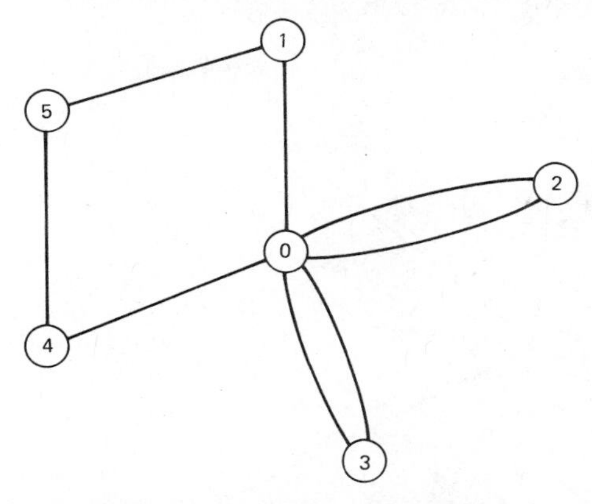

The savings list tells us to consider arc (3,5). However, city 5 is already in a tour and if we include arc (3,5) what do we do with city 4? This heuristic algorithm ignores (3,5) as 5 is in a tour and is an *interior city* of the tour, that is, it is not connected directly to city 0. Note that we can consider arcs (3,4) or (2,3) or (1,3) or (1,2) as cities 1, 3, and 4 are not interior points in any tours. The process continues until all the savings s_{ij} are processed. If a city has not been added to a tour, then it is visited by a salesman that services only that city.

This is a very simple heuristic algorithm to apply. You should write out the steps of the algorithm taking into consideration all the possibilities mentioned above. The capacity condition is considered as a side constraint. Other constraints, such as the total number of days a tour can take, can be handled in the same manner.

This *exchange algorithm* (greedy type) seems to work rather well even though it does not produce an optimal solution: it produces good solutions; it can be programmed and run on a computer rather efficiently; and it can be used to solve problems with hundreds of cities.

We next use the algorithm to solve the following four-city symmetric problem with $k = 24$ (from Larson and Odoni):

Distance Table

City	0	1	2	3	d_i
0	—				—
1	28	—			10
2	57	47	—		12
3	20	46	76	—	14
4	45	73	85	40	11

The first solution uses $n = 4$ one-city back-and-forth tours and has a total distance of $300 = 2(28 + 57 + 20 + 45)$:

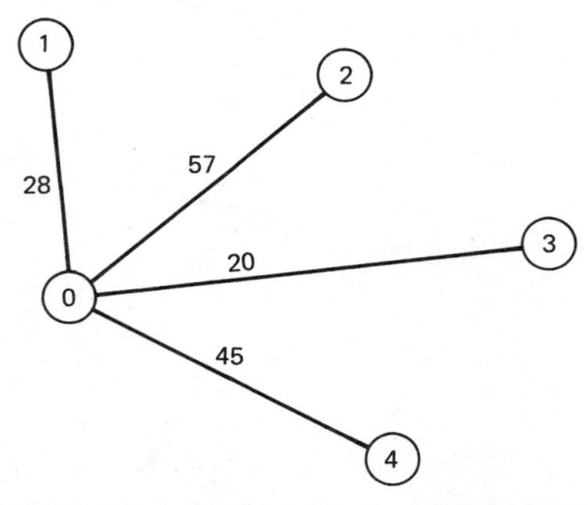

Calculating the savings by joining cities 1 and 2 we have $s_{12} = 28 - 47 + 52 = 33$. Doing this for all the other combinations, we have the following:

Savings Table

City	0	1	2	3	4
0	—	—	—	—	—
1		—	33	2	0
2			—	1	29
3				—	25
4					—

We first process the largest savings $s_{12} = 33$ through the algorithm and since $d_1 + d_2 = 22 \leq 24 = k$, we have the tours

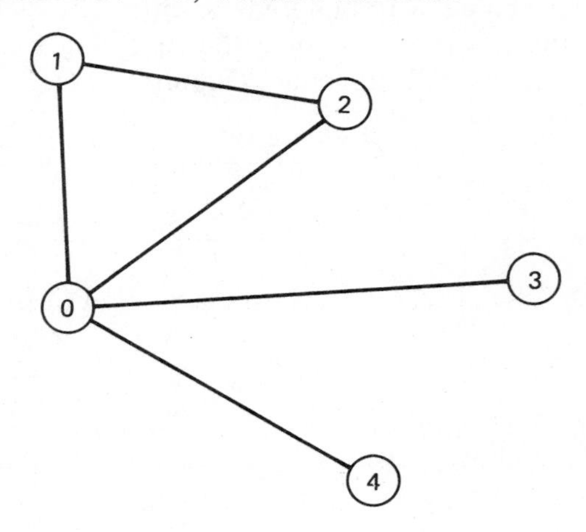

Next, $s_{24} = 29$, but city 4 cannot be combined with the tour (0,1,2,0) as $d_1 + d_2 + d_4 = 33 > 24$. Next, we cannot combine cities 3 and 4 (savings $s_{34} = 25$) as $d_3 + d_4 = 25 > 24$. Finally, we cannot combine city 3 ($s_{13} = 2$ and $s_{23} = 1$) with the tour (0,1,2,0), as $d_1 + d_2 + d_3 = 36 > 24$. We end up with the three tours (0,1,2,0) (0,3), and (0,4) with a total distance of 262, a change of 38 from the original 300. Can you do any better? Would you get a different answer if $k = 25$? Describe how you would apply the concept of sensitivity analysis to this problem. Also, solve the problem for $d_3 = 13$.

21.56. The problem of salesmen with capacity workloads is a special case of the more general problem of vehicle routing. The problem is usually stated in terms of a single depot from which vehicles with a given load (or weight) capacity k must deliver goods to n customers each having a known demand d_i. There are numerous such applications: the delivery of newspapers to distribution points, the loading of milk from dairies, the pickup of students by school buses, the delivery of oil to customers or of gasoline to filling stations, and so on. Can you think of other applications?

The following problem (from Larson and Odoni) is given in terms of refuse trucks that make pickups from nine locations. The capacity of the trucks is 23 units of trash. The distance table and the d_i units of trash at each location i are as follows:

Distance Table

Location	0	1	2	3	4	5	6	7	8	9	d_i
0	—										—
1	25	—									4
2	43	29	—								6
3	57	34	52	—							5
4	43	43	72	45	—						4
5	61	68	96	71	27	—					7
6	29	49	72	71	36	40	—				3
7	41	66	81	95	65	66	31	—			5
8	48	72	89	99	65	62	31	11	—		4
9	71	91	114	108	65	46	43	46	36	—	4

If the distance table is symmetric, Larson and Odoni suggest using the blank upper part of the table as the locations in which to place the corresponding s_{ij}'s. Solve this problem using the exchange algorithm. The solution requires two tours, (0,1,2,3,4,0) and (0,6,5,9,8,7,0), with a total distance of 397; the initial nine-tour solution had a distance of 836. Note that the solution requires two trucks or one truck that goes out twice. If $k = 16$, the exchange algorithm yields the three tours of (0,5,9,8,0), (0,2,3,4,0), and (0,1,6,7,0) with a distance of 520. This type of sensitivity analysis can be used by a city to determine what capacity refuse trucks to purchase, for example, $k = 16$ requires three trucks (or tours), while $k = 23$ requires only two tours that have less miles.

Variations of the vehicle-routing problem abound: each vehicle could have different load capacities; there could be more than one depot; time constraints could apply to a tour, that is, a tour must not take more than an 8-hour shift. Resolve the refuse-truck problem with $k = 23$, but add on a time restriction that requires that the total time t that it takes to load all the trash on a tour must be less than or equal to 4 hours. The following table lists the time t_i in hours it takes to load the trash at each location i:

Location	t_i
1	0.5
2	1.2
3	1.0
4	0.5
5	1.8
6	0.4
7	1.0
8	0.5
9	0.5

For what value of t will the original refuse tours of (0,1,2,3,4,0) and (0,6,5,9,8,7,0) still be feasible? What are the tours if $t = 3$ hours?

21.57. Another heuristic algorithm used to solve the vehicle-routing problem is the *sweep algorithm* of Gillett and Miller. This algorithm starts with any destination—say destination 1—by drawing a ray from the depot 0 to destination 1 and then moving (sweeping) the ray clockwise or counterclockwise. As the next destination is "swept," it is added to the route if the capacity restriction is not exceeded. If a destination cannot be added, it is used to initiate the next sweep, and so on. Once a 360° sweep is accomplished, the resulting clusters of destinations are used to form tours. The traveling-salesman tour for a cluster is figured out after the sweep. This algorithm requires the coordinates of the destinations and Euclidean distances. It is best to solve this problem many times by changing the starting destination. Each starting destination will produce a new set of clusters and routes. To be effective, quite a few destinations, if not all of them, should be used as the starting city, and clockwise and counterclockwise sweeps should be used. The best of all the sets of tours is selected. Try this algorithm out on your 10-city random plot. You need to impose a set of d_i and a k.

Discuss what you think are the merits and demerits of the exchange algorithm and the sweep algorithm. Construct examples for which each algorithm will yield a poor solution.

21.58. We always like to invent and work with algorithms that will find the optimal solution to a decision problem. Those like the simplex algorithm have a mathematical basis that guarantees our finding the optimum, assuming that we are diligent and correct in our numerical calculations—or that we have a computer that has a correctly programmed version of the simplex algorithm. However, there are many problems like the traveling-salesman problem that are beyond exact solution if they are large enough. For these problems, we resort to heuristic algorithms—rather simple computational procedures that take advantage of the problem's structure and are based on our insight and experience in problem solving. Heuristic algorithms yield satisfactory solutions; in most cases they are not optimal. We should not be deterred from studying a decision problem as there is no known way of finding the best solution. The process of problem formulation, combined with heuristic algorithm design, will usually provide us with an understanding that can be used to find an acceptable resolution of the problem.

Somewhat surprisingly, there is a decision problem related to networks and the traveling-salesman problem whose obvious heuristic algorithm yields the optimal solution. By an obvious heuristic we mean applying the greedy algorithm concept—do the best you can at each step and do not worry about its impact on future steps. Before we get to the problem, we need a few additional concepts from network theory.

We assume we have an undirected network with n nodes and that for any two nodes we can find a sequence of arcs that connects these nodes, that is, the network is connected. For a given set of nodes of a network, a *tree* is a set of arcs that connects these nodes without having any cycles. The first figure below is a connected network; the second is a tree that connects nodes 2, 3, and 4.

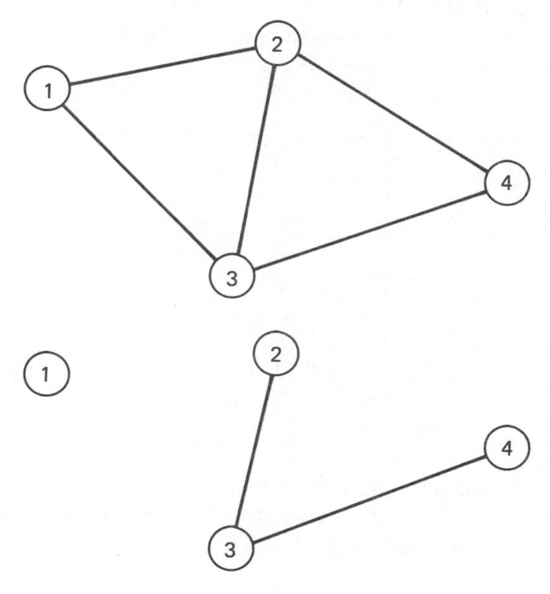

If a tree connects all the nodes of a network, it is called a *spanning tree*:

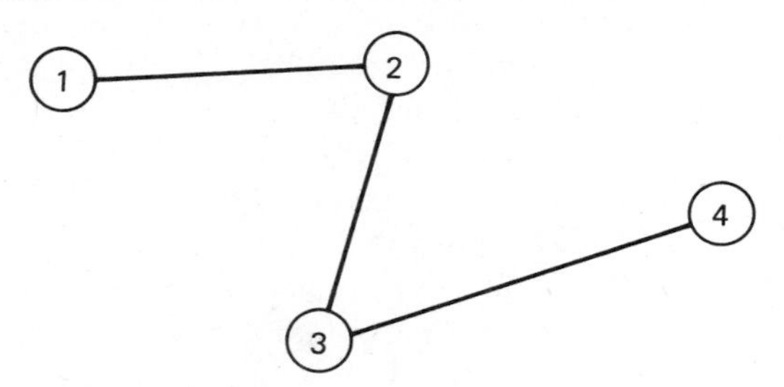

A tree has the property that there is a unique sequence of arcs joining each pair of nodes. If a network has n nodes, then a spanning tree has $(n - 1)$ arcs.

Let the arc that connects any two nodes i and j represent a distance c_{ij} (c_{ij} could be the cost of going from i to j). Here $c_{ij} = c_{ji}$. A connected network can have many different trees that span all n nodes, but we are interested in particular spanning trees that solve a decision problem related to the arc distances. This problem, one that has many applications, is to find a spanning tree whose sum of the arc distances in the tree is a minimum—the *minimal-spanning-tree* problem.

Can you think of any applications? What if the nodes are houses on a new development and you want to connect them with water, sewer, or gas lines; or the nodes are lawn sprinkler outlets that need to be connected by water pipes; or the nodes are cities that need to be connected by telephone lines or by new roads; or the nodes are pins on an electronics board that have to be connected by wire? For all these problems, we want to connect the nodes with the minimal amount of material. This means the finding of a minimal-spanning tree. How would you go about finding a minimal-spanning tree for the following network:

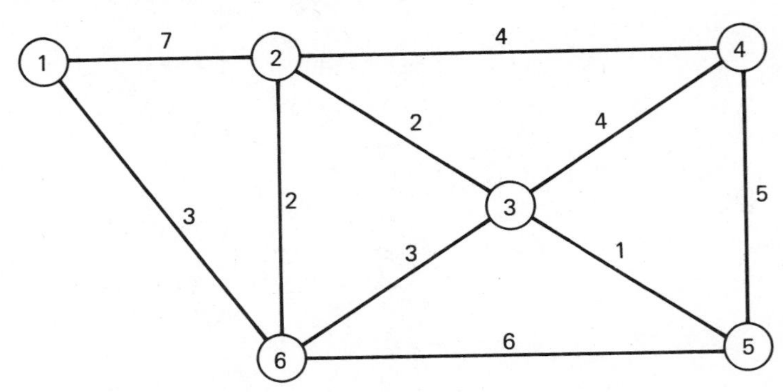

The only arcs that are allowed are shown in the figure with their respective lengths. Conceptually, we can assume that every node is connected to every other node, that is, the network is complete. For those arcs that do not actually exist, we just assign them an infinite distance. In the above figure, for example, $c_{13} = \infty$. Discuss how you would go about finding a minimal-spanning tree. What is your heuristic algorithm?

There are rather simple algorithms that solve the minimal-spanning-tree problem. They are examples of the greedy algorithm, or as it is termed by Ore, the *economy principle*.

Minimal-Spanning-Tree Algorithm 1 (Due to Kruskal)

Select the shortest arc of the network. At each successive stage, select from all arcs not previously selected, the shortest arc that completes no cycles with previously selected arcs. After $n - 1$ arcs have been selected, a minimal-spanning tree has been found. Ties in the selection process are broken arbitrarily.

For this procedure, you would arrange the arcs in a list from the shortest to the largest, and then go down the list selecting the shortest one available that does not complete a cycle. For the figure above, the list would be

Arc	Distance	Tree
(3,5)	1	Yes
(2,3)	2	Yes
(2,6)	2	Yes
(1,6)	3	Yes
(3,6)	3	No—cycle
(2,4)	4	Yes—tree complete
(3,4)	4	—
(4,5)	5	—
(5,6)	6	—
(1,2)	7	—

We would then select five arcs to form the following tree with length 12:

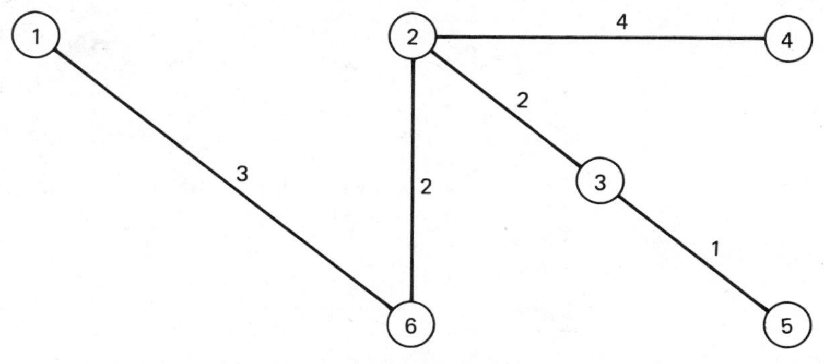

Arc (3,6) is not selected as it would complete a cycle with arcs (2,3) and (2,6) that have already been selected. If all arc lengths are distinct, then the

minimal-spanning tree is unique. Here we can obtain an alternate optimal spanning tree by substituting arc (3,4) for (2,4).

Try to prove that Algorithm 1 produces a minimal-spanning tree. The following discussion outlines the proof. Let the spanning tree produced by the algorithm be T. Denote the corresponding distance sum by $D(T)$. You should convince yourself that T is a spanning tree—it connects all nodes with $n - 1$ arcs and has no cycles. Let S be a minimal-spanning tree of the network, different from T, with distance sum $D(S)$. We assume that $D(S) < D(T)$. We want to show that $D(T) \leq D(S)$. Since T and S are not identical, there is at least one arc in T that is not in S. We find the first (the smallest distance) arc in the sequence of arcs that formed T that is not in S. Denote this arc by (k,l) and its distance by c_{kl}. If arc (k,l) is added to S, then we would have a cycle in S joining nodes k and l; the original S must contain a unique path that joins k and l, and adding (k,l) to S would then complete a cycle. This cycle must contain an arc (p,q) that is not in T as T has no cycles. If we remove (p,q) from S, we obtain another tree S_1 whose distance sum is given by $D(S_1) = D(S) + c_{kl} - c_{pq}$. But we know $D(S) \leq D(S_1)$, or $D(S) - D(S_1) \leq 0$, or, since $D(S) - D(S_1) = c_{pq} - c_{kl}$, we have $c_{pq} \leq c_{kl}$. In constructing T, arc (k,l) was the smallest arc that could be added to the previously selected arcs without producing a cycle. Also, if (p,q) is added to these arcs instead of (k,l), no cycle would be produced. Why? Thus, c_{pq} cannot be less than c_{kl} (or we would have selected it for T) and we have $c_{pq} = c_{kl}$. Then $D(S) = D(S_1)$, that is, S_1 is a minimal-spanning tree and S_1 has one more arc in common with T than S. We can continue the process (now using S_1) until we find a minimal-spanning tree S_m that is the same as T. The following figure for S might help in understanding the proof.

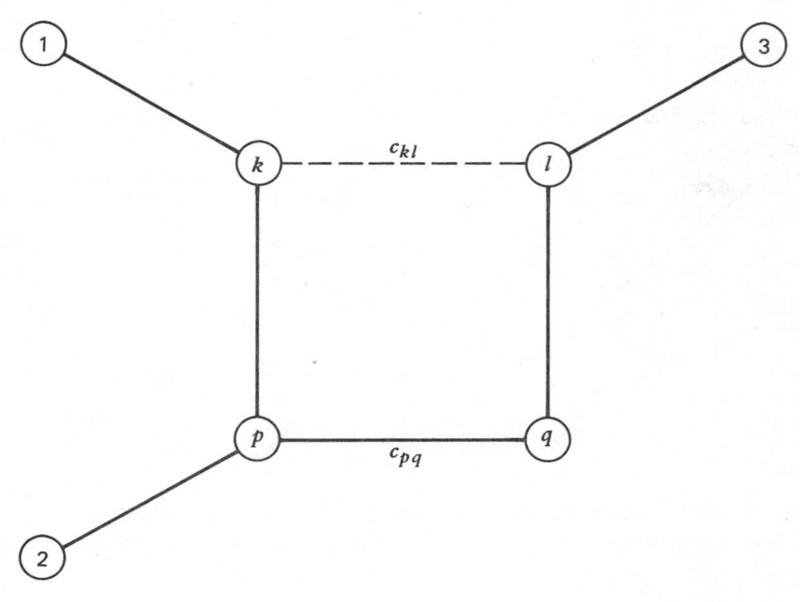

Minimal-Spanning-Tree Algorithm 2 (Due to Prim)

Select any node arbitrarily. Connect it to its nearest node. Next, find an unconnected node that is closest to a connected node and connect these two nodes. Repeat this until all nodes are connected. Ties are broken arbitrarily.

Try out Algorithm 2 on the previous example. Note that you can start anywhere. Why does this greedy algorithm work? Describe the differences between both algorithms in terms of hand computation and for use on a computer. Which is better for solution by a computer?

Minimal-Spanning-Tree Algorithm 3 (Due to Kruskal)

In his paper, Kruskal notes that if we assume that any two nodes of the network are connected (even arcs that have infinite distance), you can find the minimal-spanning tree by removing the largest arcs first, keeping the network connected, and stopping when there are only $n - 1$ arcs that connect the n nodes. These arcs form a minimal-spanning tree. Try this algorithm out on the following network (from Ford and Fulkerson):

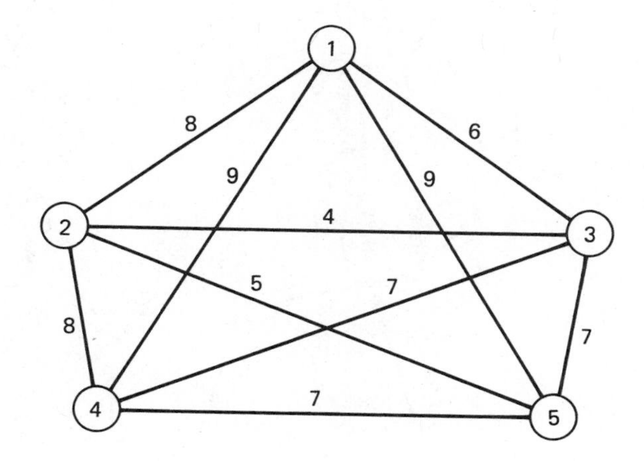

What about finding the maximal-spanning tree? Write out an algorithm and apply it to the above example.

How can you use the concept of a minimal-spanning tree as part of an algorithm to solve the traveling-salesman problem or vehicle-routing problem? Develop such an algorithm and try it out on your 10-city random problem. The following figure (from Prim) is the minimal-spanning tree that connects all the lower-48 U.S. state capitals and Washington, D.C. Using linear map distances, transform this spanning tree into a traveling-salesman tour that begins and ends in Denver. You might want to use the convex hull as an aid.

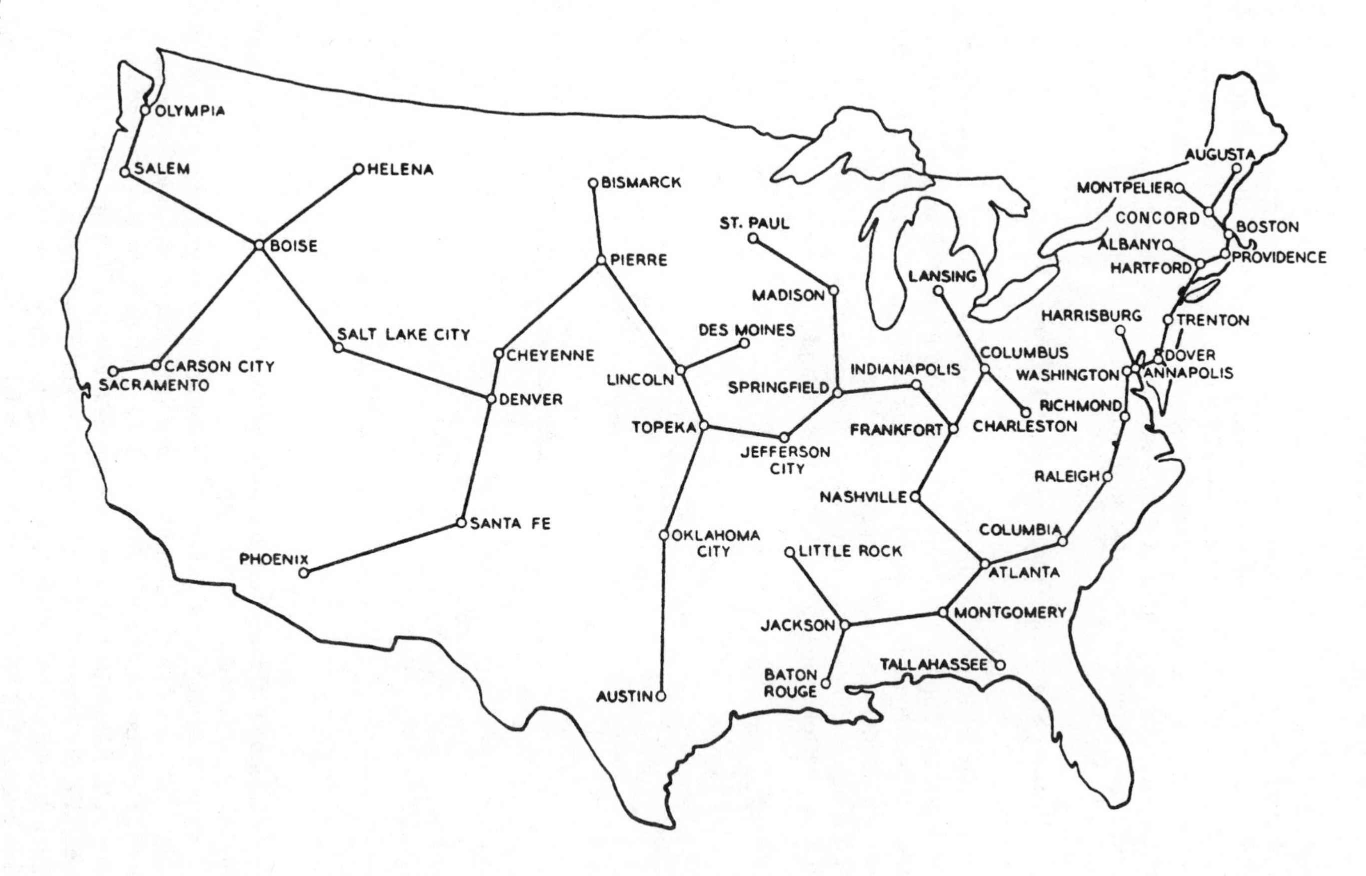
OLYMPIA
SALEM
HELENA
BOISE
CARSON CITY
SACRAMENTO
SALT LAKE CITY
PHOENIX
BISMARCK
PIERRE
CHEYENNE
DENVER
SANTA FE
ST. PAUL
MADISON
DES MOINES
LINCOLN
TOPEKA
OKLAHOMA CITY
AUSTIN
SPRINGFIELD
JEFFERSON CITY
LITTLE ROCK
JACKSON
BATON ROUGE
LANSING
INDIANAPOLIS
COLUMBUS
FRANKFORT
CHARLESTON
NASHVILLE
ATLANTA
MONTGOMERY
TALLAHASSEE
COLUMBIA
RALEIGH
RICHMOND
WASHINGTON
ANNAPOLIS
DOVER
HARRISBURG
TRENTON
HARTFORD
ALBANY
PROVIDENCE
BOSTON
CONCORD
MONTPELIER
AUGUSTA

21.59. In their paper, Dantzig et al. give a solution to a 49-city traveling-salesman problem, one city in each lower-48 state and Washington, D.C. The list of cities, the symmetric distance table (1954 road system), and the solution are given below. You might want to update the road distances (or use airline distances) and try to find the current optimal tour. [*Note:* The table distances d_{ij} have been transformed by the formula $d_{ij} = \frac{1}{17}(c_{ij} - 11)$, where c_{ij} is the actual road distance. This was done to make all numbers less than 256, which would permit compact storage of the distance table in binary notation in a computer.] The optimal tour is 1, 2, 3, . . . , 40, A, B, . . . , G, 41, 42, 1, with a total distance of 12,345 miles. Cities A through G must be traversed in that order in any optimal tour.

1. Manchester, N. H.
2. Montpelier, Vt.
3. Detroit, Mich.
4, Cleveland, Ohio
5. Charleston, W. Va.
6. Louisville, Ky.
7. Indianapolis, Ind.
8. Chicago, Ill.
9. Milwaukee, Wis.
10. Minneapolis, Minn.
11. Pierre, S. D.
12. Bismarck, N. D.
13. Helena, Mont.
14. Seattle, Wash.
15. Portland, Ore.
16. Boise, Idaho
17. Salt Lake City, Utah
18. Carson City, Nev.
19. Los Angeles, Calif.
20. Phoenix, Ariz.
21. Santa Fe, N. M.
22. Denver, Colo.
23. Cheyenne, Wyo.
24. Omaha, Neb.
25. Des Moines, Iowa
26. Kansas City, Mo.
27. Topeka, Kans.
28. Oklahoma City, Okla.
29. Dallas, Texas
30. Little Rock, Ark.
31. Memphis, Tenn.
32. Jackson, Miss.
33. New Orleans, La.
34. Birmingham, Ala.
35. Atlanta, Ga.
36. Jacksonville, Fla.
37. Columbia, S. C.
38. Raleigh, N. C.
39. Richmond, Va.
40. Washington, D. C.
41. Boston, Mass.
42. Portland, Me.

A. Baltimore, Md.
B. Wilmington, Del.
C. Philadelphia, Penn.
D. Newark, N. J.
E. New York, N. Y.
F. Hartford, Conn.
G. Providence, R. I.

Road Distances Between Cities in Adjusted Units

The figures in the table on page 316 are mileages between the two specified numbered cities, less 11, divided by 17, and rounded to the nearest integer.

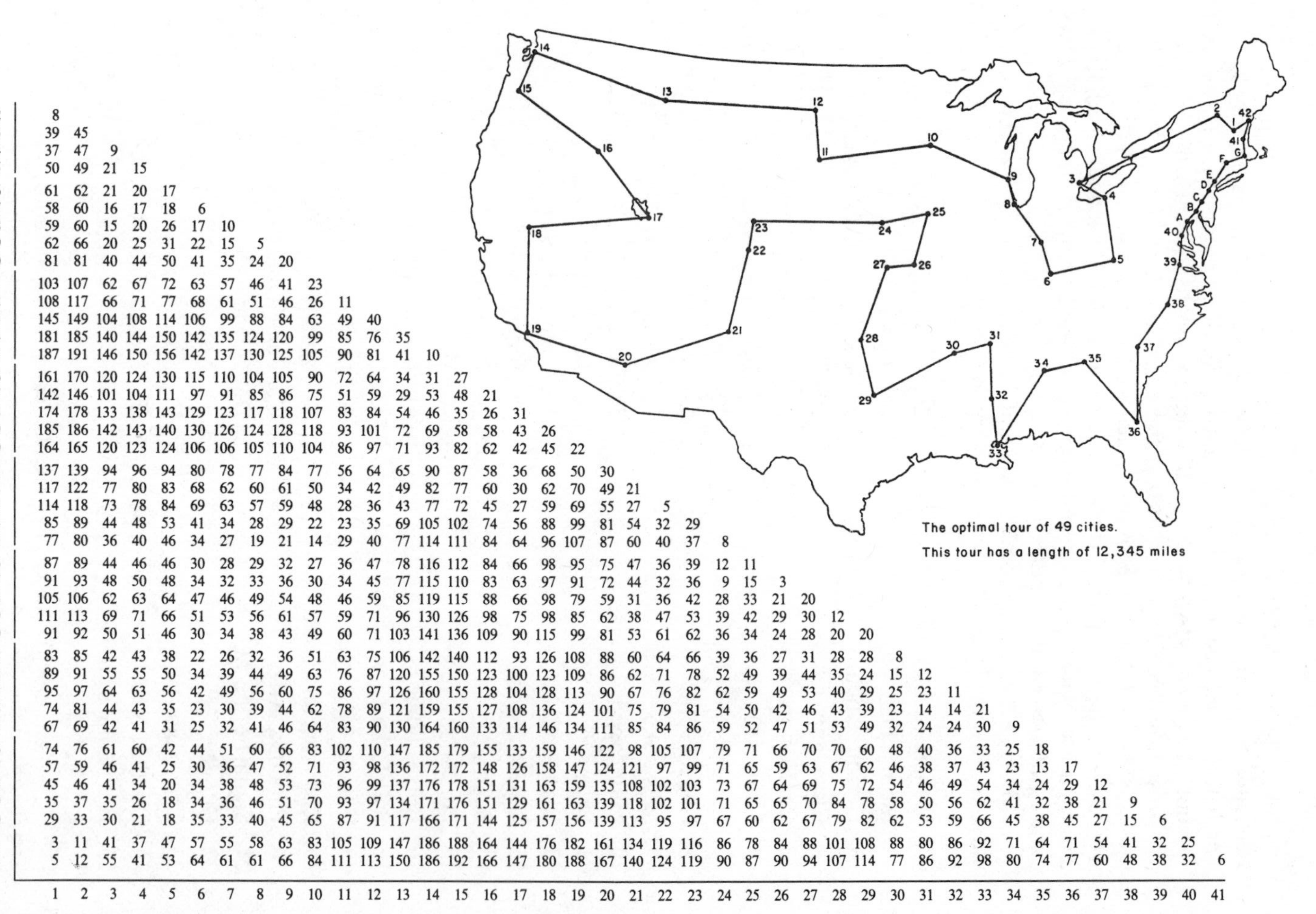

The optimal tour of 49 cities.

This tour has a length of 12,345 miles

	1	2	3	4	5	6	7	8	9	10	11	12	13	14	15	16	17	18	19	20	21	22	23	24	25	26	27	28	29	30	31	32	33	34	35	36	37	38	39	40	41
2	8																																								
3	39	45																																							
4	37	47	9																																						
5	50	49	21	15																																					
6	61	62	21	20	17																																				
7	58	60	16	17	18	6																																			
8	59	60	15	20	26	17	10																																		
9	62	66	20	25	31	22	15	5																																	
10	81	81	40	44	50	41	35	24	20																																
11	103	107	62	67	72	63	57	46	41	23																															
12	108	117	66	71	77	68	61	51	46	26	11																														
13	145	149	104	108	114	106	99	88	84	63	49	40																													
14	181	185	140	144	150	142	135	124	120	99	85	76	35																												
15	187	191	146	150	156	142	137	130	125	105	90	81	41	10																											
16	161	170	120	124	130	115	110	104	105	90	72	64	34	31	27																										
17	142	146	101	104	111	97	91	85	86	75	51	59	29	53	48	21																									
18	174	178	133	138	143	129	123	117	118	107	83	84	54	46	35	26	31																								
19	185	186	142	143	140	130	126	124	128	118	93	101	72	69	58	58	43	26																							
20	164	165	120	123	124	106	106	105	110	104	86	97	71	93	82	62	42	45	22																						
21	137	139	94	96	94	80	78	77	84	77	56	64	65	90	87	58	36	68	50	30																					
22	117	122	77	80	83	68	62	60	61	50	34	42	49	82	77	60	30	62	70	49	21																				
23	114	118	73	78	84	69	63	57	59	48	28	36	43	77	72	45	27	59	69	55	27	5																			
24	85	89	44	48	53	41	34	28	29	22	23	35	69	105	102	74	56	88	99	81	54	32	29																		
25	77	80	36	40	46	34	27	19	21	14	29	40	77	114	111	84	64	96	107	87	60	40	37	8																	
26	87	89	44	46	46	30	28	29	32	27	36	47	78	116	112	84	66	98	95	75	47	36	39	12	11																
27	91	93	48	50	48	34	32	33	36	30	34	45	77	115	110	83	63	97	91	72	44	32	36	9	15	3															
28	105	106	62	63	64	47	46	49	54	48	46	59	85	119	115	88	66	98	79	59	31	36	42	28	33	21	20														
29	111	113	69	71	66	51	53	56	61	57	59	71	96	130	126	98	75	98	85	62	38	47	53	39	42	29	30	12													
30	91	92	50	51	46	30	34	38	43	49	60	71	103	141	136	109	90	115	99	81	53	61	62	36	34	24	28	20	20												
31	83	85	42	43	38	22	26	32	36	51	63	75	106	142	140	112	93	126	108	88	60	64	66	39	36	27	31	28	28	8											
32	89	91	55	55	50	34	39	44	49	63	76	87	120	155	150	123	100	123	109	86	62	71	78	52	49	39	44	35	24	15	12										
33	95	97	64	63	56	42	49	56	60	75	86	97	126	160	155	128	104	128	113	90	67	76	82	62	59	49	53	40	29	25	23	11									
34	74	81	44	43	35	23	30	39	44	62	78	89	121	159	155	127	108	136	124	101	75	79	81	54	50	42	46	43	39	23	14	14	21								
35	67	69	42	41	31	25	32	41	46	64	83	90	130	164	160	133	114	146	134	111	85	84	86	59	52	47	51	53	49	32	24	24	30	9							
36	74	76	61	60	42	44	51	60	66	83	102	110	147	185	179	155	133	159	146	122	98	105	107	79	71	66	70	70	60	48	40	36	33	25	18						
37	57	59	46	41	25	30	36	47	52	71	93	98	136	172	172	148	126	158	147	124	121	97	99	71	65	59	63	67	62	46	38	37	43	23	13	17					
38	45	46	41	34	20	34	38	48	53	73	96	99	137	176	178	151	131	163	159	135	108	102	103	73	67	64	69	75	72	54	46	49	54	34	24	29	12				
39	35	37	35	26	18	34	36	46	51	70	93	97	134	171	176	151	129	161	163	139	118	102	101	71	65	65	70	84	78	58	50	56	62	41	32	38	21	9			
40	29	33	30	21	18	35	33	40	45	65	87	91	117	166	171	144	125	157	156	139	113	95	97	67	60	62	67	79	82	62	53	59	66	45	38	45	27	15	6		
41	3	11	41	37	47	57	55	58	63	83	105	109	147	186	188	164	144	176	182	161	134	119	116	86	78	84	88	101	108	88	80	86	92	71	64	71	54	41	32	25	
42	5	12	55	41	53	64	61	61	66	84	111	113	150	186	192	166	147	180	188	167	140	124	119	90	87	90	94	107	114	77	86	92	98	80	74	77	60	48	38	32	6

21.60. The branch and bound algorithm described in Section 21.27 was initially developed to solve the traveling-salesman problem. However, the general approach has proved of great value in solving optimization problems that have the usual linear constraints and linear objective function, but require all or some of the variables to be nonnegative integers. Although other integer algorithms have been proposed and used, the branch and bound method for integer-programming problems has been the most successful one to date (even though the number of integer variables that can be handled is only in the hundreds).

The basic idea is to solve the optimization problem (we assume minimization) as a standard linear-programming problem, that is, the integer conditions are relaxed and the variables are allowed to take on nonnegative values. The optimal value of the objective function for the relaxed problem is a lower bound for the objective function of the integer problem. (Why?) The branch and bound algorithm investigates a series of relaxed problems, using their objective-function values and noninteger solutions to guide the choice of the succeeding problems to be solved. We next describe the algorithm and illustrate it.

Branch and Bound Algorithm for Mixed-Integer Problems

First solve the associated (minimizing) continuous problem using the simplex algorithm. If the solution satisfies the integer conditions, the process stops. If not, select some integer variable x_k whose value x_{k0} in the continuous solution is not integer. For x_k to assume an integer value, we must have either $x_k \leq [x_{k0}]$ or $x_k \geq [x_{k0}] + 1$, where $[x_{k0}]$ indicates integer part of. We next construct two continuous problems as before, with the first (branch) being augmented by the condition $x_k \leq [x_{k0}]$ and the second (branch) by $x_k \geq [x_{k0}] + 1$. These problems are then solved by the simplex method. If either problem is infeasible, it (i.e., the branch) is excluded from any further consideration. If the new problems yield feasible solutions with objective-function values less than the value of any currently known integer solution, we add them to an active consideration list. If not, we can eliminate (bound) them from further consideration. (Why?) (We usually can select a reference integer solution to start the process or, at worst, assume that the best known value is $+\infty$.) Select from the active problem list the one that is associated with the lowest value of the objective function. If its (continuous) solution satisfies the integer constraints, then it is optimal. (Why?) If not, select some integer variable that has a fractional value (the one with the largest fraction is usually selected) and branch as above and solve the resulting two continuous problems. Repeat the steps, adding any feasible problems to the active list depending on how their objective-function values compare to the best known integer solution. This process will exhaust and/or eliminate all possibilities. (Can you prove why?) Branch and bound procedures can be viewed best in terms of a connected graph or tree in which the original continuous problem is the root node, with each new problem another node connected to a succeeding node by a branch defined by one of the bounding constraints on the selected integer variable x_k.

Example (The problem is from Müller-Merbach.):

$$\text{Minimize} \quad x_0 = -2x_1 - 5x_2$$

subject to

$$\begin{aligned} 2x_1 - x_2 &\le 9 \\ 2x_1 + 8x_2 &\le 31 \end{aligned} \qquad (x_j \ge 0 \text{ and integer})$$

The optimal solution is $x_1 = 3$, $x_2 = 3$; $x_0 = -21$. The branch and bound tree is the following:

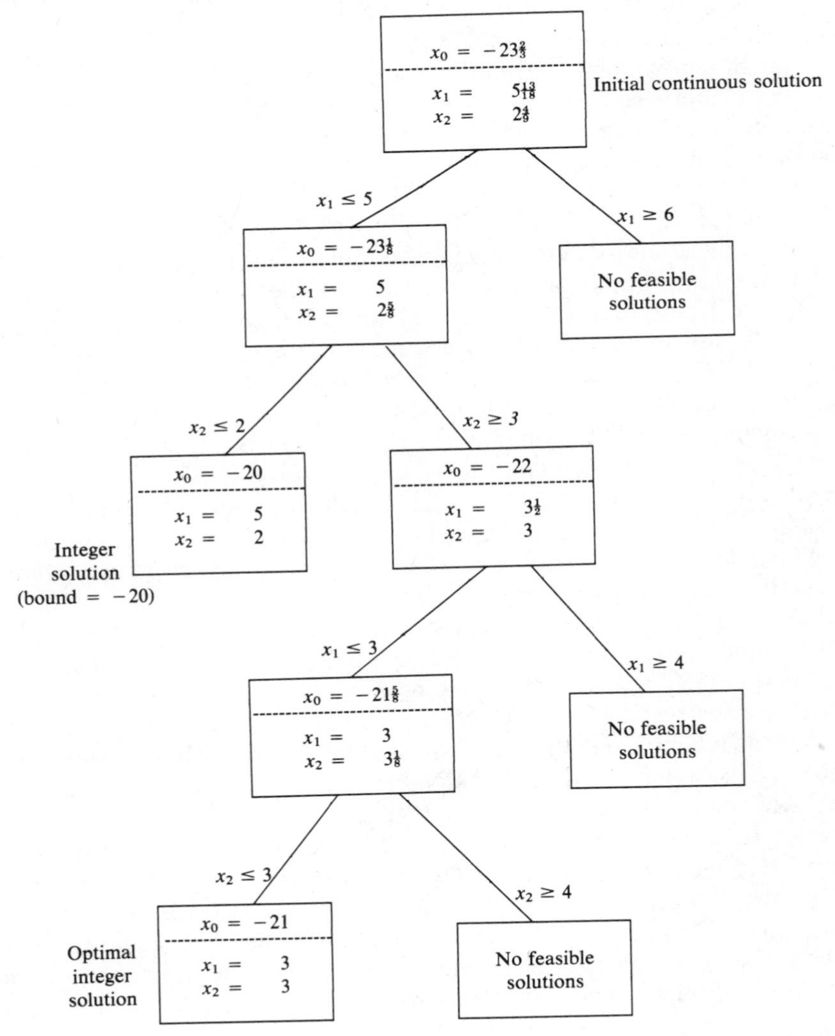

You should carry out the simplex computations that are required for the above branch and bound tree. Note that the continuous problem to be solved at each node must contain all the active variable bounding constraints that led to that node. Solve this problem graphically and indicate the node solutions on your graph.

21.61. Solve the following integer-programming problems using the branch and bound algorithm:

(a) Minimize $x_0 = 7x_1 + 3x_2 + 4x_3$

subject to

$$\begin{aligned} x_1 + 2x_2 + 3x_3 &\geq 8 \\ 3x_1 + x_2 + x_3 &\geq 5 \end{aligned} \qquad (x_j \geq 0 \text{ and integer})$$

The optimal solution is $x_1 = 0$, $x_2 = 5$, $x_3 = 0$; $x_0 = 15$. (The problem is from Garfinkel and Nemhauser.)

(b) Minimize $x_0 = -38x_1 - 20x_2 - 41x_3 - 35x_4$

subject to

$$\begin{aligned} 2x_1 + 2x_2 + 2x_3 + x_4 &\leq 32 \\ x_1 - 3x_3 + 5x_4 &\leq 2 \\ 2x_1 - 2x_2 + 5x_3 + 3x_4 &\leq 17 \end{aligned} \qquad (x_j \geq 0 \text{ and integer})$$

The optimal solution is $x_1 = 7$, $x_2 = 6$, $x_3 = 3$, $x_4 = 0$; $x_0 = 509$. (The problem is from Müller-Merbach.)

Part IV References

Balinski, M., and R. Gomory: A Primal Method for the Assignment and Transportation Problems, *Management Science*, Vol. 10, no. 3, April 1964.

Bazaraa, M. S., and J. J. Jarvis: *Linear Programming and Network Flows*, Wiley, New York, 1977.

Bellmore, M., and S. Hong: Transformation of the Multisalesmen Problem to the Standard Traveling Salesman Problem, *Journal of the Association for Computing Machinery*, Vol. 21, no. 3, July 1974.

Bodin, L., B. Golden, A. Assad, and M. Ball: Routing and Scheduling of Vehicles and Crews: The State of the Art, *Computers and Operations Research*, Vol. 10, no. 2, 1983.

Chartrand, G.: *Graphs as Mathematical Models*, Prindle, Weber & Schmidt, Boston, 1977.

Cheung, T.: Computational Comparison of Eight Methods for the Maximum Network Flow Problem, *ACM Transactions on Mathematical Software*, Vol. 6, no. 1, March 1980.

Clarke, G., and J. W. Wright: Scheduling of Vehicles from a Central Depot to a Number of Delivery Points, *Operations Research,* Vol. 12, no. 4, July–August 1964.

Cooper, L., and D. Steinberg: *Methods and Applications of Linear Programming*, Saunders, Philadelphia, 1974.

Crowder, H., and M. W. Padberg: Solving Large-Scale Symmetric Traveling Salesman Problems to Optimality, *Management Science*, Vol. 26, no. 5, 1980.

Dantzig, G. B.: *Linear Programming and Extensions*, Princeton Univ. Press, Princeton, N.J., 1963.

Dantzig, G., R. Fulkerson, and S. Johnson: Solution of A Large-Scale Traveling Salesman Problem, *Operations Research*, Vol. 2, no. 4, November 1954.

Dijkstra, E. W.: A Note on Two Problems in Connection with Graphs, *Numerische Mathematik*, pp. 269–271, Vol. 1, 1959.

Dreyfus, S. E.: An Appraisal of Some Shortest Path Algorithms, *Operations Research*, Vol. 17, no. 3, 1969.

Edmonds, J., and E. J. Johnson: Matching, Euler Tours and the Chinese Postman, *Mathematical Programming*, Vol. 5, no. 1, 1973.

Eilon, S., C. Watson-Gandy, and N. Christofides: *Distribution Management*, Griffin, London, 1971.

Ford, L. R., Jr., and D. R. Fulkerson: *Flows in Networks*, Princeton Univ. Press, Princeton, N.J., 1962.

Garfinkel, R. S. and G. L. Nemhauser: *Integer Programming*, Wiley, New York, 1972.

Gass, S. I.: *An Illustrated Guide to Linear Programming*, McGraw-Hill, New York, 1970.

Gass, S. I.: *Linear Programming: Methods and Applications*, 5th Edition, McGraw-Hill, New York, 1985.

Gillett, B. E., and L. R. Miller: A Heuristic Algorithm for the Vehicle-Dispatch Problem, *Operations Research*, Vol. 22, no. 2, March–April 1974.

Glover, F., J. Hultz, and D. Klingman: Improved Computer-Based Planning Techniques, Part 1, *Interfaces*, Vol. 8, no. 4, August 1978.

Glover, F., and D. Klingman: A Practitioner's Guide to the State of Large Scale Network and Network-Related Problems, *AFIPS Conference Proceedings*, Vol. 45, AFIPS Press, Montvale, N.J., 1976.

Golden, B.: Evaluating a Sequential Vehicle Routing Algorithm, *AIIE Transactions*, Vol. 9, no. 2, June 1977.

Golden, B., L. Bodin, T. Doyle, and W. Stewart, Jr.: Approximate Traveling Salesman Algorithms, *Operations Research*, Vol. 28, no. 3, Part II, 1980.

Hadley, G.: *Linear Programming*, Addison-Wesley, Reading, Massachusetts, 1962.

Held, M., and R. M. Karp: The Traveling-Salesman Problem and Minimum Spanning Trees, *Operations Research*, Vol. 18. no. 6, November–December 1970.

Hesse, R., and G. Woolsey: *Applied Management Science*, Science Res. Associates, Chicago, 1980.

Hillier, F., and G. J. Lieberman: *Operations Research*, 3rd Edition, Holden-Day, San Francisco, Calif., 1979.

Krekó, B.: *Linear Programming*, American Elsevier, New York, 1968.

Kruskal, J. B., Jr.: On the Shortest Spanning Subtree of a Graph and the Traveling Salesman Problem, *Proceedings American Mathematical Society*, pp. 48–50, Vol. 7, 1956.

Kuhn, H.: The Hungarian Method for the Assignment Problem, *Naval Research Logistics Quarterly*, Vol. 2, nos. 1 & 2, March–June 1955.

Kwak, N. K.: *Mathematical Programming with Business Applications*, McGraw-Hill, New York, 1973.

Kwan, M.-K.: Graphic Programming Using Odd or Even Points, *Chinese Mathematics*, pp. 273–277, Vol. 1, 1962.

Larson, R. C., and A. R. Odoni: *Urban Operations Research*, Prentice-Hall, Englewood Cliffs, New Jersey, 1981.

Lawler, E. L., and D. E. Wood: Branch-and-Bound Methods: A Survey, *Operations Research*, Vol. 14, no. 4, 1966.

Lin, S.: Computer Solutions of the Traveling Salesman Problem, *Bell Systems Technical Journal*, pp. 2245–2269, Vol. 44, 1965.

Little, J. D. C., K. G. Murty, D. W. Sweeney, and C. Karel: An Algorithm for the Traveling Salesman Problem, *Operations Research*, Vol. 11, no. 6, 1963.

Machol, R. E.: *Elementary Systems Mathematics*, McGraw-Hill, New York, 1976.

Minieka, E. *Optimization Algorithms for Networks and Graphs*, Marcel Dekker, New York, 1978.

Minieka, E.: The Chinese Postman Problem for Mixed Networks, *Management Science*, Vol. 25, no. 7, July 1979.

Müller-Merbach, H.: *Tri-Branching in Integer Programming*, Technical Report, Technische Hochschule, Darmstadt, 1982.

Munkres, J.: Algorithms for the Assignment and Transportation Problem, *SIAM Journal of Applied Mathematics*, Vol. 5, no. 1, 1957.

Murty, K. G.: *Linear and Combinatorial Programming*, Wiley, New York, 1976.

Oppenheim, M.: *Applied Models in Urban and Regional Analysis*, Prentice-Hall, Englewood Cliffs, N.J., 1980.

Ore, O.: *Graphs and Their Uses*, Random House, New York, 1963.

Padberg, M. W.: Covering, Packing and Knapsack Problems, *Annals of Discrete Mathematics*, pp. 265–287, Vol. 4, 1979.

Phillips, D. T., A. Ravindran, and J. J. Solberg: *Operations Research*, Wiley, New York, 1976.

Potts, R. B., and R. M. Oliver: *Flows in Transportation Networks*, Academic Press, New York, 1972.

Prim, R. C.: Shortest Connection Networks and Some Generalizations, *The Bell System Technical Journal*, pp. 1389–1401, Vol. 36, 1957.

Roberts, R. S.: *Graph Theory and Its Applications to Problems of Society*, SIAM Publication, Philadephia, 1978.

Rosenkrantz, D., R. Stearns, and P. Lewis: Approximate Algorithms for the Traveling Salesperson Problem, *Proceedings of the 15th Annual IEEE Symposium of Switching and Automatic Theory*, pp. 33–42, 1974.

Rothenberg, R. I.: *Linear Programming*, Elsevier North-Holland, New York, 1979.

Saaty, T. L., and R. G. Busacker: *Finite Graphs and Networks*, McGraw-Hill, New York, 1965.

Stern, H. I., and M. Dror: Routing Electric Meter Readers, *Computers and Operations Research*, Vol. 6, no. 4, 1979.

Svestka, J. A., and V. E. Huckfeldt: Computational Experience with an M-Salesman Traveling Salesman Algorithm, *Management Science*, Vol. 19, no. 7, March 1973.

Taha, H. A.: *Integer Programming*, Academic Press, New York, 1975.

Tucker, A.: *Applied Combinatorics*, Wiley, New York, 1980.

Vajda, S.: *Problems in Linear and Nonlinear Programming*, Hafner Press, New York, 1975.

Wagner, H. M.: *Principles of Operations Research*, 2nd Edition, Prentice-Hall, Englewood Cliffs, N.J., 1975.

Wardrop, J. G.: Some Theoretical Aspects of Road Traffic, *Proceedings of the Institute of Civil Engineering*, Part II, pp. 325–378, 1952.

Wiest, J. D., and F. K. Levy: *A Management Guide to PERT/CPM*, Prentice-Hall, Englewood Cliffs, N.J., 1969.

Wilson, R. J.: *Introduction to Graph Theory*, Longman Group Ltd., London, England, 1975.

PART V Games, Trees, and Decisions

Chapter 22 The Theory of Games 325

Chapter 23 The Fruits of Decision Trees 341

Chapter 24 The Analytic Hierarchy Process 355

Chapter 25 The Decision Framework, One More Time 368

Chapter 26 Part V Discussion, Extensions, and Exercises 373

22 The Theory of Games[1]

The play's the thing.

SHAKESPEARE

22.1 BASIC GAME CONCEPTS

Like linear programming, the theory of games can be considered a modern development in the field of mathematics. To the casual observer this would appear to be the only element that these areas have in common. In the general linear-programming problem, we are concerned with the efficient use or allocation of limited resources to meet desired objectives; in the theory of games we are interested in developing a pattern or strategy of play for a given game that will enable us to win as much as possible. A remarkable correspondence between these problems exists—the mathematical model of an important class of game-theory problems is identical to a linear-programming model. We shall describe some of the basic concepts of game theory and delve into its relationship to linear programming.

In general, the main concern of the theory of games is the study of the following problem: if n players, denoted by $P_1, P_2, \ldots, P_n$ play a given game, how must each player play to achieve the most favorable result? In interpreting this basic problem, we say that the term "game" refers to a set of rules and conventions for playing and a "play" refers to a particular

[1] From S. I. Gass, *An Illustrated Guide to Linear Programming*, McGraw-Hill Book Co., New York, N.Y., 1970. Reproduced with permission.

possible realization of the rules, that is, an individual contest. At the end of a play of a game, each of the players receives an amount of money, called the *payoff*. For those players that lose, the payoff is a flow of cash out of the losers' pockets into the pockets of the winning players. If all the money which is lost and won stays with the players, the game is called *zero-sum*. A game which is not zero-sum, for example, would be poker in which the house takes a percentage of each pot.

Games are also classified by the number of players and possible moves. Chess is a *two-person game* with a finite number of moves (if we include appropriate "stop rules"), and poker is a *many-person game,* also with a finite number of moves. A duel in which the duelists may fire at any instant in a given time interval is a two-person game with an infinite number of possible moves. Games are further characterized by being *cooperative* or *noncooperative*. In the former, the players have the ability to gang up on other players and work as teams, while in the latter players are concerned only with their results. Two-person games are, of course, noncooperative. We shall only discuss finite, zero-sum, two-person games; it is this type of game which bears a close relationship to linear programming.

As our first example of such a game, let us consider the now classical analysis of a real-life strategic situation—the Battle of the Bismark Sea.

22.2 HOW THE BATTLE GOT ITS NAME

In February–March 1943 intelligence reports indicated that a Japanese troop-and-supply convoy was assembling at Rabaul, New Britain, for movement to Lae, New Guinea. General Kenney, the Commander of the Allied Air Forces in the Southwest Pacific Area, was ordered by General MacArthur to intercept and inflict maximum destruction on the convoy.

The Japanese Commander—Isoruku Yamamoto—had the choice of sending the 16-ship convoy north of New Britain or south of that island. Either route required three days. These choices represented his possible strategies. Weather reports—which were available to both sides—indicated rain and poor visibility over the northern route, while it was expected to be clear in the south.

General Kenney limited himself to two possible courses of action. He could concentrate most of his search aircraft on one route or the other. The bombing force could strike the convoy on either route soon after it was found.

General Kenney's mission was to find the convoy and cause it to suffer as many days' exposure to his bombers as possible; the Japanese commander desired the minimum exposure to bombing. Each would make his decision independent of the other. What course of action should each choose? We first need to develop some additional information before the strategies can be selected.

Taking the weather, mobility of forces, and related constraints into consideration, General Kenney could construct a 2 × 2 tableau—or as we shall call it a *payoff matrix*—which indicated how many days of bombing he would have relative to how he deployed his search aircraft. The analysis goes something like this.

1. Kenney strategy—concentrate search aircraft in north
   a. *Convoy sails north.* Weather hampers search, but can get two days bombing because of proper concentration of search aircraft.
   b. *Convoy sails south.* Convoy in clear weather, but because of limited search aircraft can only expect two days of bombing.
2. Kenney strategy—concentrate search aircraft in south
   a. *Convoy sails north.* Poor visibility and badly positioned search aircraft limits bombing to one day.
   b. *Convoy sails south.* All things going properly—good weather, bulk of search aircraft—convoy would suffer three days of bombing.

Arranging the data in payoff matrix form, we have

		Japanese strategies (convoy route direction)	
		Northern route	Southern route
Kenney strategies (search aircraft concentration)	Northern route	2 days	2 days
	Southern route	1 day	3 days

To treat this military situation as a problem in game theory, we have to assume that the Japanese commander would have the same knowledge as General Kenney—the Japanese commander should be able to construct the same payoff matrix and interpret it in a similar fashion. The game is zero-sum in that a day's bombing gained by General Kenney is a day lost by the Japanese; that is, an outcome judged good by one commander is judged equally bad by the other. Part of the difficulty with game theory is obtaining the agreement between the antagonists as to the actual utility of the numbers that make up the payoff matrix.

By analyzing this simple two-strategy problem many of the features of game theory materialize. Let us generalize the concept of a payoff matrix to the concise notation of matrix algebra and represent the game as a 2×2 matrix, with the rows corresponding to the strategies of the *maximizing player* (General Kenney) and the columns to the *minimizing player* (Commander Yamamoto). We have

$$\begin{pmatrix} 2 & 2 \\ 1 & 3 \end{pmatrix}$$

as the payoff matrix.

We now join the commanders sitting in their respective headquarters staring at the numbers, trying to determine what to do. The Japanese commander, an expert in the game of Go, feels confident that he can select the best strategy for his side. The numbers in the payoff matrix cause him some concern, since it appears as if no matter what he does, his convoy will be found and subject to some damage. It is a no-win situation. The best he can hope for is one day's exposure to the bombers. Can he achieve that goal?

In comparing the columns—the northern-strategy payoffs to the southern-strategy payoffs—the Japanese commander notices that he should not select the southern strategy. If he did, he could expose himself to two days or possibly three days of bombing, while the northern-convoy route's corresponding exposure was two days or one day. The Japanese commander crosses the southern strategy from his matrix. He can even send such information direct to General Kenney, for the General, being a master strategist, would also remove the same column from his matrix. (We say that a strategy—row or column—*dominates* a second strategy if the choice of the first strategy is at least as good or better as the choice of the second strategy.)

The General now has a reduced matrix of two rows, representing his strategies, and one column, representing the fact that the Japanese must take the northern route. The reduced payoff matrix is

$$\begin{pmatrix} 2 \\ 1 \end{pmatrix}$$

The problem is now quite simple as the first row dominates the second row. As General Kenney wishes to maximize the days of bombing, he selects the northern route. And, as though calculated by the mathematicians, the Japanese actually took the northern route, General Kenney actually sent the bulk of his search aircraft to the northern route, and the Battle of the Bismark Sea occurred in the right place.

22.3 MATRIX GAMES: THE MODEL

The solution to the convoy problem is of a special kind—we say it is solved by using a *pure strategy*—here the northern route is selected by each commander, positively. A commander or player doesn't even consider the other available strategy—they don't throw dice or a coin to make a decision. The reason for this is that the matrix of the Bismark Sea problem contains what is called a *saddle point*. For a matrix, a saddle point is an element which is both the minimum of its row and the maximum of its column—two days of bombing at the intersection of the northern-route strategies is such a number. The corresponding row and column represent the optimal strategies to be used by the opposing players, and the payoff value is the value of the saddle point. The reason for this is the underlying conservatism inherent in the theory of games. Let us explain.

We assume the following 3×3 matrix is associated with some game

$$\begin{pmatrix} 3 & 5 & 6 \\ 2 & -1 & 3 \\ 0 & 7 & 4 \end{pmatrix}$$

Recall that the numbers in the payoff matrix are written in terms of the maximizing player (e.g., General Kenney), so a positive number means the row player wins, a negative number means a loss of that amount, while a zero means no money exchanges hands. The row player, player one, looks at the numbers in the first row and notes that the worst thing that could happen if the first strategy is selected is that the column player, player two, also selects the first column strategy. If this happens, player one receives three units instead of five or six units. Player one performs the same analysis on the other two rows and finds the gains to be -1 unit, that is, lose one unit, and 0 units, respectively. We write these minimum row numbers alongside the matrix

$$\begin{pmatrix} 3 & 5 & 6 \\ 2 & -1 & 3 \\ 0 & 7 & 4 \end{pmatrix} \begin{matrix} \textcircled{3} \\ -1 \\ 0 \end{matrix}$$

These numbers represent the worst things that could happen to player one for each one of the row strategies. Player one can thus select the best of these worst events and can decide to always play strategy 1—here the gain will be at least three units for every play of the game. Can player two do anything to reduce this loss per play?

Performing a similar analysis in terms of player two's frame of reference, we see that if player two selects the first column, the worst thing that could happen is that player one selects the first row; that is, player two would have to pay player one a total of three units. For columns 2 and 3 the payoffs would be 7 and 6, respectively. Adjoining these numbers to the matrix we have

$$\begin{matrix} \begin{pmatrix} 3 & 5 & 6 \\ 2 & -1 & 3 \\ 0 & 7 & 4 \end{pmatrix} & \begin{matrix} \textcircled{3} \\ -1 \\ 0 \end{matrix} \\ \begin{matrix} \textcircled{3} & \quad 7 & \quad 6 \end{matrix} & \end{matrix}$$

The best thing player two can do in this situation, besides not play the game, is to always play the first column and always lose three units. The element 3 in the matrix is a saddle point. Games with saddle points are readily solved by the above analysis. We see that if a player deviates from the pure strategy which corresponds to the location (row or column) of a saddle point, there could be a greater loss or less winnings.

The interesting games, however, are those without saddle points. If we attempt the saddle-point analysis on the game defined by the 3×4 matrix

$$\begin{pmatrix} 1 & 5 & 0 & 4 \\ 2 & 1 & 3 & 3 \\ 4 & 2 & -1 & 0 \end{pmatrix}$$

we see that the best of the worst things—this is called the *max–min*—for player one occurs for the second row, and the worst of the best things—the *min–max*—for player two occurs for the third column

$$\begin{pmatrix} 1 & 5 & 0 & 4 \\ 2 & 1 & 3 & 3 \\ 4 & 2 & -1 & 0 \end{pmatrix} \begin{matrix} 0 \\ \textcircled{1} \\ -1 \end{matrix}$$
$$\begin{matrix} 4 & 5 & \textcircled{3} & 4 \end{matrix}$$

Here we do not have a saddle point. However, player one can guarantee winning at least one unit by always playing the pure max–min strategy 2; while player two can guarantee not losing more than three units by always playing the pure min–max strategy 3. For games of this sort, the players can gain a better payoff, on the average, if they allow themselves to select among the available strategies in a probabilistic fashion. Here, player one should be able to win more than one unit, while player two's losses should be less than three units. By allowing themselves to mix the strategies and randomly select a particular one for a particular play of the game, the players can drive the expected value of the game to a number between the max–min value of one unit and the min–max value of three units; that is, each does better than the corresponding pure max–min/min–max strategies. We shall illustrate this concept, the use of a *mixed strategy*, with the well-known game of matching pennies.

To accomplish one move in this game, which here is equivalent to a play, the first player selects either heads or tails, and the second player, not knowing the other's choice, also selects heads or tails. After the choices are made known to each player, player two pays player one a unit $(+1)$ if they match or he receives a unit (-1) if they do not. The payoff of -1 represents the giving of a unit by player one to player two. The above can be summarized by the matrix:

		Player two selections	
		Heads	Tails
Player one selections	Heads	1	−1
	Tails	−1	1

We see that the matrix has no saddle point—the max–min is $+1$ and the min–max is -1.

We next define mixed strategies for each of the players. Player one would like to randomize the selection of heads or tails in that, if a fixed rule is used like "always play heads," the opponent, being a rational player (an assumption of game theory), would take advantage of such an aberration. We let x_1 be the probability that player one will select heads and x_2 the probability of selecting tails. We similarly define y_1 and y_2 for player two. A typical set of probabilities would be that player one selects the first strategy with probability $\frac{3}{4}$ and the second with probability $\frac{1}{4}$. The probabilities are interpreted in terms of the frequency of playing each strategy. For the mixed strategy $x_1 = \frac{3}{4}$ and $x_2 = \frac{1}{4}$, as player one played the game over and over again, on the average, the first strategy would be selected three-fourths of the time and the second strategy one-fourth of the time. As the probabilities for each player must sum to one, we always have $x_1 + x_2 = 1$ and $y_1 + y_2 = 1$, with the x's and y's nonnegative.

By definition, a *solution to a zero-sum game* is a pair of optimal mixed strategies—one for each player—and a number v, the value of the game, such that if player one uses an optimal mixed strategy against any strategy of player two, player one will be assured of winning at least v; and if player two plays an optimal mixed strategy against any strategy of player one, player two can be assured of not losing more than v. The main theorem of game theory shows that such optimal strategies and a number v always exists for *any* matrix. We next translate the above to a mathematical model for the matching-pennies problem.

Player one wishes to maximize the value of the game v subject to the condition that the probabilities $x_1 + x_2 = 1$ and the expected winnings by playing the strategy (x_1, x_2) against each of player two's pure strategies will be at least as great as v, that is,

$$\begin{aligned} x_1 - x_2 &\geq v \\ -x_1 + x_2 &\geq v \end{aligned}$$

The inequalities are obtained by multiplying the unknown nonnegative probabilities by the corresponding elements in the columns of the payoff matrix and summing the products, as indicated for the first column by

$$\begin{array}{l} \boxed{x_1 \times \ \ 1} \quad -1 \\ \quad + \\ \boxed{x_2 \times -1} \quad \ \ 1 \end{array}$$

Each such sum represents the *mathematical expectation* for player one as player two selects the corresponding pure strategy. Player one wants the expectation to be at least as great as the unknown value of the game, v. The value of v is not restricted as to its sign. A positive v means the game is biased to player one; a $v = 0$ means the game is *fair*; while a negative v indicates the game is biased to player two. Player one is interested in maximizing v and finding nonnegative values of x_1 and x_2 such that

$$\begin{aligned} x_1 - x_2 &\geq v \\ -x_1 + x_2 &\geq v \\ x_1 + x_2 &= 1 \end{aligned}$$

The number v is also a variable of the problem, and except for allowing v

to be negative as well as positive, the above problem is essentially a linear-programming problem. (This slight deviation can be taken care of in a number of ways. We can add a large positive number w to the game matrix so that all the elements are positive. The optimal strategies will be the same, but the new value would be equal to $v + w$, a positive number.)

A similar linear-programming problem can be constructed for the second player. This problem would be to minimize v subject to

$$\begin{aligned} y_1 - y_2 &\leq v \\ -y_1 + y_2 &\leq v \\ y_1 + y_2 &= 1 \end{aligned}$$

with $y_1 \geq 0$, $y_2 \geq 0$. The value of v will be the same. For the matching-pennies game, $v = 0$; that is, it is a fair game.

The linear-programming problems of player one and player two represent two problems which have a strong mathematical relationship in terms of primal and dual linear-programming problems. (Our game problems need a slight adjustment to be true primal and dual problems.)

By recalling the geometric approach to solving linear-programming problems described in Part III, we can develop an easy way to solve game-theory problems in which each opponent has a choice of only two strategies, that is, 2×2 games. (The procedure to be described can, in fact, be extended to solving $2 \times m$ games, where one player has only two strategies and the other has m strategies.) We shall solve the problem in terms of player one and leave it to you to solve it for player two. The matching-pennies problem to be considered, then, is the following:

$$\text{Maximize } v$$

subject to

$$\begin{aligned} x_1 - x_2 &\geq v \\ -x_1 + x_2 &\geq v \\ x_1 + x_2 &= 1 \\ x_1 &\geq 0 \\ x_2 &\geq 0 \end{aligned}$$

As the geometric approach to solving such problems is restricted to those having only two variables, we must first convert this problem, which has the three variables x_1, x_2, and v, to a problem with only two variables. This can be accomplished by using the equation $x_1 + x_2 = 1$ to represent one of the variables in terms of the other; that is, we let $x_2 = 1 - x_1$ and substitute this expression of x_2 in terms of x_1 to obtain the new constraints

$$\begin{aligned} x_1 - (1 - x_1) &\geq v \\ -x_1 + (1 - x_1) &\geq v \\ 0 \leq x_1 &\leq 1 \end{aligned}$$

or

$$\begin{aligned} 2x_1 - v &\geq 1 \\ -2x_1 - v &\geq -1 \\ 0 \leq x_1 &\leq 1 \end{aligned}$$

As this is player one's game model, we want to find the maximum value of v, recalling that v, the value of the game, is unrestricted as to its sign. We construct a two-dimensional graph with x_1 for one dimension and v for the other

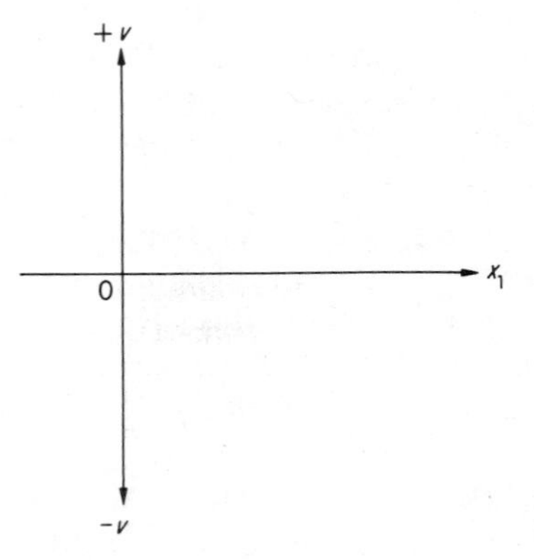

The inequality $0 \leq x_1 \leq 1$ restricts the value of x_1 to be a nonnegative number less than or equal to one. This is represented on the graph by the unbounded shaded area.

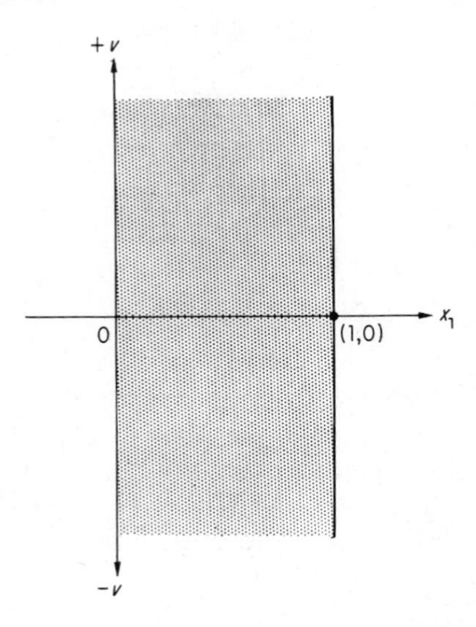

We next draw the lines $2x_1 - v = 1$ and $-2x_1 - v = -1$ to obtain the

joint solution space for the corresponding inequalities. This space is represented by the shaded area

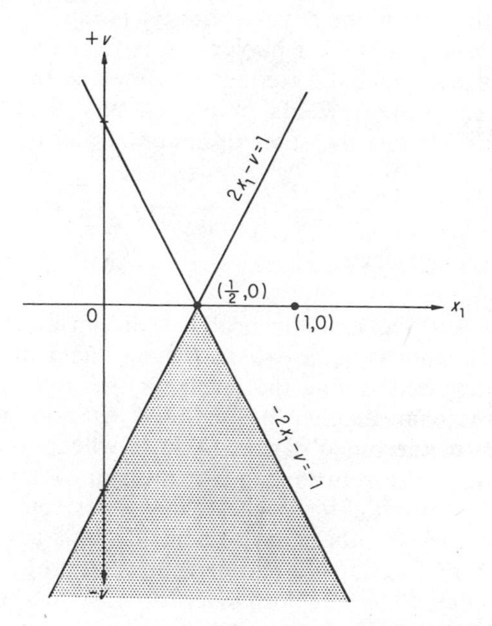

Superimposing these two graphs, we obtain the points which simultaneously satisfy the three constraints of the problem

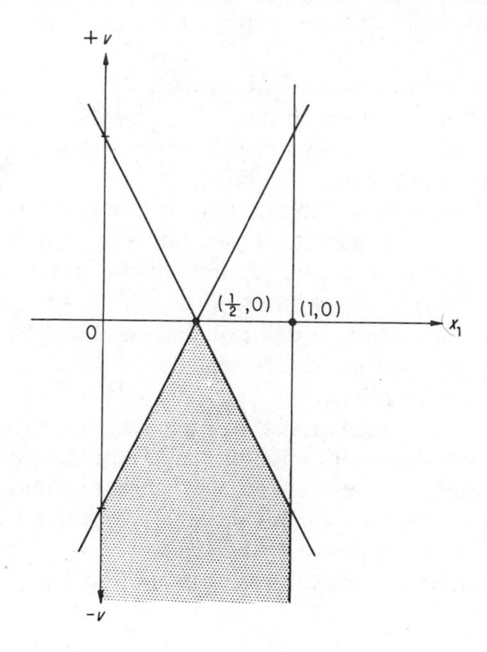

As we are looking for the point in the shaded region which has the maximum value of v, the optimal solution point is $x_1 = \frac{1}{2}$, $v = 0$. The solution is unique, as all other solutions have a negative value of v. Since $x_2 = 1 - x_1$, the optimal mixed strategy for player one is to randomize the selection of the two strategies so that the frequency of selection is $\frac{1}{2}$ for each one; $x_1 = \frac{1}{2}$, $x_2 = \frac{1}{2}$. This strategy yields an average payoff of zero; that is, the game is fair. Player two has the same optimum strategy.

22.4 AT THE CARNIVAL: THE SKIN GAME

For our last example, we join the brothers Simon at the annual Simple Manufacturing Company's picnic which is being held in conjunction with the county fair. Included among the guests is the ubiquitous Super Management Consultant team. Brother Si Simon has just rejoined the group after making the rounds of the bingo parlor, roulette wheel, and other games of chance. In fact, he really returned to borrow some money. Although his gaming ventures had emptied his pockets, his return route carried him past a new game which looked rather easy to beat. It was a card game, and after studying the rules, Si felt that he had a winning strategy. But none of his brothers, who all knew Si to be a born loser (he was vice-president in charge of returned goods), would lend him any money.

In desperation, he approached the Super Management Consultant team and explained his needs. "Tell us about this new game," they inquired, "and let us analyze it for you. If our analysis bears you out, we'll take up a collection."

"I know I can beat this game," Si replied. "Here are the rules: I play against the carnival man—we each have three cards. He has an ace of diamonds, an ace of clubs, and a two of diamonds; I also have an ace of diamonds, and an ace of clubs, but my third card is a two of clubs. We each select a card from our hands and simultaneously show it to each other. I win if the suits don't match; he wins if they do. If the two deuces are shown, there is no payoff. Otherwise, the amount of the payoff is the numerical value of the card shown by the winner. That's all there is to it. Pretty easy. All I need to do is mix up how I play my clubs and maybe throw in my ace of diamonds a few times. I'm sure to win. Oh, yes, they call this the skin game for some reason or other."

The SMC team recognized that this was a zero-sum two-person game with three possible strategies for each player. They huddled together, drawing figures on the paper picnic napkins until they reached a unanimous decision. Si should go elsewhere for his money—or better still, he should not play this game. Their analysis went like this.

The carnival man is the maximizing player one and Si the minimizing

player two. The payoff matrix is

$$
\begin{array}{cc}
 & \text{Si's strategies} \\
 & \begin{array}{ccc} \diamondsuit & \clubsuit & 2\clubsuit \end{array} \\
\text{Carnival man's strategies} \begin{array}{c} \diamondsuit \\ \clubsuit \\ 2\diamondsuit \end{array} & \begin{pmatrix} 1 & -1 & -2 \\ -1 & 1 & 1 \\ 2 & -1 & 0 \end{pmatrix}
\end{array}
$$

The carnival man would never select his first strategy—show the ace of diamonds—in that he can always get as much or better if he played the two of diamonds; that is, the payoffs for his third strategy are equal to or greater than the corresponding payoffs for his first strategy. Hence, he really plays the reduced 2×3 game

$$\begin{array}{c} \\ \clubsuit \\ 2\diamondsuit \end{array} \begin{array}{c} \begin{array}{ccc} \diamondsuit & \clubsuit & 2\clubsuit \end{array} \\ \begin{pmatrix} -1 & 1 & 1 \\ 2 & -1 & 0 \end{pmatrix} \end{array}$$

For this game Si would never play his third strategy—the deuce of clubs—in that he could do just as well, if not better, if he played his second strategy. The game finally reduces to the 2×2 game

$$\begin{array}{c} \\ \clubsuit \\ 2\diamondsuit \end{array} \begin{array}{c} \begin{array}{cc} \diamondsuit & \clubsuit \end{array} \\ \begin{pmatrix} -1 & 1 \\ 2 & -1 \end{pmatrix} \end{array}$$

Letting x_2 and x_3 be the probabilities that the carnival man plays his second and third strategies (here $x_1 = 0$), the model for this game is to

$$\text{Maximize } v$$

subject to

$$\begin{aligned} -x_2 + 2x_3 &\geq v \\ x_2 - x_3 &\geq v \\ x_2 + x_3 &= 1 \\ x_2 \quad &\geq 0 \\ x_3 &\geq 0 \end{aligned}$$

To convert it to a two-variable problem we let $x_3 = 1 - x_2$ and obtain the constraints

$$\begin{aligned} -x_2 + 2(1 - x_2) &\geq v \\ x_2 - (1 - x_2) &\geq v \\ 0 \leq x_2 &\leq 1 \end{aligned}$$

or

$$\begin{aligned} -3x_2 - v &\geq -2 \\ 2x_2 - v &\geq 1 \\ 0 \leq x_2 &\leq 1 \end{aligned}$$

The solution space in terms of x_2 and v is the shaded area.

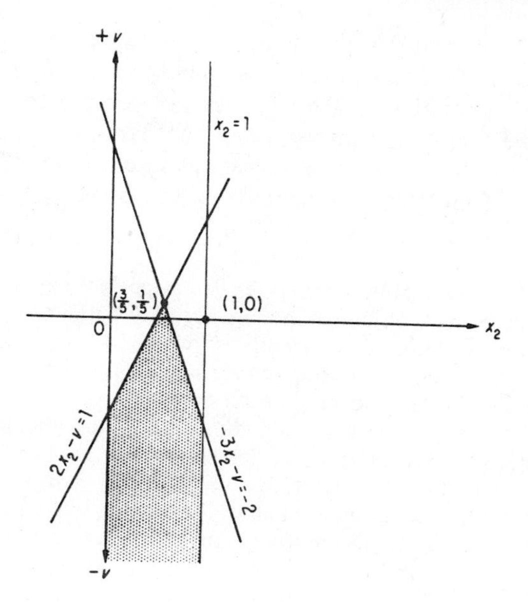

The optimum point is $x_2 = \frac{3}{5}$ and $v = \frac{1}{5}$. Thus, the value of the game is $\frac{1}{5}$ of a unit to the carnival man if he never plays his first strategy, plays the second with probability $\frac{3}{5}$, and plays his third strategy with probability $\frac{2}{5}$; that is, $x_1 = 0$, $x_2 = \frac{3}{5}$, $x_3 = \frac{2}{5}$, and $v = \frac{1}{5}$. The game is not fair; it is biased towards the carnival, and Si would continue his losing streak. But this logical analysis had no effect on Si, for "there's one born every minute."

22.5 GAMES AND LIFE

Since its inception in the 1940s by the mathematician John von Neumann and the economist Oskar Morgenstern, the theory of games has been long on theory but short on real-world applications. This is true for all variations of the game situation: zero-sum or nonzero-sum, two-person or many persons, finite number of strategies or an infinite number of strategies. Game-theory applications require some aspect of competition and strategic decision making. Military situations are an obvious place to look. Game-theory combat problems include convoy protection against submarines, submarines versus submarine combat, and helicopter versus submarine search. Competitive business situations is another area that leads to some applications: advertising campaigns, competing for market shares, bidding for contracts, or land and oil leases. Some business game-theory problems are presented in Chapter 26.

You might question the use of the game-theory model for competitive situations that are only played once: the Battle of the Bismark Sea, submarine versus submarine attack, or bidding for a particular oil lease. The idea of a randomized strategy, probabilities, and mathematical expectation seems to imply that the games are to be played more than once. The development of the mathematical theory does not require that the games be repeatable. But we feel more comfortable in applying game-theory analysis to situations that are to be played many times like the heads–tails game or the skin game. The following comments from the book by Williams address this point.

> Consider a nonrepeatable game which is terribly important to you, and in which your opponent has excellent human intelligence of all kinds. Also assume that it will be murderous if this opponent knows which strategy you will adopt. Your only hope is to select a strategy by a chance device which the enemy's intelligence cannot master—he may be lucky of course and anticipate your choice anyway, but you have to accept some risk. Game theory simply tells you the characteristics your chance device should have.
>
> You may also adopt the viewpoint that you will play many one-shot games between the cradle and the grave, not all of them being lethal games, and that the use of mixed strategies will improve your batting average over this set of games.

23 The Fruits of Decision Trees

23.1 THE RATIONAL DECISION MAKER: UTILITY FUNCTIONS

Our decision framework assumes a dispassionate decision maker whose choice between alternative solutions is based on optimizing (maximizing or minimizing) a measure of effectiveness. We all recognize that decision making in business and government (and even in personal affairs) usually involves more factors than the ones we are able to stipulate in a mathematical model. But, the model and its solution do enable us to quantify many aspects of the decision problem and thus reduce the qualitative and often unstated bases for making decisions. In a crude, but often true sense, we say that someone exhibits *rational behavior* when the person (e.g., a consumer) attempts to maximize the utility or satisfaction associated with a choice among alternatives; similarly, a rational entrepreneur desires to maximize profits.

Game theory highlights our concern with rational behavior as the game-theory mathematical model is based on the following postulates (Luce and Raiffa):

1. Each player is fully aware of the rules of the game and the utility functions (payoff functions) of each player.
2. Of two alternatives which give rise to outcomes, a player will choose the one with the more preferred outcome, or, more precisely, in terms of the utility function, he will attempt to maximize expected utility.

The main difficulty with these postulates is that we cannot specify exactly an individual's utility function. Even if we could, it would certainly change over time. For example, a gambler exhibits different patterns of betting behavior based on current winnings and losses. Contrast your behavior when you go shopping and have extra money to spend with those times when you find yourself short of cash.

Economists and game theorists have devised a set of axioms that describe rational behavior. These axioms assume that you and I, acting as a decision maker (e.g., gambler, consumer, entrepreneur), can specify the outcomes of our decisions (the alternative choices) and arrange them in order of preference. We can also define a utility for each outcome that is consistent with the ordering. The utility serves as a function that assigns a numerical figure of merit to each outcome. It represents a concept of worth or satisfaction or profit or cost. A person whose behavior satisfies these axioms is said to be a *rational decision maker,* that is, the selected decision maximizes expected utility. You should be able to work out some of the basic ideas behind these axioms. For example, if you prefer alternative outcome O_1 to O_2, and O_2 to O_3, then you should prefer O_1 to O_3.

In the usual discussions on rational behavior and utility functions, it is assumed that the decision maker is faced with an uncertain future in which alternative outcomes or prizes or events $O_1, O_2, \ldots, O_m$ can occur with known probabilities $p_1, p_2, \ldots, p_m$, respectively. Each $p_i \geq 0$ and $\sum_{i=1}^{m} p_i = 1$. This is known as a *lottery* and denoted by $L = \{(p_1,O_1)(p_2,O_2) \ldots (p_m,O_m)\}$. It states that one and only one prize will be won and the probability it will be O_i is p_i. We denote a second lottery $\overline{L} = \{(\overline{p}_1,O_1)(\overline{p}_2,O_2) \ldots (\overline{p}_m,O_m)\}$, where the O_i are taken to be the same prizes as in L, but with different probabilities $\overline{p}_i \geq 0$ and $\sum_{i=1}^{m} \overline{p}_i = 1$. How does a decision maker choose between L and $\overline{L}$? For example, let the O_i represent the outcomes (measured in profit dollars) to a company if it funds a specified set of research projects. The probability of attaining an O_i is p_i, with the probability being a function of how the company's resources are allocated to the different projects. The company's research director has developed two possible allocations and associated sets of probabilities $\{p_i\}$ and $\{\overline{p}_i\}$ that lead to the lotteries L and $\overline{L}$. On what basis should the director choose one of the allocations? Can this be done rationally? What is missing is an expression of the utility or value of L and $\overline{L}$. This can be determined by first ordering the outcomes based on the preference of the research director, with O_i being preferred to O_{i+1}. Then the director assigns a utility u_i to outcome O_i such that $u_1 = 1 \geq u_2 \geq u_3 \geq \cdots \geq u_{m-1} \geq u_m = 0$. The expected utility of L is then given by $u(L) = p_1u_1 + p_2u_2 + \cdots + p_mu_m$; and for $\overline{L}$ it is $u(\overline{L}) = \overline{p}_1u_1 + \overline{p}_2u_2 + \cdots + \overline{p}_mu_m$. The rational decision maker will choose the lottery with the maximum expected utility.

Note that the best outcome O_1 is given a utility of 1 and the worst outcome O_m is given a utility of 0. The outcomes are thus assigned a position on the interval between 0 and 1. The $(u_1, \ldots, u_m)$ represent a utility function for the decision maker. It summarizes the decision maker's complete (explicit and implicit) view of the utility of each outcome to the organization. Two decision makers in the same decision environment would probably have two different utility functions.

The difficulty in this analysis is how to determine the probabilities of the outcomes and the utility function. Can you think of a lottery situation that you face? How about selecting between two sets of courses for next semester? Denote one set by S_1 and the other by S_2. The outcomes O_i of interest are the possible grade-point averages that you might receive for the courses in S_1 and S_2. Based on the course levels, your past work, and how you feel about taking certain courses, you should be able to estimate the probabilities of attaining the various grade-point averages (GPA). Let us assume you came up with the following table:

S_1		S_2	
GPA (O_i)	Probability (p_i)	GPA (O_i)	Probability ($\bar{p}_i$)
4.0	0.3	4.0	0.0
3.5	0.3	3.5	0.5
3.0	0.2	3.0	0.5
2.5	0.1	2.5	0.0
2.0	0.1	2.0	0.0

Based on your desire to make the dean's list or the honor society, you give the following utility values u_i to the outcomes O_i:

O_i		u_i
4.0	. . .	1.0
3.5	. . .	0.9
3.0	. . .	0.8
2.5	. . .	0.5
2.0	. . .	0.0

The expected utility of S_1 is

$$u(S_1) = \sum_{i=1}^{4} p_i u_i = (0.3)(1.0) + (0.3)(0.9) + (0.2)(0.8)$$

$$+ (0.1)(0.5) + (0.1)(0.0) = 0.78$$

The expected utility of S_2 is

$$u(S_2) = \sum_{i=1}^{4} \bar{p}_i u_i = (0.0)(1.0) + (0.5)(0.9) + (0.5)(0.8)$$
$$+ (0.0)(0.5) + (0.0)(0.0) = 0.85.$$

Thus, you would select the second set of courses S_2 as 0.85 is your maximum expected utility.

Note that the expected values of the GPA for S_1 and S_2 are

$$E(S_1) = \sum_{i=1}^{4} p_i O_i = (0.3)(4.0) + (0.3)(3.5) + (0.2)(3.0)$$
$$+ (0.1)(2.5) + (0.1)(2.0) = 3.3.$$

$$E(S_2) = \sum_{i=1}^{4} \bar{p}_i O_i = (0.0)(4.0) + (0.5)(3.5) + (0.5)(3.0)$$
$$+ (0.0)(2.5) + (0.0)(2.0) = 3.25.$$

Using expected value instead of expected utility, you would choose S_1. The choice of S_2 based on expected utility is due to your assurance of getting a B or better with little risk involved.

The choice of S_2 over S_1 is, of course, a function of your probability estimates and the utility function. You might want to see how sensitive the result is to a change in the probabilities and/or utility function.

23.2 UNCERTAIN MONEY: EXPECTED MONETARY VALUE

Our interest in this chapter is not with lotteries, but with establishing a basis for the use of the term rational behavior. We shall use this concept in developing the *decision-tree model* that is used to choose among alternative actions in the face of uncertain knowledge about the future. You should contrast the decision-tree approach with our previous decision models in which we tacitly assumed that our choice of optimal solution would happen with certainty (with probability equal to one).

Before climbing our first decision tree, we need to clarify some aspects of outcomes and utility. If our outcomes are dollars, you might be wondering why we just do not let \$2.00 have a utility value that is twice as much as the utility value of \$1.00. We tried to indicate that utility functions are not that straightforward. If you have a million dollars, an extra dollar has a low utility. If you only have \$10, that extra \$1.00 has a high utility. Sometimes you might find it difficult to develop a utility function. In most situations in which money outcomes O_i occur with probability p_i, it is usually correct to dispense with the utility function and select the decision that maximizes *expected monetary value* (EMV) $= \sum_i p_i O_i$. You should be aware, however, that this might not produce a result that everyone could agree to and/or the result could be different if you used a utility function. For example, the following problem (due to Hadley) illustrates part of the dilemma (also see Sections 26.15 and 26.16).

A builder is contemplating which one of two contracts, C_1 or C_2, to undertake. The possible profit, losses, and probabilities are given in the following table:

C_1		C_2	
Profit/Loss (O_i)	Probability (p_i)	Profit/Loss (O_i)	Probability (p_i)
\$100,000	0.6	\$50,000	0.3
−\$100,000	0.4	−\$10,000	0.7

We have that the EMV of C_1 = (0.6)(\$100,000) + (0.4)(−\$100,000) = \$20,000 and the EMV of C_2 = (0.3)(\$50,000) + (0.7)(−\$10,000) = \$8000. The maximum EMV is \$20,000; the contractor should select C_1. But, depending on how the contractor feels about C_1's having a rather high risk of losing \$100,000 and the contractor's ability to absorb such a high loss, the choice of C_1 might be the wrong thing to do. Can you see how using a utility function might resolve the choice? What would you do?

If you were a decision maker in a government setting in which outcome payoffs are not measured in profit, but in how decisions affect the public, you do need to transform your preferences into some appropriate utility function. Even here, however, there could be attempts to use surrogate payoff measures. For example, a police chief has to choose between different ways to allocate the police force between patrol, detectives, undercover agents, and so on. The chief may try to do so based on estimates of how the choices will impact the crime rate or arrest rate and use the expected rate as a selection mechanism. Such surrogates are sometimes easier to understand and justify than the use of a personal utility function. But, the police chief's utility function embodies the chief's experiences and knowledge and it should be given due respect. We pay our executive decision makers for their abilities; we are buying their utility functions. Possibly, one of their decision-making mechanisms is the use of explicit utility functions.

23.3 FUTURE MONEY: EXPECTED PRESENT VALUE

Let us return to dollars and future payoffs. When it comes to money, we all have ideas about its value. But, have you ever thought of what it means to you if someone promises to give you a dollar a year from today? Of what value is that dollar to you? What if a friend agrees to pay you 50¢ today if you agree to give your friend the future dollar? Would you do it? What if your friend raised the payment to 80¢ or 90¢? What payment would make you accept?

If you would not do it for anything less than $1.00, could you find a friend or a banker that would negotiate with you? Why should they? They could invest the $1.00 at the current interest rate (say 10%) and have $1.10 instead of a $1.00 a year from now.

We see that future money, in this case, $1.10, has a current value of $1.00. At the interest rate of 10%, if you invested 91¢ today, you would then have $1.00 a year from now, that is, next year's $1.00 is worth 91¢ today. You would like to get more than that amount from your friend, but from an economic point of view, you have no reason not to agree to an exchange for 91¢. The 91¢ is termed the *discounted value, present value,* or *present worth* of a dollar available to you a year from now. Based on economic considerations alone, not with what you could do with the money if you had it, you should be indifferent to receiving 91¢ today or $1.00 in a year.

Decision situations whose monetary outcomes take place in the uncertain future usually use the *expected present value* (EPV) as the measure of effectiveness, that is, in the expected value formula $\sum_i p_i O_i$ the monetary outcomes are assumed to be discounted values. In the previous example, in which the contractor had to choose between contracts C_1 and C_2, the outcomes should really be discounted profits or losses as they occur in the future.

If a monetary payment P (gain or loss) occurs n years in the future and the interest rate i is constant over these years, it is rather easy to determine today's present value V of P. For $n = 1$, we have

$$(1 + i)V = P$$

or

$$V = \frac{P}{(1 + i)}$$

We assume that the interest is computed each year and that the principal V and interest are reinvested each year. Then, for $n = 2$, we have

$$(1 + i)[(1 + i)V] = P$$

or

$$(1 + i)^2 V = P$$

or

$$V = \frac{P}{(1 + i)^2}$$

The amount V that needs to be invested today at interest rate i so that it will be equal to the payment P received n years from today is given by the compound-interest formula

$$V = \frac{P}{(1 + i)^n}$$

For $i = 10\%$, $n = 3$, and $P = \$100$, the present value V is

$$V = \frac{\$100}{(1 + 0.10)^3} = \$75.13$$

A rational person should be indifferent to receiving \$75.13 today or \$100 three years from now as the \$75.13 can be put into the bank at 10% interest so it "grows" to \$100 in three years. The amount V is sometimes referred to as *current dollars*. In our discussions below on decision trees, we shall assume that the measure of effectiveness is in terms of EPV or EMV, as appropriate. We are assuming that there is a direct relationship between EPV and EMV and the decision maker's utility for money, that is, the monetary values of the outcomes can be used as utility values (see Section 26.16).

23.4 SOYBEANS OR CORN: A DECISION TREE

In Section 26.8, we pose as a game the problem of the farmer who has to select a strategy that calls for the planting of all soybeans, all corn, or a mixture of the two, against nature's strategies of normal rainfall, drought, or heavy rainfall. For the following discussion, we limit the farmer's strategies to the planting of all soybeans or all corn (no mixture). Here we do not assume that nature is playing the game, but that weather is a chance event and the probability of each type happening is known and is based on historical data, that is, the frequency counts. For example, the farmer knows that over the past 50 years normal conditions occurred 55% of the time (0.55 probability), drought conditions 15% of the time (0.15 probability), and heavy rainfall 30% of the time (0.30 probability). With this information, and the profit-per-acre data (reproduced below), what should the farmer plant?

		Weather			
		Normal	Drought	Heavy rain	
Farmer	Plant soybeans	$10	$ 5	$12	Profit per acre
	Plant Corn	$ 7	$ 8	$13	

A rational farmer's decision would be based on expected monetary values. If this is the case, then the farmer can use a simple analysis structure—a *decision-tree model*—to determine which strategy maximizes the EMV measure of effectiveness. The decision tree is a means of interpreting the problem's alternative solutions and data. Unlike a linear-programming model in which the many alternatives are not known or given beforehand, a decision-tree analysis requires knowledge of all possible and/or relevant alternative solutions. The decision tree can be interpreted as a graph or network in which the nodes represent decisions or actions (e.g., what the farmer does) or chance events (e.g., what nature decides to do) and the arcs or branches represent the strategy selections that combine to form each alternative. We illustrate the construction of the decision tree for the farmer's problem. It is convenient to grow the tree from left to right and use squares for decision nodes and circles for chance-event nodes. The tree begins (i.e., is rooted) by the farmer's decision node that requires a strategy selection of planting soybeans or planting corn:

Decision Chance Event

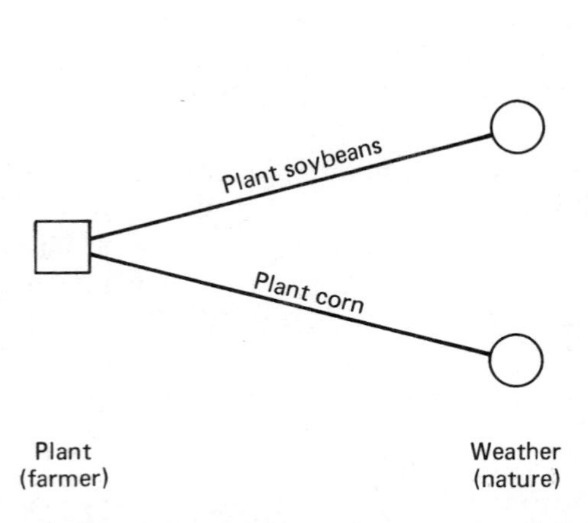

Each such decision is followed by a chance event that depends on nature. There are three weather possibilities—normal conditions, drought, heavy rainfall—with corresponding probabilities (0.55, 0.15, 0.30). These are represented on the decision tree, along with the payoffs for each decision/chance-event outcome combination:

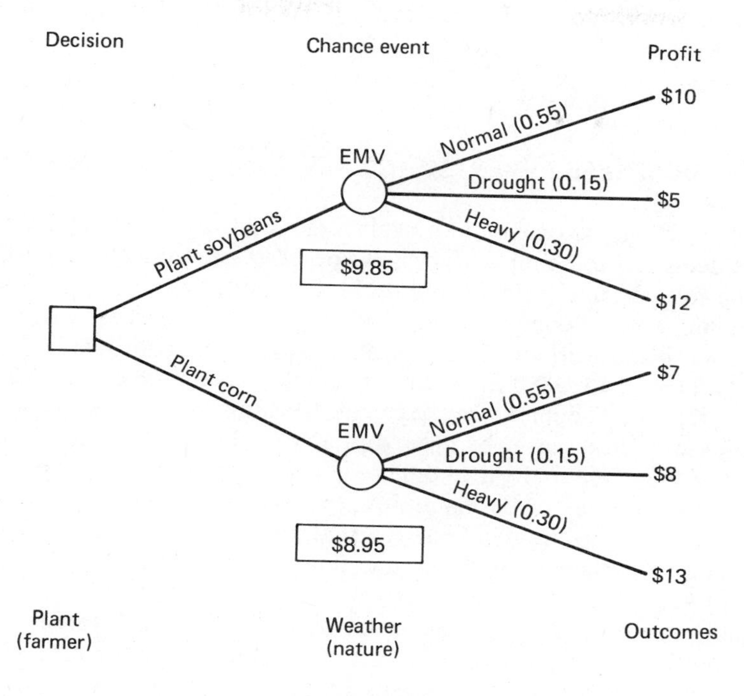

Farmer's Decision Tree

The farmer's decision tree is a graphic way of exhibiting all of the information that describes the problem. We see that there are six possible outcomes. One of them will actually occur; we will not know which one until after the crop is harvested. But, the farmer must make a decision now. The rational approach is for the farmer to calculate the expected monetary value per acre for each of the farmer's strategy choices and select the choice that maximizes the EMV. Here we have

$$\text{EMV(soybeans)} = (0.55)(\$10) + (0.15)(\$5) + (0.30)(\$12) = \$9.85$$

$$\text{EMV(corn)} = (0.55)(\$7) + (0.15)(\$8) + (0.30)(\$13) = \$8.95.$$

Thus, the farmer should plant soybeans as the EMV (soybeans) is the maximum by 90¢. Based on this analysis, should the farmer consider planting a mixture of soybeans and corn? How does this solution compare to the game-theory solution? For that problem, the farmer's optimal strategy was to plant one-sixth of the acreage with soybeans and five-sixth of the acreage with corn to achieve an expected profit (value of the game) of $7.50 per acre.

Show that if the weather probabilities were (0.5,0.5,0), that is, nature played each of the normal and drought strategies 50% of the time, the farmer would always have an expected value of \$7.50 per acre no matter what strategies the farmer employed. In this case, the (0.5,0.5,0) weather is nature's "optimal" strategy.

23.5 BUSINESS INVESTMENT: A BIGGER DECISION TREE

Decision trees are quite useful for analyzing business decision situations that require decisions to be made over a period of years. Consider the following example (based on a decision tree by McCreary).

A manufacturing company, faced with a possible increased demand for its product, has the options of buying and installing an additional production unit (at a cost of \$80,000) or putting its employees on overtime (at a yearly cost of \$52,000). The marketing department estimates there is a 60% chance that sales would increase (by 20%) and a 40% chance that sales would decrease (by 5%). Due to the lead time required to order and install new equipment, a decision has to be made now to enable the company to meet the possible increase in the demand. We first analyze the company's problem over a one-year planning horizon, and then extend the analysis to consider a second year.

Based on costs, depreciation, productivity of the work force, assumed sales, and so on, the company planners have determined the following end-of-one-year discounted profit figures for the various action chance-event combinations:

Actions	Profit (discounted) if sales rise ($p = 0.6$)	Profit (discounted) if sales drop ($p = 0.4$)
Purchase new equipment (cost: \$80,000)	\$460,000	\$340,000
Overtime (cost: \$52,000)	\$440,000	\$380,000

We can picture the decision-outcome possibilities by the following decision tree, assuming the company wants to make the decision that maximizes expected discounted net profits, that is, expected present value (EPV) minus costs.

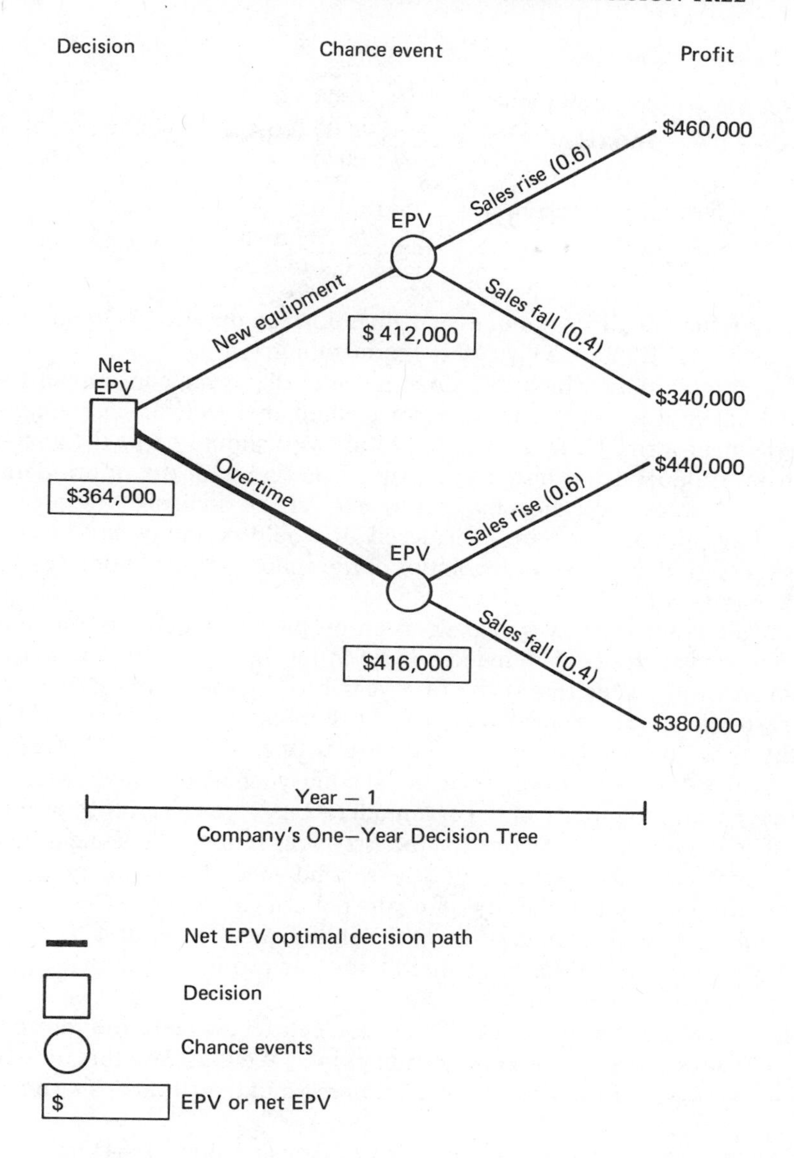

Company's One–Year Decision Tree

Using the expected present value (EPV) as the measure of effectiveness, we have

$$\text{EPV (new equipment)} = (0.6)(\$460{,}000) + (0.4)(\$340{,}000) = \$412{,}000$$

$$\text{EPV (overtime)} = (0.6)(\$440{,}000) + (0.4)(\$380{,}000) = \$416{,}000$$

The discounted net profit figures are then:

$$\begin{aligned}\text{Net EPV (new equipment)} &= \$412{,}000\\ &\underline{-80{,}000}\ \text{(cost of new equipment)}\\ &\$330{,}000\end{aligned}$$

$$\begin{aligned}\text{Net EPV (overtime)} &= \$416{,}000\\ &\underline{-52{,}000}\ \text{(cost of overtime)}\\ &\$364{,}000\end{aligned}$$

Based on the net EPV, the one-year time horizon decision is to go on overtime as its net EPV = $364,000 is the maximum.

For situations in which decisions in an earlier year can impact the decisions in future years, it is wise not to limit the decision-tree analysis to just the near term. How far into the future you should carry the analysis is unclear. It depends on how comfortable you feel with the future data estimates and some indication that the results of the analysis will not change even if additional years are considered. It would certainly pay to do some sensitivity studies on the probabilities of the chance events and the estimated discounted profits.

For the company's two-year decision problem, we need to obtain an estimate of what the sales demand will be in the second year. The marketing department indicates that if the first year had a sales increase (was a good year), they expect the second year to have a high sales increase with probability 0.5, or a moderate sales increase with probability 0.5. Even if the first-year sales were down, then the second year would have a high sales increase with probability of 0.8 or a moderate sales increase with probability 0.2. The company's decisions for the first year remain the same—buy new equipment or go on overtime. For the second year, the company could purchase two machines (if the first unit was not bought in year 1), buy a second machine, initiate overtime, or keep the status quo. These first- and second-year alternatives and the discounted two-year profits are shown in the following decision tree.

To determine the final net EPVs, we need to evaluate the intermediate net EPVs for the second-year decision points. We illustrate this for decision point (2–1). We assume that new equipment and overtime costs remain the same.

There are two branches coming out of decision point (2–1) and we must calculate the net EPV for both of them and then select the decision for (2–1) that corresponds to the maximum net EPV. We have

$$\begin{aligned}\text{Net EPV (2–1) (new equipment)} &= (0.5)(\$996{,}000) + (0.5)(\$948{,}000)\\ &\quad - \$80{,}000 = \$892{,}000\\ \text{Net EPV (2–1) (overtime)} &= (0.5)(\$1{,}012{,}000) + (0.5)(\$976{,}000)\\ &\quad - \$52{,}000 = \$942{,}000\end{aligned}$$

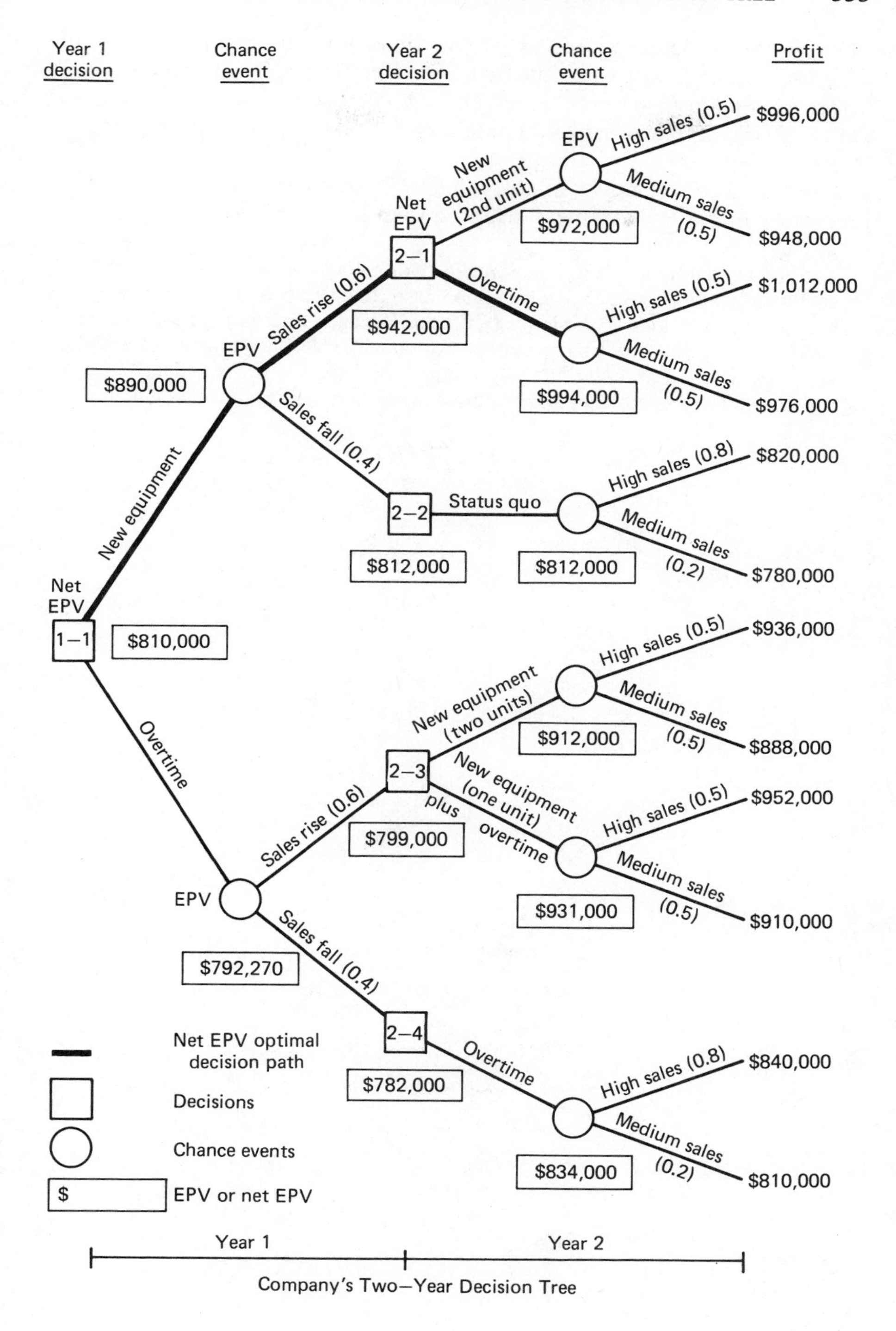

Company's Two–Year Decision Tree

Thus, the decision at (2–1) is to use overtime. We have indicated the other EPV and net EPV figures on the decision tree. They show that the net EPV at decision point (1–1) is $810,000 and the corresponding optimal decision path is to buy a unit of new equipment in the first year and go on overtime in the second year.

Decision trees are a way of analyzing well-structured, but risky, decision problems. As the management scientist John Magee notes:

> The unique feature of the decision tree is that it allows management to combine analytical techniques such as discounted cash flow and present value methods with a clear portrayal of the impact of future decision alternatives and events. Using the decision tree, management can consider various courses of action with greater ease and clarity. The interactions between present decision alternatives, uncertain events, and future choices and their results become more visible.

24 The Analytic Hierarchy Process

24.1 GOALS, CRITERIA, ALTERNATIVES: THE AHP

When we encounter monetary gambling and lottery situations, we usually have little difficulty in defining the outcomes and stating the associated probabilities. We know the payoff rules and the probabilities for a casino roulette wheel or a craps table. The probability of winning a new car in the local charity's lottery is easy to determine if we know how many tickets have been sold. For some betting situations, the probabilities might be difficult to calculate, for example, when playing the card game "21" or playing the horses. But even in these cases we have some explicit way of obtaining estimates of the probabilities (counting cards in "21"—which will get you kicked out of the casino—and the tote board dollar betting totals in horseracing). However, for decision problems that involve personal matters, or business or political outcomes, many of us find it difficult to justify a set of probabilities. There are often too many quantitative, qualitative, and judgmental aspects that limit our ability to state probabilities and associated utility functions.

In the previous example of selecting between two sets of courses based on expected grade-point average, we assumed that we knew the probabilities of obtaining the various GPAs. You might have felt uncomfortable with this assumption, but even so, you could come up with reasonable and acceptable probabilities. The same could be said about the builder who had to estimate the probabilities of gaining or losing money on the contract. In the farmer's problem, we determined the weather probabilities from historical records, a reasonable thing to do. But determining the probabilities of rising or falling sales and making an associated investment decision is another matter. Is there anything we can do for those situations in which we find it difficult to justify probabilities and utility functions? Can we, somehow, coalesce our knowledge about the decision situation and our intuition (and gut feeling) about what might happen into a nonprobabilistic analysis procedure that will guide us in making a choice among alternative outcomes? Given a number of outcomes, why not simply compare them with each other, two at a time (say, O_i vs O_j), and see if we can come up with a set of weights or priorities which order the value of the outcomes? For many difficult decision problems, usually those that are unstructured and/or involve qualitative and noncomparable measures, most of us can say we like outcome O_i better than outcome O_j and state this relationship by a numerical score. Once we had such a table of comparison scores, we would need some way of transforming them into priorities or weights that reflect the best to the worst outcome. A simple and versatile process for accomplishing such a transformation has been developed by the mathematician Thomas L. Saaty and is called the *analytic hierarchy process* (AHP). This decision model is based on structuring the problem elements in terms of how the alternative solutions (the outcomes) influence decision criteria, satisfaction of which help describe how a particular solution contributes to the accomplishment of the decision problem's main objective (measure of effectiveness). We discuss the AHP by means of examples.

For our first problem, we consider the situation in which you, as a last-semester college senior, need to select a set of five courses to complete the school year and to graduate. By that time, you would have taken most of your required major and minor courses and thus find that there is some flexibility in selecting courses for the final semester. There are still some major courses available, some hard, some easy, and some other courses that might be fun to take or others that might expose you to some new and interesting fields. But the thing that is guiding your selection is the objective of increasing your grade-point average (GPA). Looking over the schedule of classes, you determine that there are three sets of five courses, S_1, S_2, S_3 you want to consider. We can picture the relationship between your GPA objective and the sets of courses in a two-level hierarchical diagram:

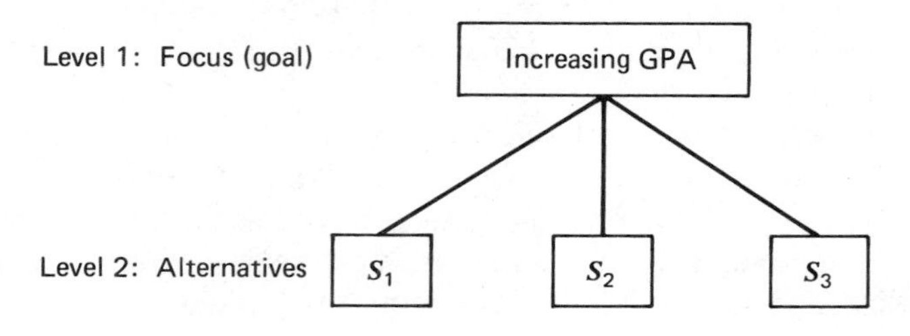

You now need to determine which of the alternatives will contribute the most to accomplishing the hierarchy's focus of increasing your GPA. This is done by comparing the sets of courses pairwise and asking the question: "In terms of increasing your GPA, which one of the sets is more important and by how much more important is it?" Here the phrase "more important" and the concept of importance should be interpreted in a generic way and comparable to preference, dominance, and similar relationships. In phrasing the AHP question, we use whichever one makes the clearest semantic sense. In this example, you should think of importance as being equivalent to preferred. An intensity of importance scale is given in Table 1—"The Pairwise Comparison Scale" developed by Saaty. For the problems that we are con-

TABLE 1. The Pairwise Comparison Scale[a]

Intensity of Importance	Definition	Explanation
1	Equal importance of both elements	Two elements contribute equally to the property
3	Moderate importance of one element over another	Experience and judgment slightly favor one element over another
5	Strong importance of one element over another	Experience and judgment strongly favor one element over another
7	Very strong importance of one element over another	An element is strongly favored and its dominance is demonstrated in practice
9	Extreme importance of one element over another	The evidence favoring one element over another is of the highest possible order of affirmation
2, 4, 6, 8	Intermediate values between two adjacent judgments	Compromise is needed between two judgments
Reciprocals	If activity *i* has one of the preceding numbers assigned to it when compared with activity *j*, then *j* has the reciprocal value when compared with *i*	

[a] From Thomas L. Saaty, *Decision Making for Leaders*, © 1982 by Lifetime Learning Publications, Belmont, CA 94002, a division of Wadsworth, Inc. Reprinted by permission of the publisher.

sidering, he has demonstrated that we need use only the whole numbers 1 to 9, with 1 indicating that the two items being compared are of equal importance and a 9 meaning that the first item is extremely more important than the second. For example (referring to the scale), if S_1 is strongly favored to S_3, then we give this comparison of S_1 to S_3 a score of 5, which means that S_1 is favored five times as much as S_3. In turn, the comparison score of S_3 to S_1 must have the reciprocal value of $\frac{1}{5}$. This seems to be a reasonable way of stating the intensity of the relationship between two items. For example, if we were comparing the physical weights of two stones A and B and concluded that A was five times heavier than B, then B would be $\frac{1}{5}$ as heavy as A. The scale 1–9 and its reciprocals enables us to capture the intensity of a relationship that we usually describe in qualitative terms: equal or indifferent (1), moderate (3), strong (5), very strong (7), and extreme (9). (You probably have filled out teacher evaluation forms that required you to rate the attributes of the teacher in these terms.) When a compromise between two consecutive terms is necessary, we employ the numbers 2, 4, 6, and 8. These scores represent our judgments on how the items compare with respect to the focus of the problem. There are no restrictions placed on the comparisons such as if S_1 is better than S_2, and S_2 is better than S_3, then S_1 must be better than S_3, although consistency in such comparisons will yield priorities with less margin of error. The comparison of an item to itself is scored a 1. It is helpful to frame the comparison question so that the answer is a whole number. In our example above, if we started out comparing S_3 to S_1, we would have just reversed them to obtain the value of 5 and then determined the required $\frac{1}{5}$. To continue with the example, we make all pairwise comparisons and arrive at the following judgmental matrix:

Comparison of sets of courses with respect to increasing GPA	S_1	S_2	S_3
S_1	1	3	5
S_2	$\frac{1}{3}$	1	2
S_3	$\frac{1}{5}$	$\frac{1}{2}$	1

The AHP determines the priorities of each alternative, that is, the importance or weight we should give each alternative, by analyzing such judgmental matrices using the advanced mathematical theory of eigenvalues and eigenvectors. We need not delve into this topic here except to note that the AHP interprets the eigenvector associated with the largest eigenvalue as the priorities that indicate the importance of each alternative in accomplishing the objective. For a matrix with n rows and columns, we approximate the required priority eigenvector as follows.[1] For each row i of the matrix, take the product of the ratios in that row and denote it by Π_i. Calculate the corresponding geometric mean P_i, where $P_i = \sqrt[n]{\Pi_i}$. Let $P = \sum_i P_i$. We normalize the P_i (i.e., transform them so that their resultant sum equals unity) by forming $p_i = P_i/P$. Each p_i is the ith priority or weight given to the ith alternative (here the set of courses S_i). The calculations for the GPA judgment matrix are shown in the following table.

GPA	S_1	S_2	S_3	Π_i	$P_i = \sqrt[3]{\Pi_i}$	$p_i = P_i/P$
S_1	1	3	5	15	2.466	0.65
S_2	$\frac{1}{3}$	1	2	0.667	0.874	0.23
S_3	$\frac{1}{5}$	$\frac{1}{2}$	1	0.1	0.464	0.12
					$P = 3.804$	

The priorities for the courses are then

	p_i
S_1	0.65
S_2	0.23
S_3	0.12

The AHP interprets this information as indicating that the set of courses S_1 will contribute the most to increasing your GPA, followed by S_2, and finally S_3. Based on the magnitudes of the p_i, you should feel pretty comfortable in choosing S_1, select S_2 with caution, and do not select S_3.

Now let's make the problem a bit more interesting. You still must select one of the three sets of courses, but you are concerned with more than just your GPA. Now your problem focus, the overall objective, is the attainment of an excellent education. But, in selecting a set of courses, there are criteria that need to be considered. The criteria deal with how the various sets of courses contribute to GPA improvement, advancing your major career field, and giving you a broader educational background. This new problem can be

[1] As discussed in Section 26.33, the geometric mean approximation to the eigenvector could cause a reversal of priorities. This is not the case for our examples here, but a more exact way of calculating the eigenvector (as discussed in Section 26.33) should be used for real-life decision problems.

pictured by the following three-level hierarchical structure:

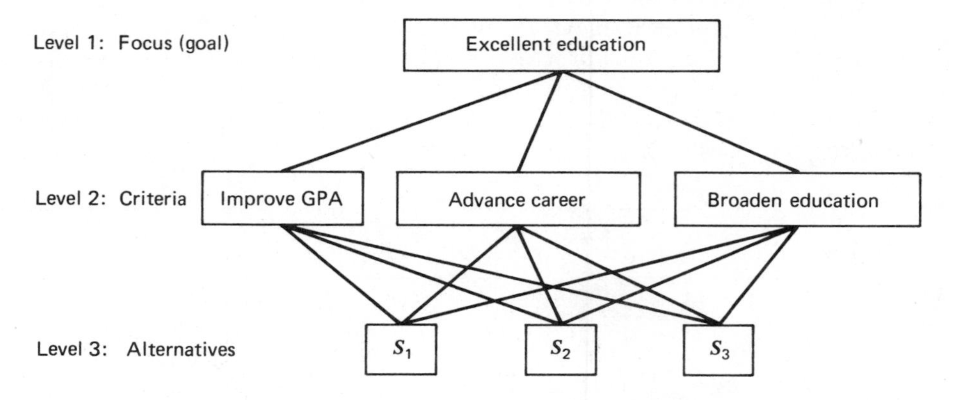

Unlike the previous two-level hierarchy example, here we first need to determine how important each criterion is in achieving the objective of an excellent education. We do this by constructing a judgment matrix as before and comparing the criteria pairwise by asking the question: "Of the two criteria, which do you consider more important in contributing to the focus and how much more important is it?" Using the pairwise comparison scale above, let us assume we determined the judgment matrix and the calculated p_i as shown next:

Comparison of the criteria with respect to the focus of obtaining an excellent education	Improve GPA	Advance career	Broaden education	Π_i	P_i	p_i
Improve GPA	1	1/5	1/2	0.1	0.464	0.11
Advance career	5	1	7	35.0	3.271	0.74
Broaden education	2	1/7	1	0.286	0.659	0.15
					4.394	

From the p_i column, we see that the advance career criterion has a much higher priority (0.74) than the other two. Each of the other criteria are about the same in their importance to the objective.

We next have to carry out the analysis of the third level of the hierarchy, that is, determine the importance of the sets of courses to each criterion. This is done by constructing three judgment matrices, one for each criterion. For example, for the criterion improve GPA, we need to ask (as was done earlier) the question: "Of the two sets of courses being compared, which is considered more important by you in improving your GPA and how much more important is it?'' We ask similar questions for the other two criteria and arrive at three judgment matrices, as given below.

GPA	S_1	S_2	S_3	Π_i	P_i	p_i
S_1	1	3	5	15	2.466	0.65
S_2	$\frac{1}{3}$	1	2	0.667	0.874	0.23
S_3	$\frac{1}{5}$	$\frac{1}{2}$	1	0.1	0.464	0.12
					3.804	

Advance career	S_1	S_2	S_3	Π_i	P_i	p_i
S_1	1	$\frac{1}{2}$	8	4	1.587	0.36
S_2	2	1	9	18	2.621	0.59
S_3	$\frac{1}{8}$	$\frac{1}{9}$	1	0.0139	0.240	0.05
					4.448	

Broaden education	S_1	S_2	S_3	Π_i	P_i	p_i
S_1	1	6	$\frac{1}{5}$	1.2	1.063	0.27
S_2	$\frac{1}{6}$	1	$\frac{1}{3}$	0.0556	0.382	0.10
S_3	5	3	1	15	2.466	0.63
					3.911	

From the GPA judgment matrix we see that the set of courses S_1 with $p_1 = 0.65$ contributes the most to achieving the GPA criterion and S_3 ($p_3 = 0.12$) the least; from the advance career matrix, we have S_2 ($p_2 = 0.59$) contributing the most to that criterion and S_3 ($p_3 = 0.05$) the least; and from the broaden education matrix, we have S_3 ($p_3 = 0.63$) being the set of courses that would contribute the most to broadening your education and S_2 ($p_2 = 0.10$) the least.

Based on the above analysis, it appears as if you did a good job in forming the three alternative sets of courses; one is better for each criterion. But their contributions to achieving the other criteria are rather varied and in some cases quite weak. We are still not in a position to pick a set of courses as we need to factor in the influence of the criteria. Based on our earlier level 2 analysis, the criteria are not equal in terms of your goal of achieving an excellent education. We determined weights of 0.11 to improve GPA, 0.74 to advance career, and 0.15 to broaden education. These weights are used to modify the corresponding level 3 weights for each set of courses. Try and think of how this should be done before reading on. We first organize the results obtained so far in the following table.

Focus: Excellent Education

Criteria		GPA	Advance career	Broaden education	Composite hierarchical priorities
Level 2 priorities		0.11	0.74	0.15	(p_i)
Alternatives					
Level 3 priorities	S_1	0.65	0.36	0.27	0.38
	S_2	0.23	0.59	0.10	0.48
	S_3	0.12	0.05	0.63	0.14

To obtain the composite hierarchical priority for an S_i, we multiply each of its level 3 priorities by the corresponding level 2 priority and sum the products. For S_1 we have

$$p_1 = (0.11)(0.65) + (0.74)(0.36) + (0.15)(0.27) = 0.38$$

and for S_2 and S_3, we have, respectively,

$$p_2 = (0.11)(0.23) + (0.74)(0.59) + (0.15)(0.10) = 0.48$$

$$p_3 = (0.10)(0.12) + (0.74)(0.05) + (0.15)(0.63) = 0.14$$

The theory of the AHP is that these final composite weights capture your explicit and implicit knowledge about each set of courses in terms of their satisfaction of the individual criteria and of your feeling as to the importance of the criteria in achieving the ultimate goal of obtaining an excellent education. The rationale for why these mathematically derived numbers can be so interpreted is discussed in Chapter 26. (Note that the final p_i sum to unity. Why?) For your decision problem of selecting the set of courses for achieving the best education, the composite p_i indicate that the set of courses S_2 would be your best choice ($p_2 = 0.48$), with set S_1 ($p_1 = 0.38$) the next best. The set S_3, although being best in terms of broadening your education, does little to improve your total education ($p_3 = 0.14$).

The ability of the scale 1–9 to express your feeling as to the intensity of a pairwise comparison is based on psychological measurement studies. In turn, the validity of the total AHP process has its roots in the quantification of physical attributes using human sensory perceptions. To demonstrate how the AHP can be used to estimate a physical attribute, consider the following geometrical problem illustrated by the figures F_1, F_2, F_3, F_4, and F_5 below.

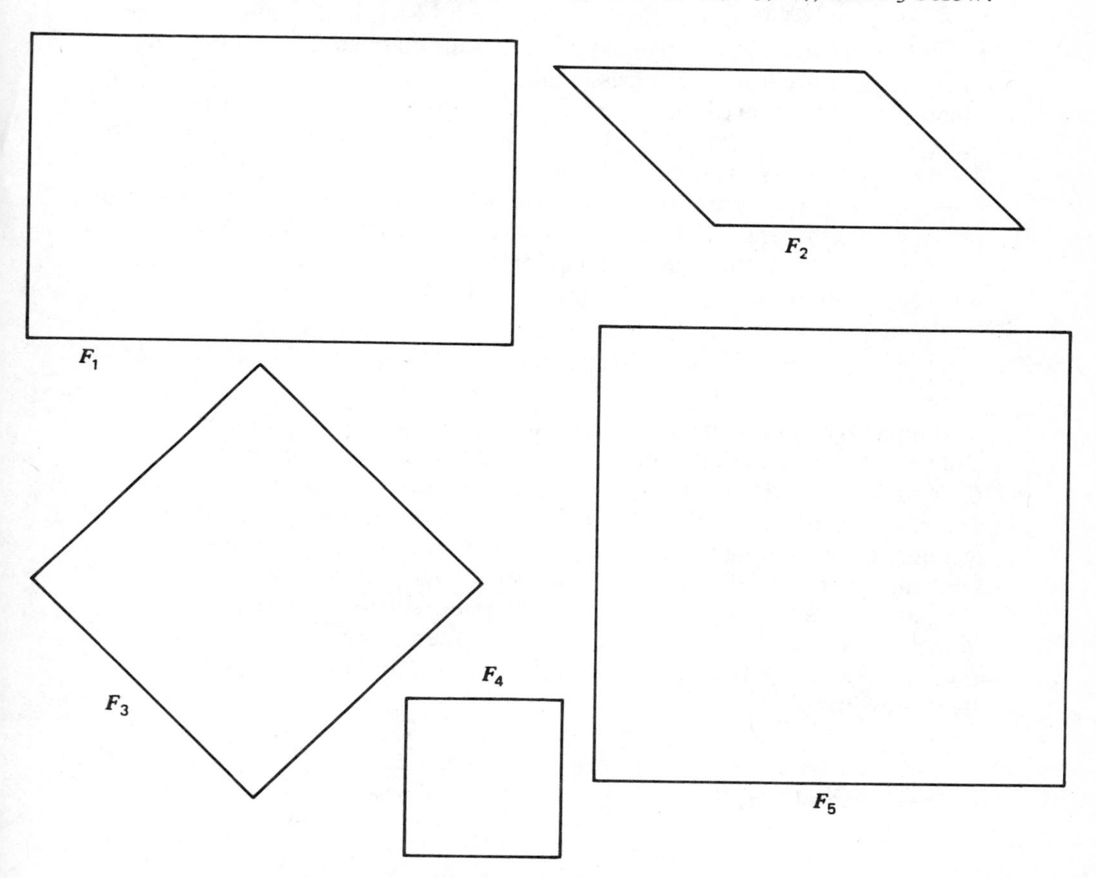

Let the unknown area of figure F_i be denoted by A_i and the total area covered by all five figures be $T = A_1 + A_2 + A_3 + A_4 + A_5$. You are asked to estimate how much of the total area is contained in each figure, that is, you are to determine the proportions A_i/T. The only measurement tool that you are allowed to use is your visual perception. To use the AHP, we visually compare each figure pairwise with the others by asking the question: "Of the two figures being compared, which one has the larger area and how much larger is it?" For example, it is clear that F_1 is larger than F_4, but is the ratio 2 to 1, 3 to 1, 4 to 1, or what? Your answer would be entered in the comparison matrix for F_1 against F_4, and the reciprocal for F_4 against F_1. Develop the comparison matrix for the figures using the intensity numbers 1 to 9 and their reciprocals, and calculate the associated priorities. Your fractional priority values should, hopefully, come close to the actual ratios of A_i/T. The answer is given in Section 26.34.

The AHP is a versatile procedure for analyzing many decision situations in which you can form a hierarchical structure involving a focus, criteria, and alternatives. More complex multilevel hierarchies can also be studied. Care must be taken in developing a hierarchy as it is difficult to compare too many criteria or alternatives; also, the alternatives need to be distinct. Saaty suggests that the comparison matrices contain no more than nine elements, and if there are more, then the hierarchy should be decomposed into more manageable parts. The AHP has been used to set priorities for such diverse applications as choosing a job, buying a car, selecting a stock portfolio, choosing a corporate research program, evaluating alternatives for improving health-care systems, and many other areas. We illustrate the AHP approach by the following examples from Saaty. You should check the computations as a means of becoming familiar with the process.

24.2 WILL YOU EVER BUY THAT NEW CAR?

You have visited a number of dealer showrooms and have reduced the possible number of cars you want to consider to three; let's call them A, B, and C. Your choice will be based on four criteria: price, running cost, comfort, and status. Note that the first two criteria are quantitative and the last two are qualitative. Using the comparison scale, you have to measure the pairwise importance of the criteria to the objective of buying the best car, and the pairwise comparison of the cars with respect to each criterion. When numerical data such as costs are available, then it should be used to determine the ratio comparisons. The hierarchical structure for this problem is the following:

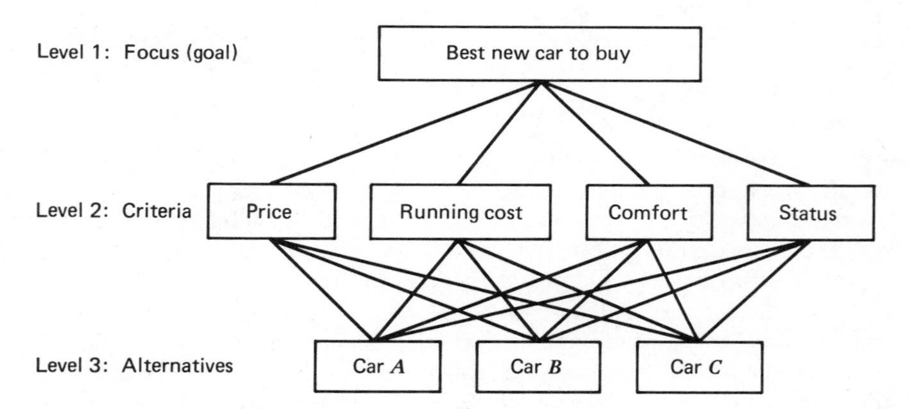

The level 2 and level 3 judgment matrices and the corresponding priorities p_i are given next (you might really want to collect data for cars you are interested in and perform an analysis along similar lines).

Level 2 Comparisons

Decision to buy a new car	Price	Running cost	Comfort	Status	p_i
Price	1	3	7	8	0.586
Running cost	$\frac{1}{3}$	1	5	5	0.277
Comfort	$\frac{1}{7}$	$\frac{1}{5}$	1	3	0.088
Status	$\frac{1}{8}$	$\frac{1}{5}$	$\frac{1}{3}$	1	0.049

We see that the criterion of price has the highest weight ($p_1 = 0.586$), followed by running cost ($p_2 = 0.277$). The other two criteria—comfort and status—have such small weights that their impact on the decision can be ignored and the problem reduced to just considering the first two criteria. However, for illustrative purposes, we shall deal with all four criteria.

Level 3 Comparisons

Price	A	B	C	p_i
A	1	2	3	0.540
B	$\frac{1}{2}$	1	2	0.297
C	$\frac{1}{3}$	$\frac{1}{2}$	1	0.163

Running cost	A	B	C	p_i
A	1	$\frac{1}{5}$	$\frac{1}{2}$	0.106
B	5	1	7	0.744
C	2	$\frac{1}{7}$	1	0.150

Comfort	A	B	C	p_i
A	1	3	5	0.627
B	$\frac{1}{3}$	1	4	0.280
C	$\frac{1}{5}$	$\frac{1}{4}$	1	0.093

Status	A	B	C	p_i
A	1	$\frac{1}{5}$	3	0.188
B	5	1	7	0.731
C	$\frac{1}{3}$	$\frac{1}{7}$	1	0.081

We summarize the above calculations and determine the composite hierarchical priorities for buying a new car in the following table.

Focus: Buying the Best New Car

Criteria		Price	Running cost	Comfort	Status	Composite hierarchical priorities (p_i)
Level 2 priorities		0.586	0.277	0.088	0.049	
Alternatives						
Level 3 priorities	A	0.540	0.106	0.627	0.188	0.410
	B	0.297	0.744	0.280	0.731	0.442
	C	0.163	0.150	0.093	0.018	0.149

The composite hierarchical priorities indicate that car B is the best buy ($p_2 = 0.442$), with car A a close second ($p_1 = 0.410$). Car C with $p_3 = 0.149$ is out of the competition.

24.3 CHOOSING THE RIGHT JOB

A graduating Ph.D. student has received three job offers from industrial organizations A, B, and C. Overall job satisfaction is the student's major objective, with the criteria to be applied being the research, growth potential, benefits, colleagues, location, and reputation of each company. Note that benefits, which includes salary, is the only criterion that has a quantitative base, the others being qualitative and difficult to measure. The corresponding three-level hierarchy and the student's level 2 and level 3 judgment matrices follow. You should think about how you would gather information so as to be able to make the pairwise comparisons of the criteria and jobs. How would you state the comparison questions?

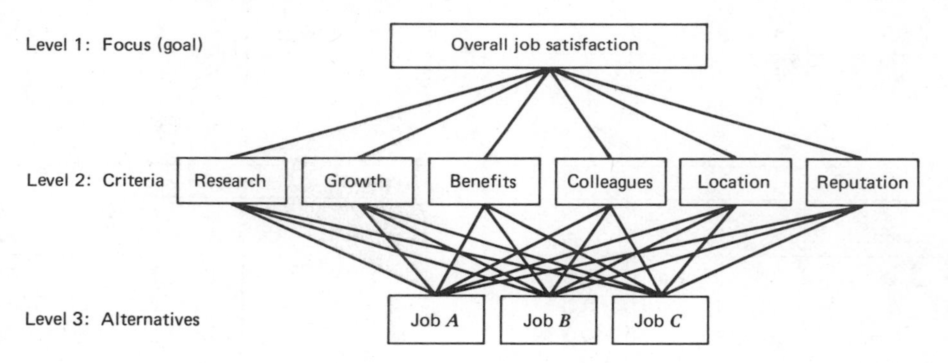

Level 2 Comparisons

Overall job satisfaction	Research	Growth	Benefits	Colleagues	Location	Reputation	p_i
Research	1	1	1	4	1	$\frac{1}{2}$	0.16
Growth	1	1	2	4	1	$\frac{1}{2}$	0.19
Benefits	1	$\frac{1}{2}$	1	5	3	$\frac{1}{2}$	0.19
Colleagues	$\frac{1}{4}$	$\frac{1}{4}$	$\frac{1}{5}$	1	$\frac{1}{3}$	$\frac{1}{3}$	0.05
Location	1	1	$\frac{1}{3}$	3	1	1	0.12
Reputation	2	2	2	3	3	1	0.30

Note that the criterion of reputation ($p_6 = 0.30$) is the most important with respect to accomplishing the goal of selecting the job that gives the best overall satisfaction; all the other criteria, except colleagues, are weighted about the same.

Level 3 Comparisons

Research	A	B	C	p_i
A	1	$\frac{1}{4}$	$\frac{1}{2}$	0.14
B	4	1	3	0.62
C	2	$\frac{1}{3}$	1	0.24

Growth	A	B	C	p_i
A	1	$\frac{1}{4}$	$\frac{1}{5}$	0.10
B	4	1	$\frac{1}{2}$	0.33
C	5	2	1	0.57

Benefits	A	B	C	p_i
A	1	3	$\frac{1}{3}$	0.32
B	$\frac{1}{3}$	1	1	0.22
C	3	1	1	0.46

Colleagues	A	B	C	p_i
A	1	$\frac{1}{3}$	5	0.28
B	3	1	7	0.65
C	$\frac{1}{5}$	$\frac{1}{7}$	1	0.07

Location	A	B	C	p_i
A	1	1	7	0.47
B	1	1	7	0.47
C	$\frac{1}{7}$	$\frac{1}{7}$	1	0.07

Reputation	A	B	C	p_i
A	1	7	9	0.77
B	$\frac{1}{7}$	1	5	0.17
C	$\frac{1}{9}$	$\frac{1}{5}$	1	0.06

We summarize the above data in the following table and calculate the composite hierarchical priorities. They indicate that job A with composite priority $p_1 = 0.40$ should be the student's first choice, with job B ($p_2 = 0.34$) a close second.

Focus: Overall Job Satisfaction

Criteria		Research	Growth	Benefits	Colleagues	Location	Reputation	Composite hierarchical priorities
Level 2 priorities		0.16	0.19	0.19	0.05	0.12	0.30	(p_i)
Alternatives								
Level 3 priorities	A	0.14	0.10	0.32	0.28	0.47	0.77	0.40
	B	0.62	0.33	0.22	0.65	0.47	0.17	0.34
	C	0.24	0.57	0.46	0.07	0.07	0.06	0.26

25 The Decision Framework, One More Time

As we reach the end of this text, we offer the following comments for emphasis and summary. We have presented our view of the important field of mathematical decision sciences and how it should be presented to the undergraduate student. The concepts, examples, and discussions presented have been limited by our interests and biases. This material is what we think is basic and important for the student to know. This is true for those students who will probably not take another quantitative, mathematical course, as well as those who will major in operations research or mathematics or engineering. No matter what your profession, your future work will certainly involve decision-making situations. You will find that the ideas in this text will serve you well. Just remember the following flow process of the decision framework.

25.1 MODELING STEPS: AN EXPANDED VIEW

Many of you will find that this simple graphical view of the decision framework is too restrictive. Many problems will involve complex model structures, will need to be solved using special computer programs, require extensive data collection and analysis, and call for a team of analysts to formulate the problem and interpret the results. You might find the following

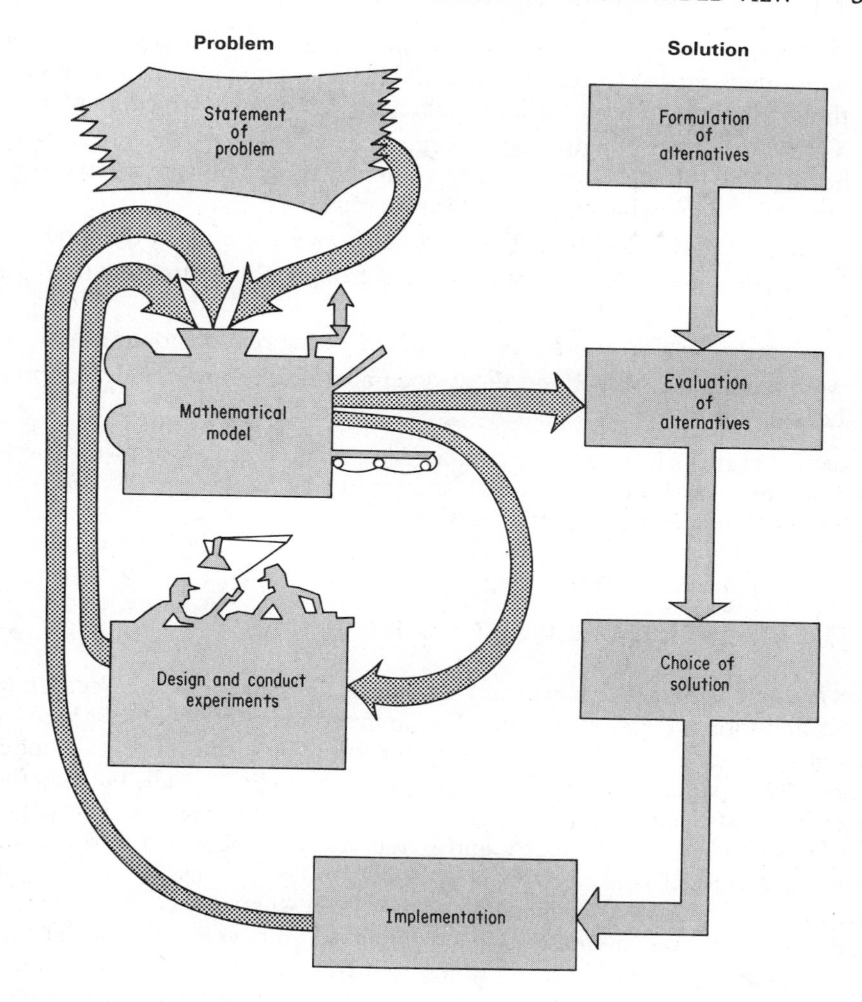

expansion of the *modeling steps* more meaningful:

- ☐ Describing the problem to be solved; defining the problem issues, study objectives, and assumptions.
- ☐ Isolating the system or process to be modeled; delineating the characteristics which can be modeled.
- ☐ Developing or adopting a supporting theory; developing a flow or logic diagram.
- ☐ Determining available data sources; formulating the mathematical model or set of models to be linked; analyzing data requirements and designing data collection procedures.
- ☐ Collecting data.

☐ Describing the program logic of the model, including basic flowcharts with input, processing, and output described; estimating parameters of the model; constructing and implementing the computer program(s).

☐ Verifying that the mathematical/logical description of the problem is correct and that the corresponding computer program(s) has been coded correctly; debugging the computer program(s).

☐ Developing alternative solutions and analyzing them using the model.

☐ Evaluating results and output obtained from the model; validating the model.

☐ Presenting results with a plan for implementing recomendations.

☐ Maintaining the model and data; documenting the total modeling process.

A view of how these steps are interrelated—the basic steps of the modeling process—is shown in the chart below (from *Guidelines for Model Evaluation,* U.S. General Accounting Office):

25.2 IS THE MODEL WHAT WE THINK IT IS?

Two important features of this chart are the *verification* and *model validation* boxes and loops. Here we assume that your problem needs to be solved by an algorithm that has been coded for a computer. In a true sense, the model is not now a paper description or what you have in your head, but it is the computer code and computer data files. Whether what you had on paper and in your head was correctly and successfully transformed into a computer-based model is always open to question. The process of *verification* includes tests that are designed to demonstrate that the computer-based model runs as you intended it to do. Even if you were the analyst and programmer, you must establish to the best of your abilities that the model as run on the computer does what you expect. This is usually accomplished by working out examples by hand that are then processed on the computer to determine if the answers are the same. You also try and get the computer model to fail by having it solve problems that are at the boundaries of the model and data assumptions. After such tests, at best, you will be able to say that the computer program will give satisfactory results for certain cases of interest; you can never know that a complex computer program is completely checked out, that is, debugged. All of your actions here must be documented so that others can get to know the limits of the computer-based model. In fact, a good analyst will document what has been done in all the modeling steps.

The second feature of importance—*model validation*—concerns itself with the total modeling process. It is an attempt to examine the correspondence of the computer-based model and its outputs to perceived reality. We

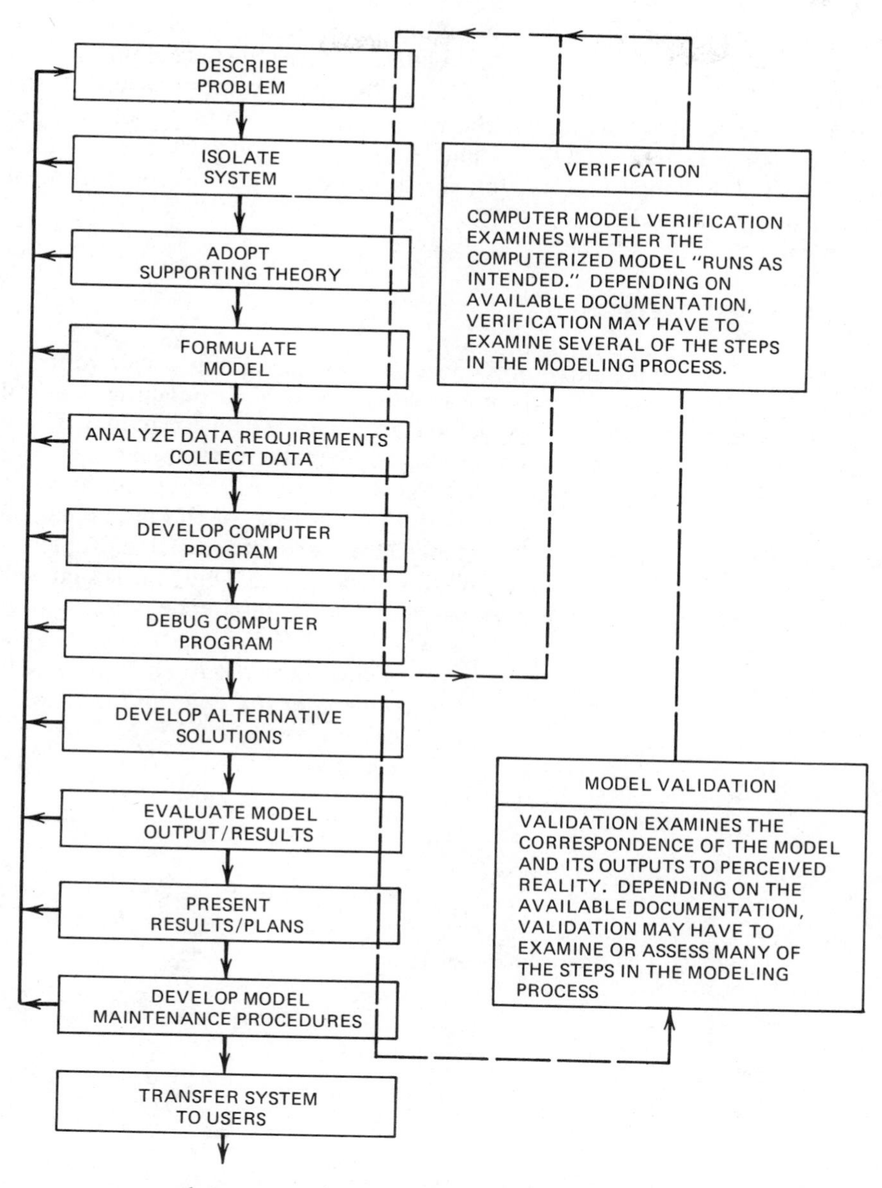

can never say that a complex model has been completely validated. Validation tests can only act as filters. If it passes this test, you must go on to the next, and so on, until you run out of tests, or demonstrate that the model has poor validity and should not be used or until you and others have had enough success with the model that your confidence in it is high enough for its use as a decision aid. You have to apply ingenuity to devise validation tests. The standard one—for ongoing systems—is to process historical input data and produce accurate historical output. This, of course, cannot be done for new or proposed systems. For these latter systems, test problems, expert opinion, and sensitivity analyses are used to obtain some measure of the computer-based model's validity. Somehow you have to put it all together into a working decision-aiding system. As you will find out, it is no small task, but the challenge and excitement of discovery and accomplishment make it worth the effort.

25.3 FINAL WORDS

Our ability to use quantitative procedures as aids in solving a wide range of decision problems has evolved into a science of decision making. Central to this development is the concept of the mathematical decision model. These models can be represented by simple logical or mathematical statements or complex mathematical/computer-based systems. By remembering how the decision framework centers on the definition and use of the mathematical model, and what you need to do to establish the worth of a decision problem's mathematical model, you will maintain objectivity in your work that will advance the science and your personal and professional endeavors.

Whatsoever thou takest in hand,
remember the end, and thou shall
never do amiss.
APOCRYPHA

26 Part V Discussion, Extensions, and Exercises

26.1. For a two-person, zero-sum game, there is an amazing relationship between the mathematical models for player 1 and player 2 that can be expressed by primal–dual linear-programming problems. This relationship is developed next. Similarly, any linear-programming problem can be transformed into a two-person, zero-sum game.

We shall assume that we are given an arbitrary payoff matrix:

$$\mathbf{A} = \begin{pmatrix} a_{11} & a_{12} & \cdots & a_{1n} \\ a_{21} & a_{22} & \cdots & a_{2n} \\ \cdots & \cdots & \cdots & \cdots \\ a_{m1} & a_{m2} & \cdots & a_{mn} \end{pmatrix}$$

From the definition of a solution to a game, the problem for player 1 is to find strategies $x_1, x_2, \ldots, x_m$ and a number v such that

$$\begin{aligned}
a_{11}x_1 + a_{21}x_2 + \cdots + a_{m1}x_m &\geq v \\
a_{12}x_1 + a_{22}x_2 + \cdots + a_{m2}x_m &\geq v \\
\cdots\cdots\cdots\cdots\cdots\cdots\cdots\cdots \\
a_{1n}x_1 + a_{2n}x_2 + \cdots + a_{mn}x_m &\geq v \\
x_1 + x_2 + \cdots + x_m &= 1 \\
x_1 &\geq 0 \\
x_2 &\geq 0 \\
&\vdots \\
x_m &\geq 0
\end{aligned}$$

Similarly, for player 2 we have

$$\begin{aligned}
a_{11}y_1 + a_{12}y_2 + \cdots + a_{1n}y_n &\leq v \\
a_{21}y_1 + a_{22}y_2 + \cdots + a_{2n}y_n &\leq v \\
\\
a_{m1}y_1 + a_{m2}y_2 + \cdots + a_{mn}y_n &\leq v \\
y_1 + y_2 + \cdots + y_n &= 1 \\
y_1 &\geq 0 \\
y_2 &\geq 0 \\
\cdot \quad & \quad \cdot \\
\cdot \quad & \quad \cdot \\
\cdot \quad & \quad \cdot \\
y_n &\geq 0
\end{aligned}$$

Since every element of **A** can be made positive by the addition of a suitable constant to all the a_{ij}, we can assume that $v > 0$. Let us divide each of the above relationships by v and let

$$x_i' = \frac{x_i}{v} \quad \text{and} \quad y_j' = \frac{y_j}{v}$$

Note that

$$\sum_i x_i' = \frac{1}{v} \sum_i x_i = \frac{1}{v}$$

and

$$\sum_j y_j' = \frac{1}{v} \sum_j y_j = \frac{1}{v}$$

Hence, by minimizing $\sum_i x_i'$, player 1 will maximize the value of the game, and by maximizing $\sum_j y_j'$, player 2 will minimize the value of the game. We can then restate the relationships in terms of equivalent linear-programming problems and obtain the following symmetric dual problems:

The primal problem. Find strategies $x_1', x_2', \ldots, x_m'$ which minimize

$$x_1' + x_2' + \cdots + x_m'$$

subject to

$$\begin{aligned}
a_{11}x_1' + a_{21}x_2' + \cdots + a_{m1}x_m' &\geq 1 \\
a_{12}x_1' + a_{22}x_2' + \cdots + a_{m2}x_m' &\geq 1 \\
\cdots\cdots\cdots\cdots\cdots\cdots\cdots\cdots\cdots \\
a_{1n}x_1' + a_{2n}x_2' + \cdots + a_{mn}x_m' &\geq 1 \\
x_i' &\geq 0
\end{aligned}$$

The dual problem. Find strategies $y'_1, y'_2, \ldots, y'_n$ which maximize

$$y'_1 + y'_2 + \cdots + y'_n$$

subject to

$$\begin{aligned}
a_{11}y'_1 + a_{12}y'_2 + \cdots + a_{1n}y'_n &\leq 1 \\
a_{21}y'_1 + a_{22}y'_2 + \cdots + a_{2n}y'_n &\leq 1 \\
\cdots\cdots\cdots\cdots\cdots\cdots\cdots\cdots&\cdots \\
a_{m1}y'_1 + a_{m2}y'_2 + \cdots + a_{mn}y'_n &\leq 1 \\
y'_j &\geq 0
\end{aligned}$$

Since every game has a solution, optimal solutions to the above problems exist and

$$\min \sum_i x'_i = \max \sum_j y'_j = \frac{1}{v}$$

The set of x'_i and y'_j which satisfies the linear-programming problems must, of course, be converted to the optimal x_i and y_j that solve the game problems.

If only the primal or only the dual is solved, the optimal strategy for the other problem is contained in the simplex tableau of the corresponding final solution. The optimal strategy corresponds to the row 0 elements of the slack vectors.

26.2. For any 2×2 zero-sum game let the payoff matrix be denoted by

$$\begin{pmatrix} a & -b \\ -c & d \end{pmatrix}$$

where a, b, c, and d are all nonnegative quantities. Prove that the optimal strategies (x_1, x_2) for player 1 and the optimal strategies (y_1, y_2) for player 2 are

$$x_1 = \frac{c + d}{a + b + c + d}, \qquad x_2 = \frac{a + b}{a + b + c + d}$$

$$y_1 = \frac{b + d}{a + b + c + d}, \qquad y_2 = \frac{a + c}{a + b + c + d}$$

and the value of the game v is

$$v = \frac{ad - bc}{a + b + c + d}$$

Thus, a game with the given payoff matrix is fair if an only if $ad - bc = 0$. (*Hint:* Develop the mathematical model for player 1 in terms of $x_1 + x_2 = 1$ and v, and for player 2 in terms of $y_1 + y_2 = 1$ and v, where $(x_1, x_2, y_1, y_2) \geq 0$. Do you have to worry about saddle points?) How does your analysis change if the 2×2 game is given by

$$\begin{pmatrix} a & b \\ c & d \end{pmatrix}$$

where the payoff coefficients a, b, c, and d can be any positive or negative numbers? What assumptions must you make? For this game, how must the payoff entries be related to ensure against a saddle-point solution?

26.3. Develop the payoff matrix for the game known as "paper-scissors-stone." The players simultaneously call out one of the three items. If both name the same object, the play is a draw. The payoffs are based on the rules that scissors cut paper (scissors win one unit), stone breaks scissors (stone wins one unit), and paper covers stone (paper wins one unit). What are the optimal strategies and the value of the game? Construct the primal and dual linear-programming problems for this game and solve the primal problem using the simplex method. Note that the solution to the dual problem is contained in the final simplex tableau of the primal problem.

26.4. Show that the following games are not fair, that is, $v \neq 0$. Why do they seem as if they might be fair? Try playing them or the skin game with an unsuspecting friend.

$$\begin{pmatrix} 2 & -3 \\ -1 & 2 \end{pmatrix}, \quad \begin{pmatrix} 2 & -3 \\ -3 & 4 \end{pmatrix}$$

26.5. Solve the following two-person zero-sum game for both players' optimal strategies:

$$\begin{pmatrix} 4 & 1 \\ 2 & 2 \\ 1 & 4 \end{pmatrix}$$

(*Hint:* Find player 2's solution by the graphical procedure. You should be able to show that player 1 should not use strategy 2 even though it might look like a good one. What happens if two units are subtracted from each payoff element?)

26.6. Solve the following games by the formulas of Section 26.2 and by graphing the corresponding linear-programming problems for players 1 and 2:

$$\begin{pmatrix} 3 & -1 \\ -2 & 3 \end{pmatrix}, \quad \begin{pmatrix} 1 & -1 \\ -1 & 1 \end{pmatrix}, \quad \begin{pmatrix} 4 & -5 \\ -4 & 5 \end{pmatrix},$$
$$\begin{pmatrix} 2 & -3 \\ -1 & 2 \end{pmatrix}, \quad \begin{pmatrix} 4 & -5 \\ -3 & 5 \end{pmatrix}$$

26.7. Discuss what happens to the optimal strategies and v if the maximum of (a, b, c, d) is added to all the payoffs in

$$\begin{pmatrix} a & -b \\ -c & d \end{pmatrix}$$

26.8. Consider the strategic situation faced by a farmer in deciding which one of two crops, soybeans or corn, to grow. The farmer's profits depend on what nature has in store for the summer in terms of a normal rainfall, drought, or heavy rain. Based on nature's "strategy," the farmer estimates the profit per acre for each crop as given in the following table:

		Nature (weather)			
		Normal	Drought	Heavy rain	
Farmer	Plant soybeans	\$10.00	\$5.00	\$12	Profit per acre a_{ij}
	Plant corn	\$7.00	\$8.00	\$13	

Formulate this decision problem as a zero-sum two-person game and solve it using the graphical procedure. How do you interpret the farmer's optimal strategies x_1 and x_2 if you allow the farmer to plant x_1% of the land with soybeans and x_2% of the land with corn? Is this a one-time game for the farmer? Is nature really player 2 who wants to minimize the value of the game? How could the farmer use the historical records that indicate the probabilities of occurrence of the three types of weather?

26.9. Consider the following strategic situation. Two local hi-fi dealers are out to capture the market for new sales. They tend to dominate the market by their volume sales and between them they account for 90% of all hi-fi sales. Their respective share of this 90% is highly volatile and depends on how they advertise. Company A has decided to initiate an advertising campaign and is contemplating two possible approaches (strategies); company B is planning to select among three possible strategies.

The grapevine (the hi-fi intelligence system) has passed this information along to each company. In fact, each company has analyzed the competitive aspects of each strategy and has determined what percentage of the market it would get, as given in the following table. The numerical entry a_{ij} represents the percentage of the market company *A* would get if it used campaign strategy *i* against company *B*'s strategy *j*. Company *B* would then expect to receive (90% $-$ a_{ij}).

		Company B			
		Campaign 1	Campaign 2	Campaign 3	
Company A	Campaign 1	30%	40%	60%	a_{ij}
	Campaign 2	60%	10%	30%	

You should be able to solve for both companies' optimal strategies. First check for a saddle point and then any dominant rows or columns. How should you interpret optimal strategies if companies *A* and *B* decide to select a new campaign (from the given sets) at the beginning of each month? Remember that the theory assumes that each player knows what the other player's strategies are and they both know the payoff elements. Do you think this application is a realistic one? (The example is from Bradley et al.)

26.10. Discuss the applicability of game theory to the following strategic situations. Find the optimal strategies and the value of the game for each payoff matrix. Remember to check for a saddle point and dominant rows or columns. (The problems are from Smith.)

(a) The management of a company and the union are preparing for the renewal of the employees' contract. Management has three strategies; the union is considering four. The average hourly wage increases for each combination of strategies is given in the following payoff matrix; the a_{ij}'s are given in terms of management being the maximizing player.

		Union				
		1	2	3	4	
Management	1	−0.40	−0.25	−0.20	−0.50	a_{ij}
	2	−0.20	0	−0.10	−0.25	
	3	−0.10	−0.10	−0.05	−0.30	

What would happen if management came up with a fourth strategy with payoffs of $(-0.15,\ -0.05,\ 0,\ -0.05)$? Should they consider it?

(b) Two countries are negotiating the price of wheat. The joint strategies, with the percentage change in the selling price (using the current contract as the base price), are given in the following payoff matrix:

		Buyer			
		1	2	3	
Seller	1	12	10	−10	a_{ij}
	2	−10	0	−25	
	3	15	−20	8	

26.11. In many strategic situations, the difficulty in using game theory is our inability to develop a proper concept of the payoff elements. Consider the following example (from Greenberg).

An environmental group and a developer are pitted against each other in gaining public support for their views of how a piece of riverfront land should be used. A referendum is to be voted on and each group has to decide on which land use to support. The developers can choose among four possibilities: (1) build a road; (2) build a barge dock; (3) build an amusement park; or (4) build oil storage tanks. The environmentalists want to select between: (1) build park and zoo or (2) preserve the land. The payoff elements represent how the selected strategies would appeal to the voters, given in terms of the developers being the maximizing player 1. For example, if the developers opt for a road and the environmentalists for a park and zoo, then the developers look bad (represented by a payoff of −15). If the developers choose an amusement park and the environmentalists choose to support a park and zoo, then the developers would get an appeal rating of 10. The assumed payoffs are given in the following table:

		Environmentalists		
		Park and zoo	Preserve land	
Developers	Road	−15	−5	a_{ij}
	Barge dock	−8	3	
	Amusement park	10	−5	
	Oil storage tanks	−30	−20	

Apply the concept of dominance to this payoff matrix to reduce it to a 2×2 game. Use the formulas of Section 26.2 to solve for the optimal strategies and the value of the game. How should you use the optimal strategies in this one-time game? As the outcome is quite sensitive to the payoff elements, you might want to vary the payoffs, one at a time, to see what happens. Adjust the formulas that solve a 2×2 game to take care of a change to each payoff element. (*Hint:* Consider the game

$$\begin{pmatrix} a + \Delta & -b \\ -c & d \end{pmatrix}$$

where Δ is a positive or negative change to the payoff a.) Can you think of another way of defining the payoff elements that would persuade the zoning commission to use your game-theory analysis? Some people might be bothered by the negative payoffs. If you add 30 to each payoff element, you have the same game without negative elements.

26.12. Once you determine the optimal strategy for playing a two-person zero-sum game, you have to make sure that you employ a device that randomizes the selection of your strategy based on the optimal frequencies. For example, in the skin game, player 1's optimal strategy is ($x_1 = 0, x_2 = \frac{3}{5}, x = \frac{2}{5}$) and the value of the game $v = \frac{1}{5}$. To play this game, we would take five pieces of paper and write "strategy 2" on three of them and "strategy 3" on two of them and mix them up in a container. We would then pick one of the papers, sight unseen, and play the corresponding strategy. Returning the paper to the container, we would then draw again for the next play of the game, and so on. You could also divide a spinner or roulette wheel so that the pointer will stop on numbers or areas such that the frequency count of 3 to 2 is maintained for the second and third strategies. Any other ideas?

26.13. A two-person, nonzero-sum game is represented by two payoff matrices that indicate the payoffs to each player as they select among the possible strategies. Such games are called *bimatrix games*. For a two-strategy game we might have the following payoff matrices:

$$\begin{matrix} & \begin{matrix} 1 & 2 \end{matrix} \\ \begin{matrix} 1 \\ 2 \end{matrix} & \begin{pmatrix} 2 & -1 \\ -1 & 1 \end{pmatrix} \end{matrix} \quad \text{Player 1's payoffs}$$

$$\begin{matrix} 1 \\ 2 \end{matrix} \begin{pmatrix} 1 & -1 \\ -1 & 2 \end{pmatrix} \quad \text{Player 2's payoffs}$$

The rows (1,2) in each matrix represent player 1's strategies and the columns (1,2) are player 2's strategies. If player 1 chooses row 1 and player 2 chooses column 1, then player 1 receives two units and player 2 receives one unit. We assume that the players do not cooperate in their selections. These type of games are usually solved in terms of *equilibrium strategies*; that is, if a player employs an equilibrium strategy, the other player cannot do better by deviating from the corresponding equilibrium-strategy choices. Every bimatrix game has at least one equilibrium point. For example, the bimatrix game

$$\begin{array}{cc} & \begin{array}{cc} 1 & 2 \end{array} \\ \begin{array}{c} 1 \\ 2 \end{array} & \begin{pmatrix} 4 & 0 \\ 0 & 1 \end{pmatrix} \end{array} \quad \text{Player 1's payoffs}$$

$$\begin{array}{c} 1 \\ 2 \end{array} \begin{pmatrix} 1 & 0 \\ 0 & 4 \end{pmatrix} \quad \text{Player 2's payoffs}$$

has the equilibrium solutions of $(x_1 = 1, x_2 = 0; y_1 = 1, y_2 = 0)$, $(x_1 = 0, x_2 = 1; y_1 = 0, y_2 = 1)$, and $(x_1 = \frac{4}{5}, x_2 = \frac{1}{5}; y_1 = \frac{1}{5}, y_2 = \frac{4}{5})$. Check them out; show that $(x_1 = 1, x_2 = 0; y_1 = 0, y_2 = 1)$ is not an equilibrium solution.

There is some concern as to whether equilibrium strategies are acceptable solutions to all bimatrix games. Consider the following problem, "*the prisoner's dilemma,*" as described by Rapoport.

> Two suspects are guilty of the crime of which they are suspected, but the D.A. does not have sufficient evidence to convict either. The state, has, however, sufficient evidence to convict both of a lesser offense. The alternatives open to the suspects, *A* and *B*, are to confess or not to confess to the serious crime. They are separated and cannot communicate. The outcomes are as follows. If both confess, both get severe sentences, which are, however, somewhat reduced because of the confession. If one confesses (turns state's evidence), the other gets the book thrown at him, and the informer goes scot free. If neither confesses, they cannot be convicted of the serious crime, but will surely be tried and convicted for the lesser offense.

In these terms, letting strategy 1 stand for "not confess" and strategy 2 for "confess" we can represent the game in terms of years in jail by the two payoff matrices

Prisoner *A* strategies	Prisoner *B* strategies: Not confess	Confess	
Not confess	1	20	*A*'s payoffs
Confess	0	5	

Prisoner *A* strategies	Prisoner *B* strategies: Not confess	Confess	
Not confess	1	0	*B*'s payoffs
Confess	20	5	

What should the prisoners do? Can they be sure that each would not confess? Note that for Prisoner *A*, the confess strategy appears to be the "rational" choice, and similarly for *B*. Why? This yields the unique equilibrium solution of each confessing and receiving a penalty of five years. Would you do that?

The original version of "the prisoner's dilemma," due to the mathematician A. W. Tucker, had the following payoff matrices:

$$\begin{array}{cc} & \begin{matrix} 1 & 2 \end{matrix} \\ \begin{matrix} 1 \\ 2 \end{matrix} & \begin{pmatrix} 0.9 & 0 \\ 1.0 & 0.1 \end{pmatrix} \end{array} \quad \text{Player 1's payoffs}$$

$$\begin{array}{cc} \begin{matrix} 1 \\ 2 \end{matrix} & \begin{pmatrix} 0.9 & 1 \\ 0 & 0.1 \end{pmatrix} \end{array} \quad \text{Player 2's payoffs}$$

Here, the payoffs are interpreted in the usual monetary sense. Again, for each player, there are strategies that are dominant in that the payoffs in the second row of player 1's matrix are greater than the corresponding payoffs in the first row; while for player 2, the second-strategy payoffs are dominant. Thus, they both should select their second strategy with probability 1 (they each make 0.1). But if they both play their first strategy, they are each better off (they each make 0.9). What if a player decides to double-cross the other? Of course, the dilemma vanishes if they can communicate.

Use this game to experiment with your friends. Set up situations in which they can communicate and situations without communication. How do the experimental results change if the payoff matrices are

$$\begin{array}{cc} & \begin{matrix} 1 & 2 \end{matrix} \\ \begin{matrix} 1 \\ 2 \end{matrix} & \begin{pmatrix} 9 & 2 \\ 10 & 0 \end{pmatrix} \end{array} \quad \text{Player 1's payoffs}$$

$$\begin{array}{cc} \begin{matrix} 1 \\ 2 \end{matrix} & \begin{pmatrix} 9 & 15 \\ -10 & -30 \end{pmatrix} \end{array} \quad \text{Player 2's payoffs}$$

A player could ask for a side payment that confuses the issue, for example, player 1 threatens to play strategy 2 unless player 2 agrees that player 1 should get 10 units. This can take place if the payoffs are in money, but not if they represent years in jail.

26.14. In the GPA example, we assumed the following utilities u_i for each GPA outcome O_i:

O_i	u_i	i
4.0	1.0	1
3.5	0.9	2
3.0	0.8	3
2.5	0.5	4
2.0	0.0	$5 = m$

A major task in developing a utility function, that is, the set of u_i, is to ensure some kind of consistency between the values. We require $u_1 = 1 \geq u_2 \geq \cdots \geq u_{m+1} \geq u_m = 0$ and if $u_i = u_{i+1}$, then we are indifferent to the outcomes O_i and O_{i+1}. This process assumes that the decision maker can rank the outcomes and select the best (the one corresponding to $u_1 = 1$) and the worst (the one corresponding to $u_m = 0$). How do we determine the values of u_i between u_1 and u_m? One of the axioms of rational behavior assumes that the decision maker can make a particular comparison between outcome O_i and outcomes O_1 and O_m which leads to the determination of a u_i. It is done in the following manner.

Consider a lottery in which there are just the two prizes O_1 and O_m. You are given the choice of receiving the outcome O_i with probability 1 or playing a lottery in which you can win O_1 with probability u_i or win O_m with probability $1 - u_i$. The axioms of rational behavior assume that you can determine a probability $0 \leq u_i \leq 1$ such that you would be indifferent to playing the lottery with this probability or receiving O_i with certainty. Playing the lottery enables you to win either O_1 (the best prize) or O_m (the worst prize). Do you think the u_i can be determined in this manner?

For example, let O_1 = \$1000, O_m = \$0, and O_i = \$500. If we select $u_i = 0.5$ then the EMV (lottery) = (0.5)(\$1000) + (0.5)(\$0) = \$500. A supposed EMV rational person should be indifferent to playing this lottery or receiving \$500 with certainty. Would you do that?! But what if we increased u_i to 0.75 or to 0.99? With $u_i = 0.99$, most of us would decide to play the lottery and chance the winning of \$1000 with little risk. Of course, if we made $u_i = 0.01$, then most of us would take the \$500 with certainty and not play the lottery. Utility theory assumes that there is a happy middle ground between these extreme positions, that is, the decision maker can select a u_i that makes playing the lottery and receiving O_i with probability 1 equivalent (the decision maker would be indifferent). Using this process and the probabilities u_i as utility values, the decision maker can determine the set of u_i that fill out the utility function between $u_1 = 1$ and $u_m = 0$.

As you can imagine, this is easier said (written) than done. You should try it out on the GPA example to develop your utility function for that situation. Compare yours with some of the other students to determine why they are different.

26.15. For the contractor example, we have the following discounted outcomes and probabilities for the two contracts C_1 and C_2:

Contract	O_i	p_i	u_i	i
C_1	\$100,000	0.60	1.0	1
C_2	50,000	0.30		2
—	0	—		3
C_2	− 10,000	0.70		4
C_1	− 100,000	0.40	0.0	5

We include the \$0 outcome as the contractor has the option of not bidding and obtaining a \$0 return. You should assume the role of the contractor and complete the u_i column. Then compute the expected utility $u(C_1)$ and $u(C_2)$ and compare them with your value of u_3, the utility of not bidding. The maximum value determines the decision to be made.

26.16. The following example (from Hadley) shows that the decision that maximizes EPV can be different from the one that maximizes expected utility.

A contractor must decide on bidding for contracts C_1 and C_2 or neither. The discounted dollar profits and probabilities are given below.

C_1		C_2	
Profit/Loss (O_i)	Probability (p_i)	Profit/Loss (O_i)	Probability (p_i)
\$50,000	0.7	\$40,000	0.6
10,000	0.1	30,000	0.2
− 20,000	0.2	− 10,000	0.2

Based on the lottery indifference approach, the contractor has furnished the following utility function for the seven possible outcomes, including not bidding:

Contract	O_i	p_i	u_i	i
C_1	\$50,000	0.7	1.00	1
C_2	40,000	0.6	0.95	2
C_2	30,000	0.2	0.80	3
C_1	10,000	0.1	0.50	4
—	0	—	0.30	5
C_2	− 10,000	0.2	0.20	6
C_1	− 20,000	0.2	0.00	7

Show that the expected utility $u(C_2)$ is larger than $u(C_1)$ and the no-bid utility u_5. Also, show that $\text{EPV}(C_1) > \text{EPV}(C_2)$. Develop a new utility function that is related to the discounted outcomes so that $u(C_1) > u(C_2)$. [*Hint:*

Assume the utility function is a function of only the monetary outcomes O_i and that this function is linear. Then the lottery that maximizes the expected discounted profit is also the one with the highest expected utility. Why? The form of the linear relationship is $u_i = aO_i + b$, where the outcomes O_i are in monetary units. You can determine the slope a and intercept b as we know the line passes through the two points $(O_1, u_1 = 1)$ and $(O_m, u_m = 0)$. Determine the linear relationship for the above contractor problem and the corresponding u_i.] (A utility function that can be expressed as $u_i = aO_i + b$ is said to be objective. As Epstein notes, "It has been stated that only misers and mathematicians truly act according to objective utilities. For them, each and every dollar maintains a constant value regardless of how many other dollars can be summoned in mutual support.")

Utility as a function of the units of outcome is most often not a linear relationship. For the contractor's seven outcomes and associated utilities, picture their functional relationship by a graph with the horizontal axis for the O_i and the vertical axis for the u_i. Do you think you could use the graph to interpolate a u_i for an O_i not in the table of outcomes? What types of continuous curves seem most appropriate for utility functions?

26.17. For a utility function $\{u_i\}$, $i = 1, \ldots, m$, show that the decision is not changed if we use the transformed utility function $U_i = au_i + b$, where $a > 0$. Thus, there is an infinity of utility functions that are related by a linear transformation. Why must $a > 0$? Note that $U_1 = a + b$ (as $u_1 = 1$) and $U_m = b$ (as $u_m = 0$). Using this result you can transform the scale of the utility function from (1, 0) to, for example, (100, 0), or (100, 50), and so on. Transform the original GPA utility function such that $u_1 = 50$ and $u_m = 10$. [Note that $50 = 40 + 10 = a + b$ and $U_i = (50 - 10)u_i + 10$.]

26.18. In developing utility functions, we all feel more comfortable when the outcomes are measured in dollars. However, many public-policy decisions have outcomes that cannot be readily compared as they are not given in dollars. For example, the mayor of a city has many options in allocating the city's budget. How should the mayor allocate a budget surplus of $1 million to the city's service departments (roads, police, fire, health, garbage collection)? What is the utility of filling all the potholes versus 10 extra police officers on patrol? The mayor's utility function might vary based on the demands of the various constituencies and how close it is to election time. Your utility function would depend on whether you lived in a high-crime area or commuted over the bumpy roads. How would you help the mayor out?

26.19. You want to keep your old car for two more years, but it needs some repair work. You can have the mechanic give it a regular tuneup plus minor replacements for $100, or spend $500 for a major engine overhaul, or delay doing anything. Based on discussions with the mechanic and the repair history of the car, over the next two years you can expect additional mechanical trouble and costs according to the following table:

Action	Major trouble	Average trouble	Minor trouble	No trouble
Tuneup (cost $100)	$1000 $p = 0.3$	$500 $p = 0.5$	$200 $p = 0.2$	0 $p = 0.0$
Overhaul (cost $500)	$1000 $p = 0.1$	$500 $p = 0.3$	$200 $p = 0.1$	0 $p = 0.5$
Do nothing (cost $0)	$1000 $p = 0.4$	$500 $p = 0.6$	$200 $p = 0.0$	0 $p = 0.0$

Use a decision tree to aid in determining which action you should take.

26.20. How could you use decision-tree analysis to help the University Construction Committee decide to expand the football-field seating by either building a 5000-seat permanent addition or to contract for 5000 temporary seats that are put up for each big game. The chance events are whether you will have a winning team for each of the next five years. What data are required? Remember to do a sensitivity analysis on critical data.

26.21. Set up and solve the following problem using a decision tree. A racing stable has a choice of running only *one* of two horses in a race. The horses named *A* and *B* have different capabilities of winning depending on the speed of the track. We classify the track speeds as dry, wet, or muddy. The weatherforecaster says that the probability of each speed on the day of the race is $\frac{1}{3}$, that is, they are equally likely. Based on past performances, if the track is dry, horse *A* has the following probabilities of finishing first, second, or third—(0.7, 0.2, 0.1), if it is wet—(0.5, 0.4, 0.1), and if it is muddy—(0.0, 0.5, 0.5). For horse *B*, the probabilities of finishing first, second, or third are (0.2, 0.5, 0.3) if the track is dry, if it is wet (0.5, 0.4, 0.1), and muddy (0.6, 0.2, 0.2). The purse (winnings) are $2000 for first place, $1000 for second place, and $500 for third place. Which horse would you run and why? (Assume you have to enter the horse a week before the race.)

26.22. For the company's one-year decision tree (Section 23.5), determine for what values of the probabilities of a sales rise and a sales drop would the company be indifferent to each action, that is, find the probabilities such that the net EPV (new equipment) = net EPV (overtime). Also perform the net EPV sensitivity analysis for sales-rise probabilities of 0.7 and 0.5. How would you present the results of a sensitivity study to the company president?

26.23. For the company's two-year decision (Section 23.5), perform a sensitivity analysis on the probabilities by changing each one by ± 0.1, as required.

26.24. (a) Call your local bank to find out what the current interest rate is on a regular savings account. Using this rate, determine the present value of a \$12,000 gift you will receive in three years. How much less than \$12,000 would you be willing to accept today?

(b) Determine present value formulas for the following situations:

1. For a payment P received n years in the future at simple interest i, that is, when interest is not compounded.
2. For a series of future payments P_t ($t = 1, \ldots, n$) at compound interest i.
3. For a series of future payments P_t ($t = 1, \ldots, n$) at corresponding compound interest rates i_t.
4. For a payment P received n years in the future when interest i is compounded k times a year.

26.25. Develop a decision tree for the decisions and chance events associated with your remaining undergraduate years. The outcome payoffs should not be in monetary terms.

26.26. The following 10-year decision tree (due to Magee) requires you to discount all the yields at a constant 10%. Note that some of the yields are over 10 years; others that come out of the second decision point at year three are for eight years. Determine the net expected present yield and the decisions the company should take.

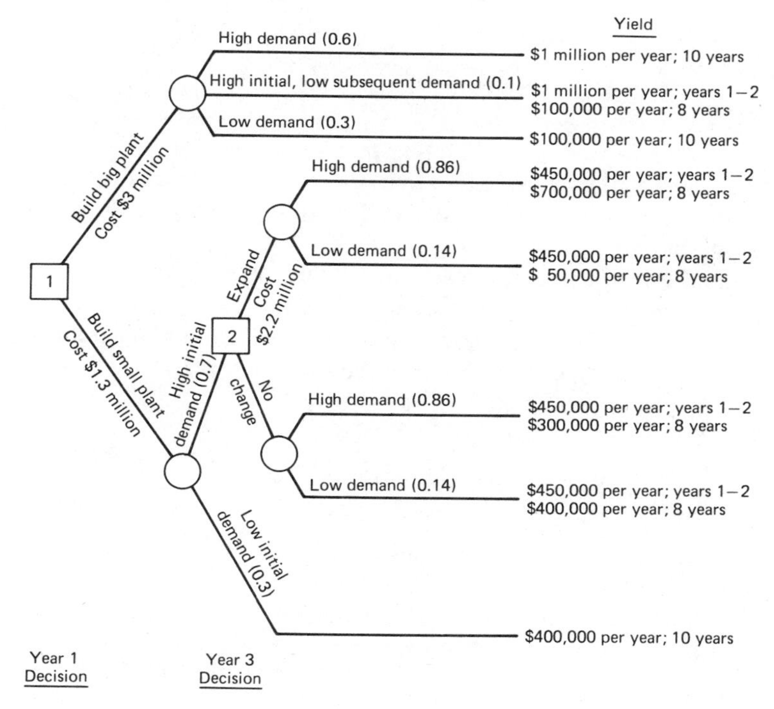

26.27. In his paper, Magee presents the following decision tree for which the decision is to either have a cocktail party outdoors or indoors, with the chance event being rain or no rain. Develop a utility function for the four outcomes. You have to determine the most favorable and worst outcomes for you. Determine the values of the probability of no rain that would make the expected utility value for the outdoor decision a maximum. You need to treat the probability p of no rain as a parameter and find the range of p that makes the expected utility of the outdoors node greater than the expected utility of the indoors node.

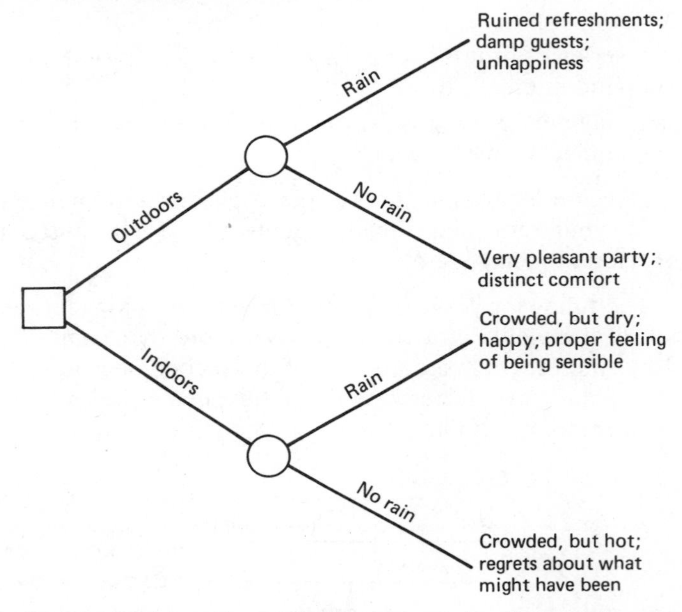

26.28. Some people are bothered by expected values as they do not seem to belong to any particular decision. The following example (due to McCreary) might help.

A company has a new process and product they want to market. For a \$1 million investment, they would expect a net cash flow of \$4 million if sales were high (probability of (0.7) and a net cash loss of \$1 million if sales were low (probability of 0.3). The simple decision tree is

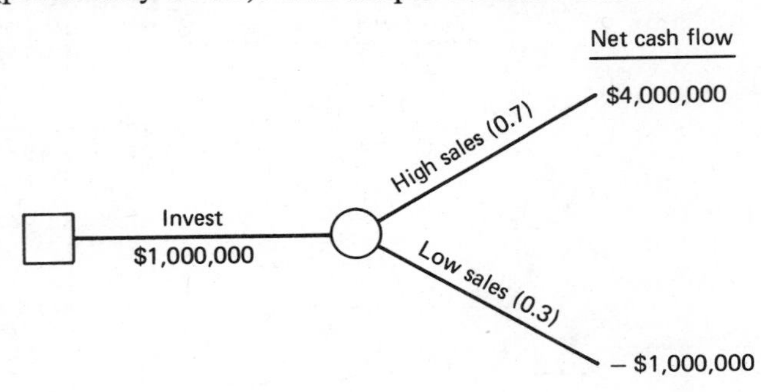

We have

$$\begin{aligned}\text{EMV (Invest)} &= (0.7)(\$4{,}000{,}000) + (0.3)(-\$1{,}000{,}000) \\ &= \$2{,}500{,}000\end{aligned}$$

What do you make of \$2.5 million expected monetary value? The company will either get \$4 million or lose \$1 million once the decision to invest is made. The \$2.5 million is difficult to understand in this context. But, what if another company wanted to buy the process from the company, how would you set a purchase figure? They wouldn't pay you \$4 million as there is some risk to the venture, and you certainly wouldn't pay them \$1 million to take it off your hands! As you think about it, the \$2.5 million might be the right purchase figure.

26.29. Discuss the construction of a decision-tree model in terms of the decision framework and modeling steps.

26.30. You might have had the idea that you can display the decision–chance activities of a two-person zero-sum game using a decision tree. Such games do have a tree structure called the *game tree*. For the heads (H)–tail(T) game we have, starting with player 1, the following tree:

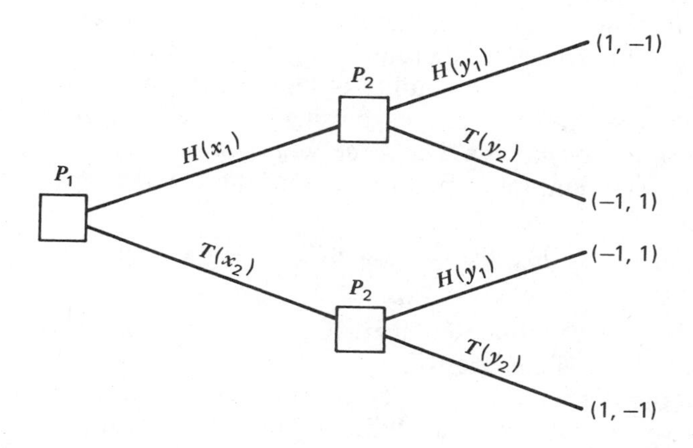

The decision points are choices by the players. The payoffs are shown as a number couple with the first number player 1's payoff and the second number player 2's payoff. The x_i's and y_j's represent the corresponding nonnegative strategy values (probabilities) with $x_1 + x_2 = 1$, $y_1 + y_2 = 1$. Is this a good way to analyze such games? What is the EMV at decision point P_1? Draw the game tree for the stone-paper-scissors game. You might also try to draw the game tree for tic-tac-toe. For the game tree, you need to instruct each player on what to do whenever a particular node occurs in the sequence of play. Can you imagine what the game tree for chess or checkers would look like?

26.31. The concept of mathematical expectation (expected value) as applied to gambling and real money has often caused problems of interpretation. Two famous situations are the following.

(a) *The Petersburg Game (St. Petersburg Paradox) Due to Daniel Bernoulli.* A single play of the Petersburg Game consists of tossing a true coin until it falls heads. If this occurs at the nth throw, the player (gambler) receives 2^n dollars. (A coin is said to be true if the probability of its falling heads is $\frac{1}{2}$ and the probability of its falling tails is $\frac{1}{2}$; we assume the toss is also accomplished without any sleight of hand.) The probability of this occurrence is the probability of the first $n - 1$ tosses coming up tails and the nth toss falling heads. This probability is just $(\frac{1}{2})^n$. The gambler would then receive \$2.00 with probability $\frac{1}{2}$, \$4.00 with probability $\frac{1}{4}$, \$8.00 with probability $\frac{1}{8}$, and so on. The mathematical expectation of a play is given by

$$2(\tfrac{1}{2}) + 4(\tfrac{1}{4}) + 8(\tfrac{1}{8}) + \cdots = 1 + 1 + 1 + \cdots$$

This expression does not sum to a finite number; the mathematical expectation is infinite. If this is so, then how much would you be willing to pay for the privilege of playing? The usual analysis implies that you would be willing to pay any sum asked, but that doesn't seem to be the right thing to do.

Bernoulli's answer to the paradox is that when we have a large sum of money, the utility of getting more is not the same as when we have a small amount of money, that is, the utility of money increases with more money but at a decreasing rate. He proposed using the base 10 logarithm as a measure of the utility of money. Thus, the worth of having d dollars is $\log d$. With this assumption, the expected value of the Petersburg Game is

$$v = (\log 2)(\tfrac{1}{2}) + (\log 4)(\tfrac{1}{4}) + (\log 8)(\tfrac{1}{8}) + \cdots$$

Which converges to a finite number. (Can you determine what v is?) Then the amount a you should be willing to play is given by $\log a = v$. Do you agree with the Bernoulli way out of the paradox? Another way of resolving the paradox is to make the game finite by stipulating that the player receives nothing if the game does not end after say m tosses. The naturalist Comte de Buffon suggested ignoring all probabilities that are rather small, say $(\frac{1}{2})^n$, where n is greater than 100. These small values correspond to "impossible" values. If we do this, then the expectation is finite. Play the game, as originally stated, 100 times to determine an average amount a you might be willing to pay for playing the game. Before you do, take a guess at what the amount will be. (The Petersburg Game was so named as Bernoulli was living in that city when he first encountered the game.)

(b) *Pascal's Wager.* The mathematician Blaise Pascal led a very religious life as he concluded that since the value of eternal happiness is infinite, and even if the probability of gaining eternal happiness is small, the expectation is still infinite.

26.32. You have probably given some thought as to why the final composite priorities derived from the analytic hierarchy process (AHP) can be interpreted as numbers that combine a decision-maker's quantitative and qualitative understanding of a complex problem situation and, as such, are correct indicators of how the alternatives are ordered from most favorable to least favorable. In his books, Saaty describes a number of ways he has found to validate this use of the AHP priorities. Some of these tests are based on experiments that deal with physical quantities, while others are derived from AHP studies whose results can be checked by independent data.

There are many methods for analyzing multicriteria decision problems. Some deal with pairwise comparisons like the AHP, while others involve more complex ordering procedures, utility functions, linear programming, and other analytical structures. Which one, if any, to use for a particular decision situation comes with experience. However, for an analysis model based on pairwise comparisons, Saaty has shown (using results from the theory of graphs) that the AHP eigenvector correctly indicates the relative importance (dominance) of each alternative with respect to the others.

Procedures like the AHP, utility functions, decision trees, and so on, are powerful ways of organizing the information of a problem and thus enable us to better understand how the elements of a problem interact. When using any decision-aiding procedure, you need to understand its assumptions and limitations and be sure that the procedure can be used for the problem at hand. An important aspect of a decision study is the related sensitivity analysis of how the results vary with changing estimates of the data and parameters. For example, in using the AHP, you will have more confidence in the final priorities if you have demonstrated that the results do not change much (if at all) when you tried out other values for your not-so-sure pairwise comparisons.

For the AHP, Saaty proposes the calculation of a consistency index that measures how consistent you were in comparing the elements in a judgment matrix. For example, were we consistent in comparing cars to each other in terms of a specific criteria? If car A was more comfortable than car B, and car B was more comfortable than car C, did we conclude that car A was more comfortable than car C? (Check it out.) When we have many factors to compare, it is difficult to maintain such consistency, and, based on intuitive and other factors, we might want to deviate from true consistency. As Saaty notes, we may not be perfectly consistent, but that is the way we tend to work. Certainly gross inconsistency would lead to invalid results. Based on certain mathematical relationships, there is a simple way to measure consistency of an AHP judgment matrix. It is derived from the properties of judgment matrices and the theory of eigenvalues and eigenvectors. A consistency measure is determined in the following manner.

Consider an AHP judgment matrix with n rows and columns, that is,

$$\begin{pmatrix} a_{11} & \dots & a_{1n} \\ \vdots & & \vdots \\ a_{n1} & \dots & a_{nn} \end{pmatrix}$$

where $a_{ij} = 1/a_{ji}$ and all $a_{ij} > 0$. Let p_i be the corresponding AHP priorities. Sum each column of the matrix and multiply each sum by the corresponding p_i. Sum the n products and denote the result by λ_{max}. The formula is

$$\lambda_{max} = p_1 \sum_{i=1}^{n} a_{i1} + p_2 \sum_{i=1}^{n} a_{i2} + \cdots + p_n \sum_{i=1}^{n} a_{in}$$

(This is an approximate way of calculating the maximum eigenvalue of the matrix.) If the matrix is consistent, then $\lambda_{max} = n$. Thus, a suggested *consistency index CI* is

$$CI = \frac{\lambda_{max} - n}{n - 1}$$

The *CI* is compared to the corresponding random consistency index *RI* from the following table:

Size of matrix (n)	1	2	3	4	5	6	7	8	9	10
RI	0.00	0.00	0.58	0.90	1.12	1.24	1.32	1.41	1.45	1.49

(The *RI* are average consistency indices for matrices whose reciprocal entries were drawn at random from the values $\frac{1}{9}, \frac{1}{8}, \dots, 1, 2, \dots, 9$.) It is suggested that if the *consistency ratio CR* = *CI*/*RI* is less than or equal to 0.10, then the results be accepted. Otherwise, the problem should be studied again and the judgment matrix revised. (Note that for our judgment matrices we should have $\lambda_{max} \geq n$.) Also, matrices of orders 1 and 2 are necessarily consistent and the *CI* and *CR* formulas are not applicable.)

We illustrate the *CI* and *CR* calculation for the level 2 judgment matrix used in deciding which car to buy.

Decision to buy a new car	Price	Running cost	Comfort	Status	p_i
Price	1	3	7	8	0.586
Running cost	$\frac{1}{3}$	1	5	5	0.277
Comfort	$\frac{1}{7}$	$\frac{1}{5}$	1	3	0.088
Status	$\frac{1}{8}$	$\frac{1}{5}$	$\frac{1}{3}$	1	0.049
Column sum	1.60	4.40	13.33	17	

$$\lambda_{max} = (0.586)(1.60) + (0.277)(4.40) + (0.088)(13.33) + (0.049)(17) = 4.16$$

$$CI = \frac{4.16 - 4}{3} = 0.0533$$

$$CR = \frac{CI}{RI} = \frac{0.0533}{0.90} = 0.06$$

As the *CR* is less than 10%, the priorities are acceptable. You should calculate the consistency ratios for the other car-buying judgment matrices and for those of the job-selection problem. The total *CI* of a hierarchy is obtained by weighting each *CI* by the priority of the element with respect to which the comparison is being made and then adding all the results. Thus, if a *CI* is high it might not have much influence due to its multiplier. This, for example, would be the case in the car-buying problem if the comfort or status matrices had a high *CI*. The hierarchy *CR* is obtained by comparing the hierarchy *CI* to the hierarchy *RI*, which is found by summing similarly weighted *RI*'s, where each *RI* corresponds to the dimension of the related individual *CI*. The hierarchy *CR* should be about 0.10.

When applying the AHP to decision problems that involve your judging qualitative factors, we should not expect to obtain a consistent judgment matrix. In fact, for such problems, strict consistency implies a rigidity in your estimates that would not allow you to factor in new information. For example, an automobile-rating magazine just came out and claimed that car *A*'s ride is smooth, car *B*'s ride is noisy, and car *C*'s ride is bumpy. You thought you had it all figured out, but now you need to combine this information with your previous impressions about each car's comfort, and redo the comparison matrix. Certainly, gross inconsistency needs to be avoided, but do not be afraid to let your judgment rule your comparisons.

If a judgment matrix has a *CR* greater than 0.10, there is a way to find out which comparisons seem to be the ones causing trouble. Denote the numbers in your original *n*th-order comparison matrix by a_{ij} (the element in the *i*th row and *j*th column) and the associated priorities by p_i. Form a new matrix in which element (*i*,*j*) is given by p_i/p_j. This resultant matrix will be consistent and produce the original p_i. (This is just a way of saying that the new matrix has a single eigenvalue which is equal to n, with a corresponding normalized eigenvector whose components are the p_i.) Take the ratio of each a_{ij} to its corresponding p_i/p_j. Those ratios that deviate greatly from the value of 1.0 probably have a_{ij} values that need to be adjusted. As we are not looking for complete consistency, Saaty suggests only adjusting the a_{ij} that corresponds to the ratio with the largest absolute deviation from 1.0. If the ratio is greater than 1.0, adjust the a_{ij} down by one unit; if the ratio is smaller than 1.0, adjust the a_{ij} up by one unit.

26.33. As we have noted, much of the AHP theory is based on the properties of reciprocal judgment matrices and their eigenvalues and eigenvectors. The scheme we used to compute the eigenvector is an approximate one and not too accurate. In fact, it has been shown that there are judgment matrices for which our approximate scheme reverses the priority rankings. Thus, we suggest that for real decision problems you use a more accurate matrix method for computing the eigenvector and the associated maximum eigenvalue. Assuming you have knowledge of basic matrix operations, then a very accurate way of computing the eigenvector of a matrix can be described as follows. Let **A** be an $n \times n$ matrix and **e** be an nth-order column vector, all of whose components equal one. Form the successive products $\mathbf{Ae}$, $(\mathbf{A}^2)\mathbf{e}$, $(\mathbf{A}^3)\mathbf{e}$, . . . , $(\mathbf{A}^k)\mathbf{e}$. For sufficiently large k, the vector $(\mathbf{A}^k)\mathbf{e}$ will be a close approximation of the required eigenvector. This eigenvector will have to be normalized to yield the desired AHP priorities. Note that the components of $(\mathbf{A}^k)\mathbf{e}$ correspond to the row sums of $\mathbf{A}^k$. You should stop the computation when the differences of $(\mathbf{A}^{k+1})\mathbf{e} - (\mathbf{A}^k)\mathbf{e}$ are smaller than a given value, say 0.0001. Let $(\mathbf{A}^k)\mathbf{e}$ be the approximate eigenvector and denote its components by y_i^k. Form the vector $(\mathbf{A}^{k+1})\mathbf{e}$ and denote its components by y_i^{k+1}. All the ratios y_i^{k+1}/y_i^k should be approximately equal and correspond to λ_{max}. (If appropriate for your class, the eigenvector computation can be programmed on a computer. In fact, you will find it instructive to program the complete AHP so that a decision maker can use the computer to input and analyze judgment matrices in an interactive manner. [See the books by Faddeeva, and Ralston and Rabinowitz for further details on computing eigenvectors and eigenvalues, e.g., the best way to organize the computation of the successive products that yield $(\mathbf{A}^k)\mathbf{e}$.])

26.34. The answer to the area problem (to two-decimal accuracy) is $p_1 = 0.27$, $p_2 = 0.09$, $p_3 = 0.18$, $p_4 = 0.05$, and $p_5 = 0.41$. The true comparison matrix is

Area	F_1	F_2	F_3	F_4	F_5
F_1	1	3	1.5	6	0.67
F_2	0.33	1	0.5	2	0.22
F_3	0.67	2	1	4	0.44
F_4	0.17	0.5	0.25	1	0.11
F_5	1.5	4.5	2.25	9	1

Here the element in the ith row and jth column $a_{ij} = A_i/A_j$, where the A_i and A_j values are the correct area values. Use the above matrix to determine more accurate values of the p_i. A consistent comparison matrix that corresponds to the p_i given above can be calculated by forming the matrix in which $a_{ij} = p_i/p_j$. Determine this matrix and show that it does yield the given p_i.

The geometric figures in Chapter 24 have the following areas in terms of measurement units: $A_1 = 150$, $A_2 = 50$, $A_3 = 100$, $A_4 = 25$, and $A_5 = 225$, with the total units $T = 550$.

26.35. Develop a utility function approach to buying a car and compare it to the AHP method. Discuss the pros and cons of both procedures.

26.36. To obtain an understanding of how the judgment matrices and the pairwise comparison scale captures your intuitive judgments, your instructor should have the class carry out the following experiment. The instructor brings in five sealed paper bags, each bag containing an object of unknown physical weight (unknown to the class, that is). The bags are numbered 1 to 5. Each member of the class has an opportunity to lift and compare the bags in any manner. Then each student prepares a judgment matrix in which bag i is compared to bag j by asking the question: "In terms of physical weight, which bag is heavier and by how much more?" You have to use the scale from 1 to 9 and the reciprocals to capture the differences between the bags. Thus if bag i to bag j is scored 5, then bag j to bag i is scored $\frac{1}{5}$. The final priorities p_i for the resultant judgment matrix should be close to the ratios of the actual weights w_i to the total weight of all the bags, $W = w_1 + \cdots + w_5$, that is, hopefully, each p_i will be approximately equal to the corresponding w_i/W. In any event, you should calculate the consistency ratio CR of your matrix. Use the actual weights to determine the true comparison matrix and show that this matrix has an eigenvector that yields the correct ratios.

26.37. The AHP has been proven of value in situations where many individuals are involved in making the decision and it is difficult to obtain agreement on the judgment matrix and the priorities. Try to have your class resolve the following situation. Your instructor has given the class three examination options: (1) a take-home examination; (2) an in-class, closed-book, three-hour examination; and (3) an in-class, closed-book, one-hour examination, plus a computer project which requires each student to formulate a linear-programming application of reasonable size, collect data, solve the problem, and write a report discussing the problem, solution, and implementation of the results. See if the class can agree on one judgment matrix that will reflect the composite class priorities. If the class cannot agree on a specific judgment number, a compromise value to use would be the geometric mean of the numbers proposed. (Why is that a good compromise?) Have each student prepare a judgment matrix, and then fight it out to obtain a class matrix. Before doing the experiment, have the students vote in secret on what is their first choice and compare the class AHP priorities with the proportions of votes for each option.

26.38. For the car-buying example, redo the analysis by removing comfort and status from the level 2 matrix. Recall that their level 2 priorities were quite small. Prorate the p_i for price and running cost so they total to 1.0.

26.39. For the following situations, determine a focus (goal), level 2 criteria, and level 3 alternatives; collect appropriate data; and determine your decisions based on the AHP analysis:

1. Form a stock portfolio by choosing three out of six possible stocks suggested to you by a stockbroker.
2. Three of your classmates are running for president of the student body. Who should you vote for?
3. During the spring break, you have a choice of going skiing, going to the beach at Fort Lauderdale, or staying home and studying. Which do you choose?
4. You have to select from among four bands to play at the class dance. Which do you choose?
5. You and your date are going out to dinner at a local restaurant. You have four choices: French, Chinese, Italian, or Tex-Mex. You are paying. Which do you choose? How does your analysis change if the dinner is Dutch treat or your date is paying for both of you?
6. A young friend has been accepted to four universities. Show your friend how the AHP can help in choosing the school to attend.

26.40. We can validate models of physical processes by seeing whether the model outputs can be put to work. For example, the answers to the museum-guard models are easy to verify; this is also true for the shipping pattern that solves a transportation problem. We saw that the answer to a diet problem is not valid for most people, but is all right for cattle and chickens. You should look back on the models described in the text to determine how you can attempt to establish their validity.

26.41. When going through life making decisions, remember that

You've got to AC-CENT-TCHU-ATE THE POSITIVE,
E-lim-my-nate the neg-a-tive,
Latch on to the af-firm-a-tive,
Don't mess with Mis-ter In-be-tween.

You've got to spread joy up to the max-i-mum,
Bring gloom down to the min-i-mum,
Have faith or pan-de-mo-ni-um li-ble to walk up-on the scene.

AC-CENT-TCHU-ATE THE POSITIVE
by Johnny Mercer and Harold Arlen

26.42. "All God's Children Got Algorhythm."

Part V References

Anonymous: *Guidelines for Model Evaluation*, U.S. General Accounting Office, PAD-79-17, Washington, D.C., January 1979.

Bell, E. T.: *Men of Mathematics*, Simon & Schuster, New York, 1937.

Boyer, C. B.: *A History of Mathematics*, Wiley, New York, 1968.

Bradley, S. P., A. C. Hax, and T. L. Magnanti: *Applied Mathematical Programming*, Addison-Wesley, Reading, Mass., 1977.

Clark, J. and J. Cole: *Global Simulation Models: A Comparative Study*, Wiley, New York, 1975.

Danskin, J. M.: A Game Theory Model of Convoy Routing, *Operations Research*, Vol. 10, no. 6, November–December 1962.

Danskin, J. M.: A Helicopter versus Submarine Search Game, *Operations Research*, Vol. 16, no. 3, May–June 1968.

Davis, M. D.: *Game Theory*, Basic Books, New York, 1970.

Epstein, R. A.: *The Theory of Gambling and Statistical Logic*, Academic Press, New York, 1967.

Faddeeva, V. N.: *Computational Methods of Linear Algebra*, Dover, New York, 1959.

Feller, W.: *An Introduction to Probability Theory and Its Applications*, Vol. 1, Wiley, New York, 1950.

Gass, S. I.: *An Illustrated Guide to Linear Programming*, McGraw-Hill, New York, 1970.

Gass, S. I.: Decision-Aiding Models: Validation, Assessment and Related Issues, *Operations Research*, Vol. 31, no. 4, 1983.

Gass, S. I.: What is a Computer-Based Mathematical Model?, *Mathematical Modelling*, Vol. 4, no. 5, 1983.

Glicksman, A. M.: *Linear Programming and the Theory of Games*. Wiley, New York 1963.

Greenberg, M. R.: *Applied Linear Programming*, Academic Press, New York, 1978.

Hadley, G.: *Introduction to Probability and Statistical Decision Theory*, Holden-Day, San Francisco, 1967.

Halmos. P. R.: The Heart of Mathematics, *The American Mathematical Monthly*, Vol. 87, no. 7, 1980.

Hayward, O. G., Jr.: Military Decision and Game Theory, *Operations Research*, Vol. 2, no. 4, 1954.

Lucas, W. F.: An Overview of the Mathematical Theory of Games, *Management Science*, Vol. 18, no. 5, Part 2, January 1972.

Luce, R. D., and H. Raiffa: *Games and Decisions*, Wiley, New York, 1957.

Magee, J. F.: Decision Trees for Making Decisions, *Harvard Business Review*, Vol. 42, no. 4, July–August 1964.

McCreary, E. A.: How to Grow a Decision Tree, *THINK Magazine*, March–April, 1967.

Neumann von, J., and O. Morgenstern: *Theory of Games and Economic Behavior*, Princeton Univ. Press, Princeton, N.J., 1947.

Owen, G.: *Game Theory*, Saunders, Philadelphia, 1968.

Raiffa, H.: *Decision Analysis*, Addison-Wesley, Reading, Mass., 1968.

Ralston, A., and P. Rabinowitz: *A First Course in Numerical Analysis*, 2nd Edition, McGraw-Hill, New York, 1978.

Rapoport, A.: *Fights, Games and Debates*, Univ. of Michigan Press, Ann Arbor, 1960.

Rapoport, A., and A. M. Chammah: *Prisoner's Dilemma*, Univ. of Michigan Press, Ann Arbor, 1965.

Saaty, T. L.: *The Analytic Hierarchy Process*, McGraw-Hill, New York, 1980.

Saaty, T. L.: *Decision Making for Leaders*, Lifetime Learning Publications, Belmont, Calif., 1982.

Shubik, M.: The Uses of Game Theory in Management Science, *Management Science*, Vol. 2, no. 1, October 1955.

Shubik, M.: On Gaming and Game Theory, *Management Science*, Vol. 18, no. 5, Part 2, January 1972.

Smith, D. E.: *Quantitative Business Analysis*, Wiley, New York, 1977.

Williams, J. D.: *The Compleat Strategyst*, McGraw-Hill, New York, 1966.

Combined References

The discussion throughout this book has been limited to problems that should be of interest to you as an undergraduate student; problems that you can challenge with your present mathematical knowledge. There are other important decision models that are not presented. You can find out about them by reading any of the operations research or management science texts, and other books and articles cited in the references. Even though we deal mainly with decision situations, the spirit of our presentation is that of the mathematician P. R. Halmos who notes that "problems are the heart of mathematics."

Ackoff, R. L.: The Development of Operations Research as a Science, *Journal of the Operations Research Society of America*, Vol. 4, no. 3, June 1956.

Ackoff, R. L.: *Scientific Method: Optimizing Applied Research Decisions*, Wiley, New York, 1962.

Anonymous: *Nutrition Labeling*, Bulletin No. 382, U.S. Department of Agriculture, April 1975.

Anonymous: *Guidelines for Model Evaluation*, U.S. General Acounting Office, PAD-79-17, Washington, D.C., January 1979.

Apostel, L.: Towards the Formal Study of Models in the Non-Formal Sciences, chapter in *The Concept and the Role of the Model in Mathematics and Natural and Social Sciences*, Gordon & Breach, New York, 1961.

Balinski, M., and R. Gomory: A Primal Method for the Assignment and Transportation Problems, *Management Science*, Vol. 10, no. 3, April 1964.

Balintfy, J. L.: Menu Planning by Computers, *The Communications of the ACM*, Vol. 7, April 1964.

Balintfy, J. L.: Linear Programming Models for Menu Planning, chapter in *Hospital Industrial Engineering*, H. E. Smalley and J. R. Freeman (eds.), Reinhold, New York, 1966.

Bartlett, J.: *Bartlett's Familiar Quotations*, 14th Edition, E. M. Beck (Ed.), Little, Brown, Boston, 1968.

Baumol, W. J.: *Economic Theory and Operations Analysis*, 2nd Edition, Prentice-Hall, Englewood Cliffs, N.J., 1965.

Bazaraa, M. S., and J. J. Jarvis: *Linear Programming and Network Flows*, Wiley, New York, 1977.

Bell, E. T. *Men of Mathematics*, Simon & Schuster, New York, 1937.

Bellmore, M., and S. Hong: Transformation of the Multisalesmen Problem to the Standard Traveling Salesman Problem, *Journal of the Association for Computing Machinery*, Vol.. 21, no. 3, July 1974.

Bland, R. G.: New Finite Pivoting Rules for the Simplex Method, *Mathematics of Operations Research*, Vol. 2, no. 2, May 1977.

Bland, R. G.: The Allocation of Resources by Linear Programming, *Scientific American*, June 1981.

Bland, R. G., D. Goldfarb, and M. J. Todd: The Ellipsoid Method: A Survey, *Operations Research*, Vol. 29, no. 6, 1981.

Bodin, L., B. Golden, A. Assad, and M. Ball: Routing and Scheduling of Vehicles and Crews: The State of the Art, *Computers and Operations Research*, Vol. 10, no. 2, 1983.

Boyer, C. B.: *A History of Mathematics*, Wiley, New York, 1968.

Bradley, S. P., A. C. Hax, and T. L. Magnanti: *Applied Mathematical Programming*, Addison-Wesley, Reading, Mass., 1977.

Bross, I. D.: *Design for Decision*, Macmillan Co., New York, 1953.

Bullock, A., and O. Stallybrass (Eds.): *The Fontana Dictionary of Modern Thought*, Harper & Row, New York, 1977.

Charnes, A., W. W. Cooper, and A. Henderson: *An Introduction to Linear Programming*, Wiley, New York, 1953.

Chartrand, G.: *Graphs as Mathematical Models*, Prindle, Weber & Schmidt, Boston, 1977.

Cheung, T.: Computational Comparison of Eight Methods for the Maximum Network Flow Problem, *ACM Transactions on Mathematical Software*, Vol. 6, no. 1, March 1980.

Churchman, C. W., R. L. Ackoff, and E. L. Arnoff: *Introduction to Operations Research*, Wiley, New York, 1957.

Chvátal, V.: A Combinatorial Theorem in Plane Geometry, *Journal of Combinatorial Theory*, Vol. 18, no. 1, February, 1975.

Chvátal, V.: A Greedy Heuristic for the Set-Covering Problem, *Mathematics of Operations Research*, Vol. 4, no. 3, 1979.

Clark, J., and J. Cole,: *Global Simulation Models: A Comparative Study*, Wiley, New York, 1975.

Clarke, G., and J. W. Wright: Scheduling of Vehicles from a Central Depot to a Number of Delivery Points, *Operations Research*, Vol. 12, no. 4, July–August 1964.

Cleland, D. J., and W. R. King,: *Systems Analysis and Project Management*, McGraw-Hill, New York, 1968.

Cooper, L., and D. Steinberg: *Methods and Applications of Linear Programming*, Saunders, Philadelphia, 1974.

Crowder, H., and M. W. Padberg: Solving Large-Scale Symmetric Traveling Salesman Problems to Optimality, *Management Science*, Vol. 26, no. 5, 1980.

Danskin, J. M.: A Game Theory Model of Convoy Routing, *Operations Research*, Vol. 10, no. 6, November–December 1962.

Danskin, J. M.: A Helicopter versus Submarine Search Game, *Operations Research*, Vol. 16, no. 3, May–June 1968.

Dantzig, G. B.: *Linear Programming and Extensions*, Princeton University Press, Princeton, N.J., 1963.

Dantzig, G., R. Fulkerson, and S. Johnson: Solution of A Large-Scale Traveling Salesman Problem, *Operations Research*, Vol. 2, no. 4, November 1954.

Davis, M. D.: *Game Theory*, Basic Books, New York, 1970.

Dijkstra, E. W.: A Note on Two Problems in Connection with Graphs, *Numerische Mathematik*, pp. 269–271, Vol. 1, 1959.

Dreyfus, S. E.: An Appraisal of Some Shortest Path Algorithms, *Operations Research*, Vol. 17, no. 3, 1969.

Edmonds, J., and E. J. Johnson: Matching, Euler Tours and the Chinese Postman, *Mathematical Programming*, Vol. 5, no. 1, 1973.

Eilon, S., C. Watson-Gandy, and N. Christofides: *Distribution Management*, Griffin, London, 1971.

Emshoff, J. R., and R. L. Sisson: *Design and Use of Computer Simulation Models*, Macmillan Co., New York, 1970.

Epstein, R. A.: *The Theory of Gambling and Statistical Logic*, Academic Press, New York, 1967.

Faddeeva, V. N.: *Computational Methods of Linear Algebra*, Dover, New York, 1959.

Feller, W.: *An Introduction to Probability Theory and Its Applications*, Vol. 1, Wiley, New York, 1950.

Fisk, S.: A Short Proof of Chvátal's Watchman Theorem, *Journal of Combinatorial Theory*, Series B24, 374, 1978.

Flanders, H., and J. J. Price: *Elementary Functions and Analytic Geometry*, Academic Press, New York, 1973.

Ford, L. R., Jr., and D. R. Fulkerson: *Flows in Networks*, Princeton University Press, Princeton, N.J., 1962.

Gale, D.: *The Theory of Linear Economic Models*, McGraw-Hill, New York, 1960.

Gannon, M. J.: *Management: An Organizational Perspective*, Little, Brown, Boston, 1977.

Garey, M. R., and D. S. Johnson: *Computers and Intractability*, Freeman, San Francisco, 1979.

Garfinkel, R. S., and G. L. Nemhauser: *Integer Programming*, Wiley, New York, 1972.

Gass, S. I.: *An Illustrated Guide to Linear Programming*, McGraw-Hill, New York, 1970.

Gass, S. I.: Comments on the Possibility of Cycling with the Simplex Method, *Operations Research*, Vol. 27, no. 4, July–August, 1979.

Gass, S. I.: Decision-Aiding Models: Validation, Assessment and Related Issues, *Operations Research*, Vol, 31, no. 4, 1983.

Gass, S. I.: What is a Computer-Based Mathematical Model?, *Mathematical Modelling*, Vol. 4, no. 5, 1983.

Gass, S. I.: *Linear Programming: Methods and Applications*, 5th Edition, McGraw-Hill, New York, 1985.

Gass, S. I., and R. L. Sisson (Eds.): *A Guide to Models in Governmental Planning and Operations*, Sauger Books, Potomac, Md., 1975.

Gillett, B. E., and L. R. Miller: A Heuristic Algorithm for the Vehicle-Dispatch Problem, *Operations Research*, Vol. 22, no. 2, March–April 1974.

Glicksman, A. M.: *Linear Programming and the Theory of Games*, Wiley, New York, 1963.

Glover, F., J. Hultz, and D. Klingman: Improved Computer-Based Planning Techniques, Part 1, *Interfaces*, Vol. 8, no. 4, August 1978.

Glover, F., and D. Klingman: A Practitioner's Guide to the State of Large Scale Network and Network-Related Problems, *AFIPS Conference Proceedings*, Vol. 45, AFIPS Press, Montvale, N.J., 1976.

Golden, B.: Evaluating a Sequential Vehicle Routing Algorithm, *AIIE Transactions*, Vol. 9, no. 2, June 1977.

Golden, B., L. Bodin, T. Doyle, and W. Stewart, Jr.: Approximate Traveling Salesman Algorithms, *Operations Research*, Vol. 28, no. 3, Part II, 1980.

Goode, H. W.: An Application of a Highspeed Computer to the Definition and Solution of the Vehicular Traffic Problem, *Operations Research*, Vol, 5, no. 6, December 1957.

Gordon, G.: *Systems Simulation*, Prentice-Hall, Englewood Cliffs, N.J., 1969.

Greenberg, M. R.: *Applied Linear Programming*, Academic Press, New York, 1978.

Greenberger, M., M. A. Crenson, and B. L. Crissey; *Models in the Policy Process*, Russell Sage Foundation, New York, 1976.

Hadley, G.: *Linear Programming*, Addison-Wesley, Reading, Mass., 1962.

Hadley, G.: *Introduction to Probability and Statistical Decision Theory*, Holden-Day, San Francisco, 1967.

Hadley, G., and T. M. Whitin: *Analysis of Inventory Systems*, Prentice-Hall, Englewood Cliffs, N.J., 1963.

Halmos, P. R.: The Heart of Mathematics, *The American Mathematical Monthly*, Vol. 87, no. 7, 1980.

Hayward, O. G., Jr.: Military Decision and Game Theory, *Operations Research*, Vol. 2, no. 4, 1954.

Held, M., and R. M. Karp: The Traveling-Salesman Problem and Minimum Spanning Trees, *Operations Research*, Vol. 18, no. 6, November–December 1970.

Hesse, R., and G. Woolsey: *Applied Management Science*, Science Research Associates, Chicago, 1980.

Hille, E., and S. Salas: *First-Year Calculus*, Blaisdell, Waltham, Mass., 1968.

Hillier, F., and G. J. Lieberman: *Operations Research*, 3rd Edition, Holden-Day, San Francisco, Calif., 1979.

Hogan, W. W.: Energy Modeling: Building Understanding for Better Use, *Proceedings of the Second Lawrence Symposium on Systems and Decisions*, Lawrence Energy Laboratory, Livermore, Calif. 1978.

Honsberger, R.: *Mathematical Gems*, Vols. I and II, The Mathematical Association of America, Washington, D.C., 1973 and 1976.

Isaacs, R.: On Applied Mathematics, *Journal of Optimization Theory and Applications*, Vol. 27, no. 1, January 1979.

Jewell, W. S.: A Classroom Example of Linear Programming, Lesson No. 2, *Operations Research*, Vol. 8, no. 4, July–August 1960.

Kac, M.: Some Mathematical Models in Science, *Science*, Vol. 166, no. 3906, November 1969.

Kemeny, J. G., A. Schleifer, Jr., J. L. Snell, G. L. Thompson: *Finite Mathematics*, 2nd Edition, Prentice-Hall, Englewood Cliffs, N.J., 1972.

Khachiyan, L. G.: A Polynomial Algorithm in Linear Programming, *Dokl, Akad. Nauk SSSR*, pp. 1093–1096, Vol. 224, (English Translation in *Soviet Math.—Dokl.*, pp. 191–194, Vol. 20), 1979.

Klee, V., and G. J. Minty: How Good is the Simplex Algorithm?, chapter in *Inequalities III*, O. Shisha, (Ed.), pp. 159–174, Academic Press, New York, 1972.

Koopmans, T. C. (Eds.): *Activity Analysis of Production and Allocation*, Cowles Commission Monograph 13, Wiley, New York, 1951.

Krekó, B.: *Linear Programming*, American Elsevier, New York, 1968.

Kruskal, J. B., Jr.: On the Shortest Spanning Subtree of a Graph and the Traveling Salesman Problem, *Proceedings American Mathematical Society*, pp. 48–50, Vol. 7, 1956.

Kuhn, H.: The Hungarian Method for the Assignment Problem, *Naval Research Logistics Quarterly*, Vol, 2. nos. 1 & 2, March–June 1955.

Kwak, N. K.: *Mathematical Programming with Business Applications*, McGraw-Hill, New York, 1973.

Kwan, M.-K: Graphic Programming Using Odd or Even Points, *Chinese Mathematics*, pp. 273–277, Vol. 1, 1962.

Larson, R. C., and A. R. Odoni: *Urban Operations Research*, Prentice-Hall, Englewood Cliffs, N.J., 1981.

Lawler, E. L., and D. E. Wood: Branch-and-Bound Methods: A Survey, *Operations Research*, Vol. 14, no. 4, 1966.

Lee, C.: *Models in Planning*, Pergamon Press, Elmsford, N.Y., 1973.

Lin, S.: Computer Solutions of the Traveling Salesman Problem, *Bell Systems Technical Journal*, pp. 2245–2269, Vol. 44, 1965.

Little, J. D. C., K. G. Murty, D. W. Sweeney, and C. Karel: An Algorithm for the Traveling Salesman Problem, *Operations Research*, Vol. 11, no. 6, 1963.

Lucas, W. F.: An Overview of the Mathematical Theory of Games, *Management Science*, Vol. 18, no. 5, Part 2, January 1972.

Luce, R. D., and H. Raiffa: *Games and Decisions*, Wiley, New York, 1957.

Machol, R. E.: *Elementary Systems Mathematics*, McGraw-Hill, New York, 1976.

Magee, J. F.: Decision Trees for Making Decisions, *Harvard Business Review*, Vol. 42, no, 4, July–August 1964.

March, J., and H. Simon: *Organizations*, Wiley, New York, 1958.

McCreary, E. A.: How to Grow a Decision Tree, *THINK Magazine*, March–April 1967.

Minieka, E. *Optimization Algorithms for Networks and Graphs*, Marcel Dekker, New York, 1978.

Minieka, E.: The Chinese Postman Problem for Mixed Networks, *Management Science*, Vol. 25, no. 7, July 1979.

Miser, H. J.: Operations Research and Systems Analysis, *Science*, Vol. 209, July 4, 1980.

Müller-Merbach, H.: *Tri-Branching in Integer Programming*, Technical Report, Technische Hochschule, Darmstadt, 1982.

Munkres, J.: Algorithms for the Assignment and Transportation Problem, *SIAM Journal of Applied Mathematics*, Vol. 5, no. 1, 1957.

Murty, K. G.: *Linear and Combinatorial Programming*, Wiley, New York, 1976.

Neumann von, J., and O. Morgenstern: *Theory of Games and Economic Behavior*, Princeton Univ. Press, Princeton, N.J., 1947.

Nicholson, T.: *Optimization Techniques*, Vol. I, Longman Press, London, 1971.

Oppenheim, M.: *Applied Models in Urban and Regional Analysis*, Prentice-Hall, Englewood Cliffs, N.J., 1980.

Ore, O.: *Graphs and Their Uses*, Random House, New York, 1963.

Owen, G.: *Game Theory*, Saunders, Philadelphia, 1968.

Padberg, M. W.: Covering, Packing and Knapsack Problems, *Annals of Discrete Mathematics*, pp. 265–287, Vol. 4, 1979.

Paull, A. E.: Linear Programming: A Key to Optimum Newsprint Production, *Pulp and Paper Magazine of Canada*, Vol. 57, no. 1, January 1956.

Pearl, J.: *Heuristics*, Addison-Wesley, Reading, Mass., 1984.

Phillips, D. T., A. Ravindran, and J. J. Solberg: *Operations Research*, Wiley, New York, 1976.

Polya, G.: *How to Solve It*, 2nd Edition, Doubleday, New York, 1957.

Potts, R. B., and R. M. Oliver: *Flows in Transportation Networks*, Academic Press, New York, 1972.

Prim, R. C.: Shortest Connection Networks and Some Generalizations, *The Bell System Technical Journal*, pp. 1389–1401, Vol. 36, 1957.

Raiffa, H.: *Decision Analysis*, Addison-Wesley, Reading, Mass., 1968.

Ralston, A., and P. Rabinowitz: *A First Course in Numerical Analysis*, 2nd Edition, McGraw-Hill, New York, 1978.

Rapoport, A.: *Operational Philosophy*, Harper & Brothers, New York, 1954.

Rapoport, A.: *Fights, Games and Debates*, Univ. of Michigan Press, Ann Arbor, 1960.

Rapoport, A., and A. M. Chammah: *Prisoner's Dilemma*, University of Michigan Press, Ann Arbor, 1965.

Richmond, S. B.: *Operations Research for Management Decisions*, Ronald Press, New York, 1968.

Roberts, R. S.: *Graph Theory and Its Applications to Problems of Society*, SIAM Publication, Philadelphia, 1978.

Rogers, E. M.: *Physics for the Enquiring Mind*, Princeton Univ. Press, Princeton, N.J., 1960.

Rosenkrantz, D., R. Stearns, and P. Lewis: Approximate Algorithms for the Traveling Salesperson Problem. *Proceedings of the 15th Annual IEEE Symposium of Switching and Automatic Theory*, pp. 33–42, 1974.

Rothenberg, R. I.: *Linear Programming*, Elsevier North-Holland, New York, 1979.

Rubinstein, M. F.: *Patterns of Problem Solving*, Prentice-Hall, Englewood Cliffs, N.J., 1975.

Saaty, T. L.: *The Analytic Hierarchy Process*, McGraw-Hill, New York, 1980.

Saaty, T. L.: *Decision Making for Leaders*, Lifetime Learning Publications, Belmont, Calif., 1982.

Saaty, T. L., and R. G. Busacker: *Finite Graphs and Networks*, McGraw-Hill, New York, 1965.

Samuelson, P. A.: *Economics*, 5th Edition, McGraw-Hill, New York, 1961.

Sasieni, M., A. Yaspen, and L. Friedman: *Operations Research: Methods and Problems*, Wiley, New York, 1959.

Shor, N. Z.: Cutoff Method with Space Extension in Convex Programming Problems, *Kibernetika*, pp. 94–95, Vol. 13, (English translation in *Cybernetics*, pp. 94–96, Vol. 13), 1977.

Shubik, M.: The Uses of Game Theory in Management Science, *Management Science*, Vol. 2, no. 1, October 1955.

Shubik, M.: On Gaming and Game Theory, *Management Science*, Vol. 18, no. 5, Part 2, January 1972.

Silver, E. A., R. V. V. Vidal, and D. de Verra: A Tutorial on Heuristic Methods, *European Journal of Operational Research*, pp. 153–162, Vol. 5, 1980.

Sisson, R. L.: Introduction to Decision Models, Chapter 1 in *A Guide to Models in Governmental Planning and Operations*, S. I. Gass and R. L. Sisson (Eds.), Sauger Books, Potomac, Md., 1975.

Smith, D.: *Linear Programming Models in Business*, Polytech Publishers, England, 1973.

Smith, D. E.: *Quantitative Business Analysis*, Wiley, New York, 1977.

Stern, H. I., and M. Dror: Routing Electric Meter Readers, *Computers and Operations Research*, Vol. 6, no. 4, 1979.

Stewart, J. Q.,: Concerning "Social Physics," *Scientific American*, Vol. 178, May, 1948.

Stigler, G. J.: The Cost of Subsistence, *The Journal of Farm Economics*, Vol. 27, no. 2, 1945.

Strauch, R. E.: *A Critical Assessment of Quantitative Methodology As A Policy Analysis Tool*, P-5282, The Rand Corporation, Santa Monica, Calif., August 1974.

Svestka, J. A., and V. E. Huckfeldt: Computational Experience with an M-Salesman Traveling Salesman Algorithm, *Management Science*, Vol. 19, no. 7, March 1973.

Taha, H. A.: *Operations Research*, Macmillan Co., New York, 1971.

Taha, H. A.: *Integer Programming*, Academic Press, New York, 1975.

Townsend, R.: *Up the Organization*, Fawcett Publications, Greenwich, Conn., 1970.

Tucker, A.: *Applied Combinatorics*, Wiley, New York, 1980.

Vajda, S. : *Problems in Linear and Nonlinear Programming*, Hafner Press, New York, 1975.

Wagner, H. M.: *Principles of Operations Research*, 2nd Edition, Prentice-Hall, Englewood Cliffs, N.J., 1975.

Wagner, H. M.: *Principles of Management Science*, 2nd Edition, Prentice-Hall, Englewood Cliffs, N.J., 1975.

Walker, W. E., J. M. Chaiken, and E. J. Ignall (Eds.): *Fire Department Deployment Analysis*, The Rand Fire Project, North-Holland, New York, 1979.

Wardrop, J. G.: Some Theoretical Aspects of Road Traffic, *Proceedings of the Institute of Civil Engineering*, Part II, pp. 325–378, 1952.

Westlake, J. R.: *A Handbook of Numerical Matrix Inversion and Solution of Linear Equations*, Robert E. Krieger Publishing Co., Huntington, New York, 1975.

Wiest, J. D., and F. K. Levy: *A Management Guide to PERT/CPM*, Prentice-Hall, Englewood Cliffs, N.J., 1969.

Williams, J. D.: *The Compleat Strategyst*, McGraw-Hill, New York, 1966.

Wilson, A. G.: *Urban and Regional Models in Geography and Planning*, Wiley, New York, 1974.

Wilson, R. J.: *Introduction to Graph Theory*, Longman Group Ltd., London, England, 1975.

Index

How index-learning turns no student pale,
Yet holds the eel of science by the tail

ALEXANDER POPE

Abstract decision model, 20–21, 38–39
Accounting price, 152
Activity-analysis model, 101–106
Acyclic network, 259–266
Adjacent extreme point, 151, 168–169
Advertising problem, 377–378
AHP, 355–367, 391–396
Airline-crew-scheduling problem, 124
Air pollution, 20
Algorithm, 19, 37
Alternative solutions, 3, 10, 26, 28, 39
Analogy, 25, 51–52, 57
Analytic Hierarchy Process (AHP), 355–367, 391–396
Area problem, 363, 394–395
Artificial basis, 182
Artificial variable, 182
Assignment problem, 50–51, 95–99, 120–122, 243–249, 288–290
Assignment problem algorithm, 243–249
Assignment table, 97
Attack bomber problem, 263–264
Automobile manufacturer problem, 214, 266–267
Automobile replacement problem, 53–56, 385–386

Bakery problem, 113–114, 202
Barrier line, 148
Basic feasible solution, 164
Basic variable, 164
Battle of the Bismark Sea, 326–329
Best decision, 16
Best solution, 15
Biased game, 332
Bimatrix game, 380
Bipartite graph, 230
Bipartite network, 287
Blending problem, 72
Branch and bound algorithm, 223, 278–280, 317–319
Breakfast problem, 68–72, 156–159
Business investment problem, 350–354, 387

Carrying cost, 40
Caterer problem, 74–87, 113, 290–294
Chinese postman problem, 282
Clever Hans, 268
Cocktail party problem, 388
Competitive equilibrium, 155
Complete elimination, 159, 166
Complete enumeration, 49
Computer programming, 29*n*.
Connected graph, 282
Connected network, 231
Conservation of flow, 208
Consistency index, 391–393
Consistency ratio, 392–393
Constraints, 16
Continuous linear-programming problem, 119
Contract-awards problem, 294–298
Contractor bidding problem, 345, 383–385
Controllable variables, 17, 39
Convex hull, 270
Convex-hull algorithm, 270–271
Convex polyhedron, 146, 152
Convex polytope, 146, 149
Convex set, 145
Cooperative game, 326
Cost coefficients, 107
Cost function, 43
Counterintuitive results, 16
CPM, *see* Critical-path method (CPM)
Critical path, 262
Critical-path method (CPM), 262
Current dollars, 347
Cut, 212
Cycling, 179–180

Decision-aiding model, 11, 16
Decision analyst, 24
Decision framework, 4, 6, 23–28, 368–372
Decision maker, 3, 6–8
Decision making, 1–8, 10, 27
Decision problem, 3, 5–6, 18–20, 23
Decision problem model, 21
Decision science, 23
Decision space, 21, 38–39
Decision-tree model, 344, 347–354, 386–389
Decision variables, 17, 20, 37–39, 107
Declaration of model independence, 21
Degenerate basic feasible solution, 179
Degree of a node, 282
Descriptive model, 15
Deterministic model, 18
Diet problem, 68–73, 112–113, 156–158
Directed arc, 208
Discounted value, 346
Dominant strategy, 329
Dressing problem, 29–30
Duality theorem, 153
Dual problem, 152, 158, 189–190

Economic order quantity (EOQ), 42
Economy principle, 311
Eigenvalue, 358, 391–394
Eigenvector, 358, 391–394
Ellipsoid algorithm, 201

Empirical bound, 201
Engine overhaul problem, 86–87
Enumeration, 143–145
Equilibrium, 155, 158, 202
Equilibrium strategies, 380–381
Euler circuit (tour), 282–283
Exchange algorithm, 302–307
Expected monetary value, 344–345
Expected present value, 345–347, 386–387
Expected utility, 342
Exponential-time algorithm, 201
External model, 16
Extreme point, 145, 164, 189
Extreme-point solution, 164, 201, 231

Fair game, 332
Farmer planting problem, 347–350, 377
Feasible programs, 30
Feasible solution, 45
Feedback loop, 28
Feed-mix problem, 72
Fictitious destination, 241
Fictitious origin, 240
Fire department problem, 121–122
First basic feasible solution, 181, 241
Fixed cost, 40
Flow-augmenting path, 251
Force formula, 24
Formulation, 65, 115
Furniture manufacturing problem, 100–106, 113, 147, 202

Game theory, 325–340, 373–382
Game tree, 389
Gauss-Jordan elimination, 159
GPA problem, 359–362
Graph theory, 281
Gravity model, 24–26, 50–52, 58
Greedy algorithm, 120, 132, 269, 305, 309–310
Guggenheim Museum, 133

Hamiltonian circuit, 274–275
Heuristic algorithm, 44, 51, 58, 223, 269, 302–313
Heuristic reasoning, 50
Heuristics, 19
Highway traffic problem, 265–266
Hirshhorn Gallery, 133
Holding cost, 40
Holistic model, 18
Horse racing problem, 386
Hungarian algorithm, 243–249, 289
Hungarian theorem, 246

Iconic model, 14, 34
Ill-structured problem, 7
Implementation, 26, 28
Independent set of points, 245
Indirect cost, 235
Information, 11
Insertion algorithm, 270
Integer-programming model, 118, 123, 220, 317–319
Integer restrictions, 119
Integer solution, 38, 44, 98, 317–319
Intermediate nodes, 207
Internal model, 16
Intuition, 27
Inventory ordering policy, 40
Inventory problem, 39–43
Investment problem, 108
Iteration, 168

Job selection problem, 366–367
Joseph in Egypt, 133
Judgment matrix, 358, 392

Kanban, 40*n*.
König-Egérváry theorem, 246
Königsberg Bridge problem, 281–282

Law of diminishing returns, 50
Law of universal gravitation, 24
Linear equations, 37
Linearity assumption, 34, 43
Linear-programming model, 37, 43, 64, 107
Linear programs, 31
Linear relationships, 31
Logical model, 15, 26
Longest-route problem, 262–263

Mad Hatter, Inc., 74
Mailbox pickup problem, 283
Many-person game, 326
Marginal analysis, 152
Marginal profit, 153
Marginal value, 153, 199
Marriage problem, 99, 119–120
Matching pennies game, 331–336
Mathematical expectation, 332
Mathematical model, 3, 12, 15, 26–28
Mathematical-programming system, 193
Mathematical programs, 29
Matrix game, 329–336
Max-flow Min-cut theorem, 212
Maximal-flow algorithm, 250–253
Maximal-flow network problem, 208–212, 250–254

Maximizing player, 328
Measure of effectiveness, 18, 20–21, 26, 28, 30, 38, 43, 45, 49, 107
Measure of efficiency, 13
Mental model, 16
Menu planning, 112, 118
Metaphor, 57
Minimal-cost network-flow problem, 212–214, 290
Minimal-spanning tree, 310
Minimal-spanning tree algorithm, 311–313
Minimal-spanning tree problem, 310–313
Minimizing player, 328
Missile targeting problem, 124
Mixed-integer problem, 317–319
Mixed strategy, 331
Model, 11–22
Modeling steps, 368–370
MPS, 193
Multicriteria objective function, 56
Multiple optimal solutions, 151, 179–180, 196, 239
Multi-traveling salesmen problem, 299–302
Museum design problem, 4, 6, 8–10, 53, 124–133

National Bureau of Standards (NBS), 298
Nearest-neighbor algorithm, 269–270
Negotiating problems, 378–379
Network-flow problems, 207–217, 251–266
New car problem, 364–366
Node-labeling algorithm, 215–217, 250–253
Nonbasic variable, 164
Noncooperative game, 326
Nonlinear relationship, 34, 49–50
Nonnegative quadrant, 143
Nonnegativity restrictions, 37
Nonprogrammed decision, 8
Normative model, 15
Northwest-corner rule, 227–229, 284–285
Nut mix problem, 65, 109–112

Objective function, 18, 21, 30, 107
Objective utility function, 385
Opportunity costs, 152
Optimal decision, 16
Optimality test, 171
Optimal solution, 15, 21, 26
Optimal variables, 18
Optimization, 19
Optimization algorithm, 45
Output variables, 17

Pairwise comparison scale, 357
Paper-scissors-stone game, 376
Partial enumeration, 49
Pascal's wager, 390–391
Path, 282
Payoff, 326
Payoff matrix, 327
PEANUTS, 59
Personnel assignment, 95–99
Personnel classification, 95
PERT, 262–263
Petersburg Game, 390
Pill peddler, 158, 190–191
Pivot column, 167
Pivot element, 167
Pivot row, 166
Police scheduling problem, 115–116, 202
Polynomial-time algorithm, 201
Prediction, 19
Predictive model, 15
Prescriptive model, 15
Present value, 346
Present worth, 346
Primal-dual problems, 192, 333, 373–375
Primal problem, 152, 158
Prisoner's dilemma, 381–382
Probabilistic model, 18
Problem solving, 5
Procedural model, 15
Production planning problem, 214, 298–299
Program evaluation review technique (PERT), 262–263
Programmed decision, 7
Programming problems, 29
Proportionality assumption, 34
Pure strategy, 329

Quadratic equation, 49–50

Rational behavior, 341, 383
Rational decision maker, 341–342
Rational decision making, 8
Ratio test, 171, 179
Reference system, 14, 21, 24
Refuse collection problem, 20
Relationships, 16–17
Relaxation, 119, 317
Result variables, 17
Retail-sales gravity model, 51–52
Right-hand-side elements, 107
Routing problem, 302–308

Saddle point, 329
St. Petersburg Paradox, 390
Science of decision making, 6, 8
Scientific method, 23

SEAC, 298
Sensitivity analysis, 55, 193–200, 285–286, 307, 391
Set-covering problem, 122–124, 132
Set-partitioning problem, 123
Setup cost, 40
Shortest-route algorithm, 216–217, 254–255, 260–261
Shortest route problem, 214–217, 255–261
Simple Furniture Company, 101, 147, 161, 163, 194
Simple lot size formula, 42
Simplex algorithm, 163–172
Simplex tableau, 171
Simplex transportation algorithm, 224–242
Single-variable problem, 139–141
Sink node, 207
Skin game, 336–339
Slack variable, 161
Snow removal problem, 282–283
Social physics, 52
Solution space, 140, 143
Source node, 207
Spanning tree, 310
Spike's choice, 59
Stochastic model, 18
Straight-line relationship, 31
Street cleaning problem, 282
Structured problem, 7
Suboptimization, 67
Subtour, 222
Super Management Consultants, 77, 101, 218, 336
Surplus variable, 160
Sweep algorithm, 308
Symbolic model, 15
System, 9
System environment, 9, 18

Tea party problem, 74–87, 290–294
Technological coefficients, 107
Theory of games, 325–340, 373–382
Tour, 221
Tour-improvement algorithm, 271–272
Transportation network, 207
Transportation problem, 20, 32–38, 43–49, 50, 56, 58, 191–192, 224–242, 283–288, 290
Transshipment points, 207
Transshipment problem, 213–214
Traveling-salesman problem, 20, 218–223, 267–281, 299–316
Tree, 309–310
Triangle inequality, 271
Trim problem, 88–94, 113, 117–118
Truck delivery problem, 123
Truck leasing (replacement) problem, 264
Two-arc exchange algorithm, 272
Two-person game, 326
Two-phased simplex algorithm, 183–187
Two-variable problem, 142–146

Unbounded solution, 145, 152, 163, 175–177
Uncontrollable variables, 17, 20, 38–39
U.S. Department of Agriculture RDAs, 112
Urban gravity model, 25, 50
Urban planning, 24–26, 51–52
Utility function, 341–344, 382–385
Utility variables, 17

Validation, 14, 19, 24, 28, 370–372, 396
Valid model, 19
Value variables, 17
Variables, 16–17
Vehicle-routing problem, 302–308
Verification, 370–371

"What if" questions, 19
Wilson's formula, 42
Worst-case estimate, 201

Zero-sum, 326
Zero-sum game solution, 332